Neurothemis tullia tullia (Drury, 1773)
(Source: B.K. Tyagi)

Dr Bastiaan Kiauta

Odonatologist and Polymath

His Life, Works and Universe

Related Books

Title	Author
• 21st Century Biology and Agriculture: Textbook Series (Vol. 1-5)	*A. Raman*
• A Manual of Practical Entomology *3rd Ed*	*M.M. Trigunayat*
• A Manual of Practical Zoology: Biodiversity, Cell Biology, Genetics & Developmental Biology (Part 1)	*M.M. Trigunayat*
• Agricultural Insect Pests and their Control *2nd Ed*	*V.B. Awasthi*
• Agriculture Herbicide Study Manual	*Arnold P.Appleby*
• बागवानी फसलों में समेकित रोग–कीट रोग प्रबंधन	*S.K. Singh*
• Beekeeping: A Compressive Guide to Bees and Beekeeping	*D.P. Abrol*
• Biodiversity Conservation and Systematics	*Prithipal Singh*
• Biofertilizer Technology	*S. Kannaiyan*
• Changing Faunal Ecology in the Thar Desert	*B.K. Tyagi*
• Commercial Beekeeping**	*Dharm Singh*
• Competitive Nematology**	*N.G. Ravichandra*
• Dynamics of Insect Behaviour	T.N. Ananthakrishnan
• Ecological Entomology: Insect Life in Odd Environments	T.N. Ananthakrishnan
• Emerging Crop Pest Problems: Redefining Management Strategies	*P.P. Reddy*
• Entomology Ecology & Biodiversity	*B.K. Tyagi*
• Entomology in the Doon Valley	*B.K. Tyagi*
• Handbook of Insects: Photographic Guide for Identification**	*V.B. Awasthi*
• Insect Mite and Vertebrate Pests and their Management in Horticultural Crops	*P.P. Reddy*
• Insect Neuroendocrines	*V.B. Awasthi*
• Insect Physiology (21st Century Biology and Agriculture: Textbook Series)	*K.P. Sanjayan*
• Introductory Plant Nematology	*P.P. Reddy*
• Medical Entomology	*B.K. Tyagi*
• Odonata Biology of Dragonflies	*B.K. Tyagi*
• Organic Evolution	*B.L. Chaudhary*
• Pests of Forest Importance and their Management	*B.K. Tyagi*
• Pranishastra Pryogik Manual (Part 1)	*M.M. Trigunayat*
• Research Methodology in Zoology	*P.S. Narayana*
• Research Methods in Plant Sciences: Allelopathy (Vol. 1-5) (Set)	*S.S. Narwal*
• Scientist: A Unique Species and the Backbone of a Nations Economic and Intellectual Prosperity	*B.K. Tyagi*
• The Invincible Deadly Mosquitoes	*B.K. Tyagi*
• Theory and Practice of Integrated Pest Management	*A.K. Dhawan*
• Vector-Borne Diseases: Epidemiology and Control	*B.K. Tyagi*
• Vertebrate Pests in Agriculture: The Indian Scenario	*S. Sridhara*

Dr Bastiaan Kiauta

Odonatologist and Polymath

His Life, Works and Universe

B.K. Tyagi

Emeritated Scientist 'G' (Director) & Director in-Charge
Centre for Research in Medical Entomology (ICMR), Madurai, TN, India;
Advisor, SpoRIC, VIT University, Vellore, Tamil Nadu, India
&
President, Indian Dragonfly Society

SCIENTIFIC PUBLISHERS

Published by
SCIENTIFIC PUBLISHERS (INDIA)
5 A, New Pali Road, P.O. Box 91
Jodhpur 342 001 (INDIA)
E-mail: info@scientificpub.com
Website: http://www.scientificpub.com

ISBN: 978-93-88043-57-1
eISBN: 978-93-88043-80-9

Printed in India

Dedicated to

My teachers, in and outside classroom, especially
Dr M.A. Lieftinck, Prof. Dr Janny M. van Brink, Prof. Dr Minter J. Westfall, Prof. Dr Philip S. Corbet, Prof. Dr T.N.Ananthakrishnan, Prof. Dr P.K. Sen-Sarma
and
Prof. Dr S.K. Sangal,
and
My beloved Parents

Foreword

"Verba volant, littera scripta manet."
"Words fly away, written letters remain."

The monograph in front of you is not merely a biographic book. It is a book of deepest admiration and respect, devoted to one of the greatest odonatologists of our time – Prof. Dr Boštjan Kiauta.

Born in a small green country on the sunny side of the Alps, in Slovenia, Prof. Kiauta began to develop a keen interest in entomology in his earliest childhood. And not long thereafter, in 1953, as a 16-years old boy, he already published his first two articles on butterflies, shortly followed by the two earliest odonatological accounts, in 1954. Was it due to a lucky twist of the destiny, or already then due to a deeper visionary inclination that put dragonflies forever in the forefront of his life and activities, is actually not so important. What matters, however, was firm and faithful devotion of Prof. Kiauta to an odonatological scientific trajectory. From this early decision onwards, and now already for well over six decades, he has remained deeply devoted to all aspects of the studies of strikingly beautiful flying pearls we odonatologists all admire.

The name of Prof. Kiauta is known to almost every dragonfly enthusiast, in every continent of the world, wherever the modern odonatology has its home. However, his expertise goes so far beyond that of a specialized odonatologist. He is an entomologist in the broadest sense of the word, an editor, geneticist, speleologist, mountaineer, environmentalist, an animal lover and advocate of their rights, a passionate collector and an exceptional connoisseur of Slovenian literature and history, and even a knight.... Above all, he is a great man, and a true mentor and teacher. As the knowledge is only multiplied by sharing, by his multifaceted activities, Prof. Kiauta is among the greatest facilitators of the enormous advances that odonatology has seen in the last decades. His exceptional work is amply and respectably presented in the present publication.

Being from Slovenia, my early interest in odonatology was sparked by Prof. Kiauta's numerous accounts on the dragonfly fauna of our country that he published between 1954 and 1969. It did not last long when upon my completion of high school in 1991, I finally accumulated the courage to hand write a letter to the "far-off" Netherlands, to be followed by hundreds of others in the years to come. Back then, unfortunately being completely unaware of the enormous amount of everyday work and obligations of Prof. Kiauta, my letters were full of questions and requests for advice and assistance. Nevertheless, Prof. Kiauta always answered all of them promptly, offering invaluable advice and guidance, with due commendation, as well as occasional fair criticism, but always in a polite, suggestive manner. Much to my today's embarrassment, I followed with more requests over the following years when my focus shifted also towards Oriental dragonfly fauna. Again, in the times far before different modern electronic options, it was Prof. Kiauta who would help to obtain literature that was unavailable to me, whenever needed, thereby always rescuing "dangerously stalled studies" with copies of hundreds and hundreds of pages. I am sharing this part of my personal experience with Prof. Kiauta to express my deepest gratefulness for his benevolent help and mentoring. But more importantly, Prof. Kiauta is known as such to the multitude of odontologists around the world – as a global odonatological authority and mentor in the broadest sense of the word, to whom it is never difficult to help anyone.

Nowadays, in the times of computers and all kinds of exquisite modern tools in the fields of publishing and editorship, everyday communications and research, it is very hard to imagine how Prof. Kiauta succeeded to accomplish all that the present book summarizes "in only one man's lifetime" and with the use of "only a type-writer and snail mail". What was needed is his visionariness, infinite knowledge, dedication, organizatinon, discipline, and endurance... Yes, all of this for sure, however, as acknowledged by prof. Kiauta on every occasion, his great contributions to the field were only possible with the enormous understanding, support and help of his wife Marianne.

Therefore, dear Boštjan and Marianne, your works, as also exemplified by this book, speak for themselves. As most of your legacy is published – systematically, with meticulous accuracy, quality, dedication and love – in the sense of introductory Latin proverb it will remain in the service of odonatology and global odonatological family forever, I thank you sincerely and wish you both many years to come, with a humble request for cooperation and further sharing of your invaluable knowledge also in the future.

Dr Matjaž Bedjanič

Braslovče, SLOVENIA; December 31, 2017

Preface

"Who comprehends his trust, and to the same
Keeps faithful with a singleness of aim;
And therefore does not stoop, nor lie in wait
For wealth, or honour, or for worldly state;
Whom they must follow, on whose head must fall,
Like showers of manna, if they come at all."

—William Wordsworth

The life of Professor Dr Bastiaan Kiauta, an utterly indefatigable odonatological genius *nonpareil*, truly reflects the spirit radiated from these famous lines by William Wordsworth. Dr Kiauta is one of those rare scientists who by their energetic character not only practically educate the budding researchers in habits of industry, but by the example of diligence and perseverance which they set before them, largely influence the scientific activity in all directions and contribute in a great degree to form the national character. The national progress, it is aptly said, is after all the sum of individual industry, energy and uprightness and Dr Kiauta is a glorious example of focused odonatology research for the entire scientific fraternity in our own lifetime. In recognition of his lifelong and unparallel dragonfly research work, Her Majesty – the Queen of the Netherlands, felt herself honoured in having bestowed upon him on 26th April 2002, the highest civilian award of the country: *Knighthood in the Order Oranje-Nassau* (vide infra).

Professor Dr B. Kiauta, April 2002. (Source: B. Kiauta archive)

Throughout history, the greatest scientific minds have not only demystified the world with their discoveries, but helped shape how we live in it with their ingenious inventions and innovations. From Isaac Newton to Charles Darwin to Albert Einstein and many more brilliant minds, there have been an endless group of famous scientists who have made major advances in the field of science. In the vast field of odonatology there may indeed be many examples of great scientists, or for that matter – odonatologists, but there are certainly absolutely inevitably very few who have influenced in all directions the minds of so many students, scholars and researchers across the world as is impacted by Dr Kiauta - singularly the most glorious example almost synonymous to the word 'dragonfly'. Glittering like the North Star in our odonatological constellation, he is, without doubt in my mind, the world's foremost odonatological authority today, thus belonging to the same league of extraordinary odonatologists to which in past the all-time greats like Edmond de Selys-Longchamps, W.F. Kirby, R.J. Tillyard and Philip S. Corbet etc. had been associated! He is a complete scientist, with an exceptionally wide range of interests including general insect phylogeny and evolution, general insect cytotaxonomy and karyotypic evolution, insect speciation under extreme environmental conditions, such as prevailing in high mountains and in caves, insect population biology and population genetics, history of odonatology with emphasis on Slovenia, and cytogenetics relative to the Turbellarian theory of Cnidaria and invertebrate evolution and phylogeny. Major insect orders of particular interest to him are Odonata, Ephemeroptera, Trichoptera, Coleoptera, Collembola etc., though the odonates (Dagonflies) have always been his first love! The range and depth of his knowledge are astonishing, and sense of recalling scientific details and events is extraordinarily unparalleled.

During more than four decades (1971-2013) Dr Kiauta has managed the world's truly first and only worldwide dragonfly association with the largest number of membership ever for a single insect order, the Societas Internationalis Odonatologica (SIO), and served as the Executive Editor of *Odonatologica*, one of the finest scientific entomological journals to date, not to forget in least the time consuming *Odonatological Abstracts* covering 19,871 works, published between 1971 and 2013. He has sacrificed his today so that the odonatologists can have a better tomorrow with the odonatological information. He even influenced foundation of many odonatological societies the world over and, besides the SIO Regional Office in Southern Asia (now South Asian Council of Odonatology, SACO), he has been also instrumental in seeding the birth of the Indian Dragonfly Society, the first ever formal and indigenous association of odonatologists in the country, in 2017.

Why is it necessary to have this biobibliography? The great scientists have always provided inspiration by their findings, philosophy, and understanding of the world around us. They have inspired generations of young explorers, eager to learn more about the world and motivated toward betterment with constant efforts in the quest for knowledge. Dr Kiauta is truly a complete scientist; an odonatologist extraordinaire and a polymath, and his life story provides a great stimulus to the budding scientists in all walks of life, not just the odonatologists alone, in understanding the value of hard work, discipline and commitment. Even for several years following his emeritation he, now an octogenarian, has been actively engaged in odonatological research projects and prublishing results of his research work prolifically. He is a great teacher and a global odonatological mentor, with his colleagues and students, taught directly in classroom or indirectly through correspondence, are today spread in all the continents of the world and, like a responsible protégé, feel truly grateful to the lessons learnt from him. We have hard learnt in the school of life that it is energetic individualism which produces the most powerful effects upon the life and action of others, and really constitutes the best practical education. Even though Albert Einstein, the Nobel Laureate and a great physicist, has extolled the following lines in a different context (of Mahatma Gandhi), I am nevertheless inclined to recount it in the case of Dr B. Kiauta, 'the odonatological phenomenon':"*Generations to come, it may well be, will scarcely believe that such a man as this one ever in flesh and blood walked upon this Earth.*"

Biographies of great scientists, but especially of good men, like Dr Kiauta, are invariably most instructive and useful, as helps, guides, and incentives to others. His biography teaches simple living, high thinking, and energetic action for our own and the science's good. His life furnishes valuable examples of the power of self-help, of patient purpose, resolute working, and steadfast integrity, issuing in the formation of truly noble and inexorable character of commitment, exhibit in language and correspondence not to be confused, what it is in the power of each to accomplish for himself or herself; and eloquently illustrate the efficacy of self-respect and self-reliance in enabling budding odonatologists to work out for themselves an honourable competency and a solid reputation, like the Shakespeare's famous adage goes:

"This above all,—To thine own self be true;
And it must follow, as the night the day,
Then canst not then be false to any man."

The object of the book briefly is to inculcate the lessons that young and budding scientists, in particular, must work hard in order to enjoy,—that nothing creditable can be accomplished without application and diligence,—that the researcher must not be daunted by difficulties, but conquer them by patience and perseverance,—and that, above all, he must seek elevation of character, without which capacity is worthless and worldly success is naught. If the author has not succeeded in illustrating these lessons, he can only say that he, and not the protagonist, has failed in his object. I nevertheless hope and wish with my clear conscience and inner enlightenment that the untold story of life and works of Professor Dr B. Kiauta, as narrated here in a simple yet authoritative manner, will inevitably provide a much needed stimulus to the younger generation of odonatologists, entomologists and zoologists to imbibe in them the unflinching spirit of self-help, hard work, and commitment – as solidly epitomized with a phlebotomist's precision in each and every task he undertakes.

"*Books reach much farther than spoken words.*"

Prof. Dr B.K. Tyagi

Jodhpur; January 20, 2019

Acknowledgement

It is unimaginable to encompass in a small composition like this a vast and diverse life attribution dedicated single-mindedly to the cause of evolution of odonatology, the science of dragonflies, without the firsthand help of its central figure, the protagonist, Professor Dr Bastiaan Kiauta, ably supported by his benevolent wife, Mrs Marianne Kiauta. I conceived the idea of writing a biography on the life and works of Dr Kiauta in 2003 and sought Dr Kiauta's permission which he gave only after great persuasion and convincing. He, however, provided me with every bit of information that I sought, sometimes on weekends by emails or phone calls! To them I am thankful from the core of my heart and will remain indebted lifelong for helping me to accomplish this task for the sheer reason of bringing a highly powerful stimulus to younger generation of odonatologists, in particular, besides an undeniable spirit of assiduous perseverance, focused attention, and selfless application in the global advancement of odonatology through, among many works, establishment of Societas Internationalis Odonatologica (S.I.O.) and its mainstream international journal, *Odonatologica*.

A large number of odonatologist colleagues from all over the world have most generously provided several different materials (e.g., copies of air-mailed letters and emails, photographs, national offices' newsletters, international symposia reports and anecdotes etc.) and I particularly feel grateful to Dr K. Inoue (*Past* President, International Odonatological Foundation, Societas Internationalis Odonatologica, S.I.O.; 5-9, Fuminosato 4-chome, Abeno-ku, Osaka 545, Japan), Dr Matti Hämäläinen (Helsinki University, Finland), Dr Milen Martinov (Plant Health & Environment Laboratory, Diagnostic and Surveillance Services, Ministry for Primary Industries, Auckland, New Zealand), Dr Martin Schorr (Waldfrieden 25, 54314, Zerf, Germany), Dr Stephen Butler (UK), Dr Robert A. Cannings, Curator of Entomology (Emeritus), Royal British Columbia Museum, Canada), Dr L.M. Mola (Universidad de Buenos Aires, Argentina), Dr A.B.M. Machado, (Universedade Federal de Minas Gerais, Brazil). Much earlier in 1990 when Prof. Peter Miller (Oxford University, UK) visited India, and my home in Jodhpur, we talked about many odonatological interests of Dr Kiauta, and I am indeed highly grateful for his extensive knowledge associated with Prof. Kiauta, *Odonatologica* and the SIO.

My heartfelt thanks go to Dr Matjaž Bedjanić for contributing a highly delightful and inspirational 'Foreword' for this book.

As always I thank my wife, Ajita, who was all supportive and motivating to me throughout the period of completing Dr Kiauta's biography.

I owe much to Mr Pawan Kumar Sharma, CEO of the Scientific Publishers (India), Jodhpur, for his trust in me and for the patience and help extended to me throughout the process of writing this monograph.

Contents

SECTION I: ORIENTATION

1. The Family History and Early Childhood
2. The under school-age period and the World War-II time (1941-1945)
3. In the "Entomology School" of uncle Cyril Azman (1943-52)
4. Period of the early friendship with Mathias Gogala (1943-1953)
5. Breakthrough into the community of adult entomologists (1949-1952)
6. Decision on dragonflies (1952)
7. The first steps in odonatology (1953)
8. Bastiaan's first encounter with Professor Jovan Hadzi (1953)

1

The Family History and Early Childhood

कर्मण्येवाधिकारस्तेमाफलेषुकदाचन।

माकर्मफलहेतुर्भूर्मातिसङ्गोऽस्त्वकर्मणि।

Transcription:

Karmanye vadhikaraste Ma Phaleshu Kadachana,
Ma Karmaphalaheturbhurma Te Sangostvakarmani

Translation:

You have the right to work only but never to its fruits.
Let not the fruits of action be your motive, nor let your attachment be to inaction.

Srimadbhagwadgita, Chapter 2, Verse 47

Slovenia, though in an area less than a tenth the size of the United Kingdom, as seen presently, is extraordinarily blessed with natural beauty, mainly comprising towering mountains, picture-perfect lakes, vast caves, elegant cities and fast-running rivers. The small country Slovenia also boasts a bounty of tasty dishes. Some are influenced by its neighbours Austria, Italy and Hungary while others are home-grown, but all are prepared with the freshest ingredients in season. Complementing all that goodness on the plate, the local wine, beer, fruit brandy and even water are also of the highest quality. Ljubljana (say *lyoob-lya-nah* – the 'j's are mostly silent), Slovenia's biggest city, also happens to be one of Europe's greenest and most liveable capitals (Fig. 1.1).

Fig. 1.1. Present day Ljubljana seen in a bird's-eye view (Image by Tomo Jeseničnik; www.slovenia.info)

It's conveniently set up for visitors too – most of the city's top attractions are clustered in a compact pedestrianized area around a bend in the pretty Ljubljanica River, perfect for casual wandering. Castle Hill in the center of the town attracts with a mesmerizing view. Between Castle Hill and the Ljubljanica river is the city's Old Town, defined by three 'squares', which are actually more like narrow, cobbled streets. Mestni trg is dominated by the Town Hall and, in front, the Robba Fountain, decorated with three titans representing the Sava, Krka and Ljubljanica rivers. Stari trg, further south, is the true heart of the Old Town, lined with 19th century shop fronts, quiet courtyards and cobblestone passageways. Gorni trg, the southwestern extension of Stari trg, is the quietest and most evocative section of the Old Town, with some buildings dating to the medieval period. At Ljubljana's heart is the highly photogenic Prešernov trg, just a short walk from the Old Town across the three-pronged Triple Bridge. The square is flanked by beautiful historic architecture, including the salmon-pink Franciscan Church of the Annunciation, and holds a monument to Slovenia's greatest poet, France Prešeren, after whom the square is named. Triple Bridge was designed by Slovenia's master of early-Modern minimalism, Jože Plečnik, an architect and urban planner who almost single-handedly transformed the city, creating elegant new structures and remodeling almost everything else, including the river banks.

Ljubljana, with such an extraordinary ecological setting, has been the birth place and/or seat of works of many a great scientist, nature lovers, archeologists, artists, musicians, engineers, philanthropists and public figures etc. There is a natural fragrance of grandness in everything it has be it museums, public libraries, wild life or even underground rocks embedded in multitudes of caverns all over. Craving for higher education and witty discussions were always a way of life in Ljubljana. Bastiaan Kiauta, one of the world's greatest odonatologists (meaning a scientist dedicated to working on various different research aspects of biology of the insects called damsel- and dragonflies), was born here on 20th January, 1937, who would later on, all along his life, continue to ignite minds, especially young, all over the world and light up the desire of thousands of odonatologists to research in possibly the largest ever assemblage of 'family members' in a given insect order, i.e., Odonata, in the entire history of entomology!

Bastiaan Kiauta providentially has a great legacy of academic and research to emulate from his glorified family roots, on both sides of parents (Fig. 1.2). Kiauta's family, on father's side, originates from Carnia (NE Italy). Its original language is Ladinic/Rheto-Romance, as spoken currently in Carnia, elsewhere in Friuli (Italy) and in canton Grissons (Switzerland). The original site of the family is the village of Cavazzo carnico. The earliest record so far known is from 1347, when the family was called Chiavalute (see G.B. CORGNALI, Schedario antroponimico, deposited in Biblioteca civica V. Joppi, Udine) and since 1521 as Chiautta (Archivio di Stato di Udine, No. 4884). During mid-19th century, his great-grandfather, Giuseppe Chiautta, moved from Cavazzo carnico (then Austria, now Italy) to Landol near Postojna (then Austria, now Slovenia). On 23rd August 1909, the Landol branch of the family amended their surname to the phonetic spelling, Kiauta (see Status animarum, Sv. Martin near Landol).

Bastiaan's great-grandfather was a trader and vast estate owner, the grandfather, Giuseppe/Josip Chiautta/Kiauta, was a civil servant with the Austrian Ministry of Justice. The father, Milan Chiautta/Kiauta was a lawyer in Ljubljana (Slovenia).

Fig. 1.2. Bastiaan's parents. Left: father Milan Kiauta, right: mother Marija (born Stupica) Kiauta. (Photo: Studio Holynski; dated before World War II).- (Source: B.Kiauta archive)

The mother, Marija born Stupica, was the daughter of Fran Stupica, a wholesale iron trader, industrialist and owner of a large landed property in Ljubljana and in Vir (both Slovenia). Her mother, Irma (born Ažman) was of south Tyrolean extraction (formerly Austria, now Italy). Among her brothers, three were acknowledged naturalists, *viz.* Cyril (Professor of Ancient Oriental Languages, Ljubljana), Jacob (artist-sculptor, Novo mesto, Slovenia) and Method (book trader, London). She graduated in Romance and classical languages (Latin, Greek) from the Universities of Ljubljana (Slovenia) and the Sorbonne in Paris (France).

Bastiaan has one brother and two sisters; brother Thomas Kiauta (born 1948), D.Sc., is a neurologist (Fig. 1.3). He began his career first as Professor in Medical Faculty of the University of Ljubljana, then became a specialist neuro-surgeon (Ljubljana), and finally took over as Director of a Hospital in Ljubljana. Older sister Marija (born Kiauta) Javoršek (born 1939) graduated in Romance Languages and Literatures from the University of Ljubljana. She worked initially as the Librarian in the Department of French in the National and University Library of Ljubljana, and subsequently continued with the same position at the Slovenian Academy of Sciences and Arts, Ljubljana. Upon her retirement she became a free-lance translator of Old-French poetry. She is holder of several highest national awards for her literary lifework. Younger sister Andreja (born Kiauta) Grošelj (born 1945) completed the study of piano at the High-School of Music in Ljubljana and graduated in Classical Languages and Cultures (Latin, Greek) at the University of Ljubljana. She began her career as a Lector in Latin and Greek at the University of Ljubljana and is now retired.

Fig. 1.3. The Kiauta family photograph from Piran, Mediterranean coast, Slovenia; taken by Milan Kiauta, on 21 August 1956. – From left to right, front row: sister Andreja (born 1945, later a classical philologist), father Milan (1909-1987, lawyerand barrister), mother Marija (1911-2005, Romanist & classical philologist), brother Thomas (born 1948, later a neuro surgeon); standing: Bastiaan and his sister Marija (born 1939, later a Romanist).- (Source: B. Kiauta archive)

Dr Kiauta is married to Maria Antonetta Johanna Elisabeth Brink (born 1948), a biologist (University of Utrecht). Since 1971 she helped him as his assistant in the Department of Genetics, University of Utrecht, and is now retired. Additionally, she studied Sanskrit and Oriental Languages and Cultures (University of Utrecht) and Tibetology (University of Leiden). She is also a noted haiku poet and the author of 5 books in this field. As a strong helping hand to Dr Kiauta in his various undertakings, her help in the editorial matters of *Odonatologica* and in the management of SIO during 1972-2013 could not be exaggerated. She has a mother's heart for animal care and, since 2008, she is running an asylum for old and discarded animals (horses, goats, rabbits etc. (*cf.* SITAR, 2012a).

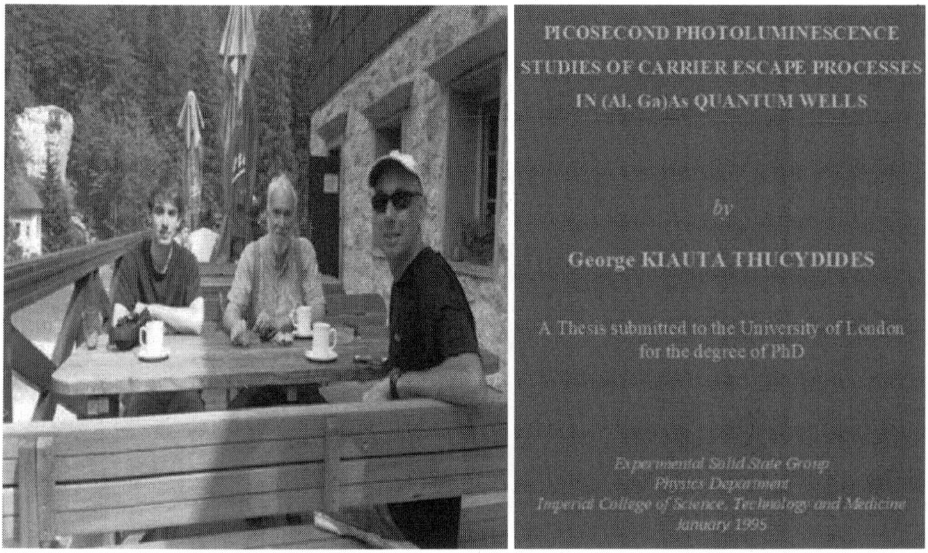

(a) **(b)**

Fig. 1.4. (a-b):(a) A photograph of part of B. Kiauta's family. From left to right: Nicholas Kiauta, the oldest grandson, aged 18, now a student of Mathematics and Philosophy of the University of British Columbia in Vancouver, Canada.; – Prof.Dr B. Kiauta.; – his son, Adjunct Professor of Electronics Dr George Kiauta, of the Simon Fraser University of Vancouver. – [Photo: Jure Javoršek, July 2016]; (b)Title page of doctoral dissertation (1995) of George Kiauta-Thucydides, the son of Dr B. Kiauta. - (Source: B. Kiauta archive)

Dr Kiauta has a son, George Kiauta, who graduated in Theoretical Physics at the University of Oxford (United Kingdom), and later earned his PhD (1995) from the University of London on the dissertation titled "*Picosecond photoluminescence studies of carrier escape processes in (Al, Ga) as quantum wells*" (Fig. 1.4). He was first employed as a Scientific Collaborator with the Experimental Solid State Group, Physics Department, Imperial College of Science, Technology and Medicine, University of London, and now is the Principal Engineer with Nokia, Vancouver (Canada) and Adjunct Professor of Electronics, Simon Fraser University of Vancouver. He is living in Vancouver (Canada), with his wife Colette and 3 sons: Nicholas (born 1998), Anthony (2000) and Alexander (2001), all of whom live in Vancouver (Canada).

In Ljubljana, the Kiautas had a two-storied house in an open yet much vegetated area. It was commonplace to hear the melodious bird chirpings in the continuous ruffles of foliage of tall trees surrounding the house. One could always also see a number of butterflies, and a few dragonflies, in the sprawling lawns that naturally boasted of many water bodies and flowering plants. Rarely though a dragonfly would also defy its natural outdoor flight and make a strange guest inside the house. Mother Marija would love to see the elegant dragonfly quietly perching on wall or some object in the room. The dragonfly would then make a few short movements in the room and, with a stunning speed, fly out of the window. She was so enchanted with the beautiful dragonfly that she would without fail daily keep the windows open so that the unusual guest, a dragonfly, could come in! Bastiaan Kiauta was born here (Fig. 1.5).

Fig. 1.5 The parental home in Ljubljana, Slovenia, where Bastiaan was born and spent his early childhood.- (Source: B. Kiauta archive)

In the times of his childhood and early youth, Ljubljana was a small city, of Austrian appearance and of Central European spirit (Fig. 1.6). In the city centre prevailed the architecture of the century transition period. The municipal borders are continuously changing but at that time the city had a population of some 80,000 inhabitants at most, possibly less. Among these, there were some old Austro-Hungarian "Austrians", all speaking Slovenian, and some Russian immigrants from the October Revolution times. They mostly stuck together, missed the command of Slovenian and their colloquial language was mostly French. The University had a few foreign professors and some foreign instructors (lectors) of various languages. The immigration of settlers from the Balkans commenced not until some years after WW-II. On the street, the Ljubljana slang generally prevailed. In a "decent society" it was considered unpolished and its use in the school was of course forbidden.

Fig. 1.6. Ljubljana during early 20th century (ljubljanapastandpresent.blogspot.com; accessed 31st Oct., 2016).

With its numerous monumental buildings in various classical styles, such as, *e.g.*, the University, Academy of Sciences and Arts, Philharmonic Society, Opera, National Library, National Museum, various churches, gymnasiums and many more, Ljubljana was at that time often called "small Vienna", since there was hardly any other city that could exhibit such a diverse and prolific academic prosperity to be able to compare.

Bastiaan's parents inoculated into their children early in life the discipline of experience hard earned from daily happenings, however trifle in nature. After all, practical wisdom is only to be learnt in the school of experience. Precepts and instructions are useful as far as they go, but, without the discipline of real life, they remain the nature of theory only. To be worth anything, character must be capable of standing firm upon its feet in the world of daily work, temptation and trial, and able to bear the wear-and-tear of actual life. Bastiaan grew up in such an open atmosphere of free thinking and discussion.

As a child, Bastiaan strolled about the surroundings, with parents, afoot, later also by tram and railway. There were hardly any personal cars and lorries were utmost scarce. Somewhere under the skyscraper, there was parked a single taxi, whereas some classical "Viennese" coaches, with their "fjakars" dressed up in black and with silk hats, were parked in front of the Central Railway Station, near the Franciscan church and in the skyscraper area. The tram was the sole means of public transportation, horse spans represented almost the sole means of goods transport. Some services, such as firewood sawyers, vegetable transporters and street organ grinders, used smaller cars, with pony or mule spans. There were very few bicycles, only one of Bastiaan's classmates possessed (a junior one). Smaller burdens were transported on wheel-carts, rarely on bicycles. For assistance with heavy burdens, travellers' luggage and when moving from one home to a new address, the citizens used to call on porters, stationed near the Central Railway Station and near the Central Post Office. They had gray uniforms and caps with a red ribbon, showing their registration number.

Everywhere in the city centre one could find the professional shoe blacks, who cleaned the shoes within a few minutes to an incredible brightness and without taking them off the client's foot. Professionally cleaned shoes resisted for some time the humidity, dust and mud. Even as a primary school child Bastiaan often made use of this service, after returning from his "insect field work". His mother paid for it. Bastiaan thus learnt at a very early age the value of work and the respect to it. No work, after all, is either small or big, if carried out with dignity, honesty and dedication.

In the capacity of park wardens, the Tivoli Park Management contracted a number of pensioners. They did not wear a service uniform, but had a distinct cap, with a green ribbon. In case of a trespass, they friendly warned the parents, but reprimanded the children strongly and severely. They did not fine, nevertheless the children had respect and were frightened of them, avoiding a confrontation whenever possible. They knew all. Walking in the grass, particularly during the hay-making season and, generally, walking outside paths was forbidden. A sole exception was made in the autumn, when it was allowed to collect horse-chestnuts (even by climbing up on the trees), used by housewives for preparing a substitute for soap, which was during the war not readily available in the store and then only on consumer cards. Walking with butterfly net was drawing wardens' attention, as a sign of threatening transgression.

In Tivoli, only (small) junior bicycles could be used, all other mechanized traffic was forbidden, except in the Lattermann Avenue up to the Sleighing Grounds and up to the Tivoli Castle. The Park Management had a smaller cart with two donkeys for their own use.

On Tivoli pond, boating was practiced in summer and skating in the winter. There was no fishing, but the wardens did not object to the capture of a small fish, a mussel, a few insect larvae or any other small animals for aquarium. The vegetation on the banks was sparse, the water could be accessed everywhere easily. Due to the boating, there was no floating vegetation. Prior to the construction of Children Playgrounds (1943), a wet meadow extended on the eastern side of the pond.

Walking and insect collecting was completely free anywhere in the Tivoli woods, on the slopes of Rožnik, Golovec, Šišenski and Ljubljana Castle hills. The Ljubljanica river could be mostly hardly approached and save for the higher animals the collecting there and in all other streams was free. The smaller streams had natural courses.

●●●

2

The Under School-Age Period and the World War-II Time (1941-1945)

"The childhood shows the man,
As morning shows the day."
John Milton, 1671

Home is the first and most important school of character. It is there that every human being receives his best moral training. It is there that one imbibes those principles of conduct which endure through manhood, and cease only with life. It is mainly in the home that the heart is opened, the habits are formed, the intellect is awakened, and character moulded for good! The training of any man can not fail to be powerfully influenced by the moral surroundings of his early years. The child's character is the nucleus of the man's; all after-education is but superposition; the form of the crystal remains the same. Bastiaan's parents not only inoculated into him a strong sense of humility, honesty and hard work but also an indomitable power of endurance and patience, which he nurtured all along his later life.

Bastiaan's father, Milan Kiauta, was of Friulian (Rheto-Romanic) extraction and moved upon the completion of gymnasium in Kranj, where one of his schoolmates was the later Nature Conservancy Lawyer, Dr Tone Dremelj, to Ljubljana, with the objective to matriculate in Medical Faculty. Since he could not expect any financial support from home, he first had to find a job with the insurance bank "Slavija". He understood immediately that a regular job could not be combined with medical studies, therefore he matriculated instead in Biology; botany attracted him already during the high school times. The leading professors at that time were Dr Fran Jesenko (botany) and Dr Jovan Hadži (zoology). Despite his concluding the first year very successfully, probably with exceptionally good results, he had to discover that due to compulsory laboratory training and the obligatory field trips (particularly botanical ones), biology, too, could not be combined with the obligations of a regular job. After a careful consideration and upon the advice of the Head of his Slavija department, Dr Mikiška (Life Insurances), he decided to switch over to the Faculty of Law. Initially the subject did not attract him very much, but the order that prevailed there fitted very well his character and his natural inclination. The contacts with the two Biology professors and with most of his former biology colleagues he cultivated throughout his life and even as a student he gained the friendship of Professor Fran Kidrič, the famous historian of the older Slovenian literature. His interests in nature, mountaineering and also generally in history he mediated also to Bastiaan.

Bastiaan's mother, Marija (born Stupica), had in her paternal line the roots in the ancient Slovenian farmer tradition, though born in a merchant and fast estate owner family of her time. Graduated in Romanic languages and literatures, she studied in Ljubljana and in Paris and she was lifelong actively working in late medieval Latin, but she never had an official job. All her life she devoted to her family, as it was habitual for married wives and mothers of her time. She had an intimate and very special relationship with nature and knew much on animals and plants, which Bastiaan did never hear at University. Bastiaan's love for plants and animals thus sprang up much earlier at a tender age of four with the teaching of mother at home, reminding me of what George Herbert said, *"One good mother is worth a hundred schoolmasters!"* With her, it was always a combination of knowledge and love for the subject. Her power over the abysmal looking subjects was amazingly absolute, and he looked up to her for everything. Therewith she exercised much influence on Bastiaan and she was always understanding and tolerant to all of Bastiaan's nature escapades and the animals he was bringing home. Despite the mess and extra work this caused her, she never pulled faces about that.

Fig. 2.1. In the Rožna dolina Garden, with Marija Kozamernik, the later Head of the Microbiology Laboratory of the Golnik Hospital; 10th May, 1942 (Photograph by Milan Kiauta, Ljubljana). (Source: B. Kiauta archive)

When still a child, Bastiaan's father, Milan Kiauta, taught him how to identify the plants, using the 1941 edition of Dr Angela Piskernik's "*Key for identification of the flora of Slovenia.*" Butterflies, beetles and other insects Bastiaan was identifying with Schwentner's picture booklets, since in the early 1940s nothing else was available in this field in bookstores. The first flower he identified "on his own" was *Forsythia.* He could not read at that time as yet, therefore father read text of the key and Bastiaan had to decide on the alternative – after father's explanation of diagnostic characters to him. This remarkable early home teachings put the seed of love in Bastiaan's mind for plants and insects.

More than by plants, however, Bastiaan was attracted by the insects, which were in abundance in his surroundings and fascinated him with their flight and colours. This must have been around 1942.

Bastiaan's mother was daily conducting his two years younger sister, Marija, and himself to the Tivoli Park (Fig. 2.1). There they were often encountering the long ago retired Slovenian classical author, Dr Ivan Pregelj, who was in his younger years teaching in gymnasium first his father in Kranj and later also his mother in Ljubljana. At almost every encounter Pregelj was telling Bastiaan on the life of spiders and occasionally demonstrating him, among other things, their descending on the threat they produce themselves. Rarely he had also destroyed an animal with a match (cigar-lighter?). Bastiaan was certainly younger than five years then, but the details from his narrations he (now an octogenarian!) still recollects vividly, though the spiders were never attracting his particular attention. He could never ascertain what had Pregelj to do with them; there are no cues in his literary works and his daughter, Dr Bazilija Pregelj, does not make any reference to spiders in her memorial work. Bastiaan's father also never referred to this, either. Nevertheless, it is perhaps due to Dr Pregelj that Bastiaan never saw in spiders anything "bad" and, at variance with many other children, he was never frightened of them, which reminds me of the famous maxim by Dr A.P.J. Abdul Kalam, renowned rocket scientist and the President of India:

"Learning gives creativity
Creativity leads to thinking
Thinking provides knowledge
Knowledge makes you great"

The Second World War had by this time intensified much and its catastrophic impact on the living of civilians was becoming clearly discernible. Bastiaan's world was small; it was closed by barbed wire fence, encircling Ljubljana. The impassable border between Italy and Germany ran close near Podutik and on the Sava river bridge in Ježica. Tivoli park represented almost all of the "free nature" within his reach and until he was about 10 years old it was attainable solely in the company of his parents, mostly the mother.

The first war years did not bring young Bastiaan much emotion and did not interfere with his "natural science" interests. Despite the introduction of consumer cards, they did not suffer a shortage of anything. The far-seeing father had taken care also of acquisition of adequate stocks of cloth and leather and, above all, of a very substantial store of school necessities, that served them well also during the first post-war years. Bastiaan was particularly disturbed by frequent street blockades, followed by domiciliary search, conducted by the Italian army. This always passed without fatal consequences for his family, though father occasionally (more or less friendly) disputed with one or the other Italian Officer the denial of national rights to the Friulian nation in Italy. Mother kept the conversation mostly centred on her own and other children, to which many Italian soldiers were rather sensitive. Once an Italian policeman ("carabiniere") told her frankly that pointing to Bastiaan and his sister, he has similar children at home and he was rather forced to serve so far away from his family. By and large, the policemen showed utmost respect to the family.

With the onset of the 1943/44 winter and with the German occupation, the war painfully hit also Ljubljana. The German soldiery was much harder and more brutal than the Italians, the previously incidental air-raid warnings turned to become daily and the probability of actual bombing attacks became realistic. The school authorities responded to this development by changing and shortening the hours of daily lessons. These started at 7 am, whereafter, Bastiaan and his sister immediately went with mother or both parents either to Tivoli or to their garden in Rožna dolina, both localities at some distance from the city. Around 10 am roared the first air-raid warning, wherewith the walking on the street was prohibited. Ljubljana experienced actually but a few bombings. Heavy air-cannonading of railway station and, above all, the catastrophic bombing of Vič and Trnovo, resulted in a tremendous devastation and in numerous civilian casualties, also among persons known to Bastiaan. His grandfather hardly escaped death. The lifted dust covered the sun and the fire roared all over the place. Bastiaan passed the bombing in the bomb-proof shelter of an inn ("Pri Katrci"), close to their garden. This shattered his nerves greatly. In consequence, whenever they were at home, he ran to the shelter in the cellar at the first sound of an air-raid warning, taking with him the teddy-bear, Sivček (pron. "Sywchek" meaning "The little grey") and, if he had the time, he grasped also the book on cock-chafers (May bugs) and the set and glass-boxed moth *Actias selene*, a gift from his parents. Sivček is still living with him and also the two entomological objects he still has.

The two years of German occupation were hard also from the economic point of view and from that of personal security, but even worse were the first post-war years. All shops and private enterprizes were seized by the (communist) State and, as a rule, their owners jailed. Every moment somebody disappeared in prison or escaped abroad, also among Bastiaan's schoolmates and their families. With the closing down of Slavija Insurance Co., father left without work for some time and mother had to work miracles to be able to bring adequate dishes daily on the table. In Tivoli, the traditional wardens were replaced by militia-men and by partly Serbian speaking soldiers, who, by twos and also on horseback, supervised the visitors. Everywhere, even on children's playing grounds, there lay cast off weapons and ammunition. Consumer cards remained in use for a long time after the "liberation", the number and kind of goods in the stores were limited and many things Bastiaan needed for his "entomology" became either unavailable or they were manufactured as rudely as to be useless. As the oldest and fast growing child Bastiaan also soon started feeling the shortage of clothes, but he did not mind that.

From about 1941, these were the years of Bastiaan's ever increasing interest in nature; since 1943 almost exclusively in entomology: in Tivoli, on the Rožnik and Šišenski hill slopes and in Rožna dolina. In Tivoli they were often accompanied by two of his mother's lady-friends from her university years, with their seven children. Among these was Michael Poljanec, a later agricultural science man and meritorious for his marking of mountain paths for the Slovenian Alpine Society. The two of them strolled day after day about Tivoli and the adjacent woods, catching land- and aquatic insects, searching for bumblebee colonies and much more.

During that time Bastiaan was also attracted by rocks. In Ljubljana there could not be found in the nature much else than various forms of Gröden sandstone, a flint and perhaps some interesting stone in the road rubble. He was also familiar with conglomerate, as used for the rimming of garden paths. His grandmother ("pama"), Irma Stupica, arranged with the stone-cutter firm, "Kunovar", in Žale that he was allowed to search among their rubbish for any stones he liked. There he enriched his collection by various ornamental rocks, such as Podpeč limestone, dolomit and tuf, a number of foreign marmors and by numerous igneous rocks, e.g., porphyry, granite, basalt and tonalite. In school he acquired by bargaining from two schoolmates pieces of pyrite, limonite, barite, a large calcite crystal and some more minerals. In exchange Bastiaan was giving small, coloured bulbs, as used for illumination of Christmas little crib, which he bought from a shop close to his home – from the pocket-money he had earned by dish-drying. The minerals were identified by his older cousin, Zvone Bahar, then already a student of chemistry. He also possessed Žurga's book, *From the life of stones* (1935). Bastiaan's fascination by limestone rocks and by the Sava pebble-stones grew enormously when father took him for the first time (1945) to the Šmarna Gora hills. Soon after conclusion of the primary school, his interest in "mineralogy" rather decreased. It was the "entomology" that was left over.

From about the age of four (1941) Bastiaan was much interested in butterflies and moths, but also in dragonflies, beetles and grasshoppers. He was not building a collection of specimens, because he did not know (neither he was thinking about) how to kill and prepare them for the cabinet, though he was familiar with insect collections in the National Natural History Museum. He was using a junior butterfly net as available in some toyshops. With father's help he was identifying them, based on the respective Schwentner booklets. Some butterflies and moths they also bred out from the incidentally found caterpillars. Among these were *Pieris brassicae*, *Aglais urticae*, *Inoachis io* and also *Aceronthia atropos* and *Saturnia pyri,* but at that time he knew only their Slovenian vernacular names. Also some dragonflies (for example *Coenagrion* spp., *Aeshna* sp. and *Libellula depressa*) emerged from larvae kept with some vegetation in large glass jars. He did not feed them, as he did not know what they eat. Initially he did not recognize the larvae as a dragonfly – until the adults emerged!

Saturnia pyri was a surprise. Bastiaan found its huge and exceptionally beautiful caterpillar on a fruit tree. It soon made a cocoon and remained in the chrysalis stage for months, throughout the winter. The large moth emerged suddenly in the spring. For an unexpected surprise was good also *Aceronthia atropos*, whose caterpillar he found in the potato field. When it transformed, the chrysalis bent and squealed on touch. He put it into a large box on a cupboard in his parents' bedroom, where it emerged after some time. Homebred butterflies, moths and dragonflies he was always setting free in the garden. Most of them took off immediately, but some dragonflies did not. These he was hanging up on bush leaves. Thus began a new era of interaction between Bastiaan and dragonflies, which has survived a long, tumultuous seventy years' span! The seed of an outstanding future research on dragonflies was sown, just as Plato, the great Greek philosopher and thinker, professed: "*The beginning is the most important part of the work.*"

●●●

3

In the "Entomology School" of Uncle Cyril Ažman (1943-1952)

"Might I give counsel to any young man, I would say to him, try to frequent the company of your betters. In books and in life, that is the most wholesome society; learn to admire rightly; the great pleasure of life is that. Note what great men admired; they admired great things; narrow spirits admire basely and worship meanly."

—W. M. Thackeray

Bastiaan's "entomological destiny" was decided upon in 1943, at the age of six, when his grand-uncle Professor Dr Cyril Ažman moved from Novo mesto to Ljubljana (Fig. 3.1). Cyril Ažman was born on the 11th July, 1881 in Šentjurij-pod-Kumom (Slovenia), in a large family with 11 children. Father Jacob Ažman (1847-1920) was a tax accountant by profession. Cyril attended primary school in Kamnik (Slovenia) and high school in Ljubljana, where he lived in Alojzijevišče. He studied at the Papal University in Rome and stayed in the Austro-Hungarian Institute,"Collegium germanicum". There on 24th July, 1905 he completed his Ph.D. at the Faculty of Philosophy and on 16th July, 1909 at the Faculty of Theology. Dr. Ažman was sharp in his erudition, interlocutor and knowledge of technical details. Although rigorous, he had a great sense of humor and was very popular and everyone who knew him, respected him. He was a naturalist of the classical stamp, a thorough expert on almost the entire Slovenian fauna and flora, had rich entomological collections and occupied himself also with the study of biology of numerous animals, above all insects and birds. In his unusually large library he kept almost complete Slovenian biological literature, from Scopoli onwards, including the reprints and typewritten transcripts of related papers, published in various periodicals, at home and abroad, and a huge collection of foreign scientific books, many of them purchased from the Imperial Austrian Library. Due to his heart disease he could not easily walk, therefore, he was spending most of the summer days in the garden (where he came by coach) and later in the garden of the Franciscan monastery. Initially Bastiaan's grand-uncle was living with his sister, Irma Stupica, but later, after the "liberation" the authorities dislodged her from her apartment, he found a shelter with the Franciscans in Ljubljana. In the garden, uncle instructed Bastiaan on the names of the insects he was bringing him, was telling on their life and development and on numerous interesting peculiarities.

The evenings Bastiaan was spending with his books and grand-uncle's microscope.

So, Bastiaan knew from grand-uncle that wasps are not really predators, but are rather catching various insects solely to feed their carnivorous larvae. Every time a wasp brings food to a larva, the latter excretes a droplet of a sweet substance, which the wasp sips up. By late summer, when all larvae had emerged, the wasps turn over to any sweet vegetarian food. When later in the season such food sources are gone, the wasps die on starvation before the low temperatures set in. This forced event in the biology of wasps left an indelible imprint on the young mind of Bastiaan and he was greatly moved by the thought of hunger and consequent death of the unfortunate insects. The result was: until to-day, both Mrs and Prof. Bastiaan Kiauta are keeping on a garden table a cup with some fruit, honey and jam – to feed the starving wasps and hornets. There are always many of them about the table; they often fight each other, but they are not stinging. Both of them were never stung by a single wasp or hornet. They do so solely when irritated by an attempt to drive them away.

Fig. 3.1. Professor Dr Cyril Ažman (Original source: Archive B. Kiauta, photo taken in 1944; supplied courtesy Matjaž Bedjanić, Editor of Erjaveciawhere it was published in 2005).

Bastiaan's junior butterfly net soon appeared to be inadequate and uncle Cyril presented him with the frame of a professional, folding entomological net, a product of the Viennese firm Wagner & Wenger. Following his instructions and after the model shown in the *Slovenian butterflier* of J. Bučar, his mother stitched a net from white mosquito-net material. Bastiaan then also received a killing jar from uncle, and ether and chloroform by father from an apothecary's shop. Bastiaan soon had many stretching boards, manufactured by grand-uncle Rado (Jakob) Ažman, at his disposal, while father made a few insectaries, as used for the breeding of caterpillars. The instrument for the preparation of these Bastiaan did never possess.

In addition to various other entomological utensils, uncle Cyril presented him also with four highly important entomological works, viz., in mid-1940s, *Slovenian butterflier*, by J. Bučar, soon afterwards: *Die Grossschmetterlinge und Raupen Mitteleuropas*, by K. Lampert and *Fauna germanica die Käfer* (5 vols), by E. Reitter and finally, in 1952: *Die Geradflügler Mitteleuropas*, by R. Tümpel. In this way he was well provided with literature for the identification of all insect orders he was interested in – and much more. Of course, uncle Cyril kept communicating to him also personally most of the specific names of butterflies, moths, beetles, dragonflies and grasshoppers. As far as he had in the primary school and lower gymnasium some difficulties with German technical terms, uncle and both parents, occasionally also grandmother and aunt Marjana Korber helped him.

From 1943 on Bastiaan commenced gathering a (small) cabinet collection. Father helped him with the preparation of butterfly, moth and dragonfly specimens. He had at most anything like 60 butterfly and moth species, perhaps some 30 beetles, at least some 20 dragonfly species, possibly more, pertaining to the genera *Calopteryx*, "*Agrion*" (all Coenagrionidae), *Platycnemis*, *Lestes*, "*Aeschna*", *Anax*, "*Gomphus*", *Cordulegaster*, "*Cordulia*" and *Libellula,* and about 10 grasshopper taxa, some of the latter in penultimate instars only. A few dragonfly larvae he also kept in alcohol.

The post-mortem darkening in many dragonflies was of course a disappointment. On uncle's suggestion Bastiaan had tried, with father's help, remedy by inserting in the abdomens some cottonwool, impregnated in boric acid, but the improvement was slight. The acetoning, that gives excellent results, he started not until some three decades later.

For the sex of insects Bastiaan did not bother much and he was unable to discern it either, unless in cases of a pronounced sexual dimorphism, such as, for example, in demoiselle dragonflies *Calopteryx* and in brimatone butterfly *Genopteryx rhamni* (by wing colour), in some *Pieris* whites (by wing spots), in broad-bodied chaser dragonfly *Libellula depressa* (by abdomen colour), in saturniid moths (by antennae), in some grasshoppers (by presence of ovipositor) and so on. Males and females they called in ancient Slovenian as "mandelci" and "babce" – the expressions that do not appear in current dictionaries.

During the war years Bastiaan often accompanied his grandmother to the gardener Franc Juvan in the Botanic Gardens and to the Herzmanski Gardening on the Glinščica stream, where she was ordering plants for her garden. On such occasion he caught in the Botanic Gardens, quite incidentally and surprisingly easily, a large brown dragonfly. He took it to uncle Cyril and he told him, it was the brown hawker (*Aeshna grandis*), one of Slovenian largest dragonflies and the only one having brown body and wings. During the war time he could not encounter this species again nor catch it. He must have collected a mélange of species on the Glinščica stream, but *Calopteryx* was most memorable from the Ljubljanica and Gradaščica rivers even though he had never seen those in Tivoli or in Rožna dolina.

Initially, in uncle's "school", Bastiaan was most interested in butterflies, which he was collecting mostly in Tivoli and its environs, first invariably in parents' company. He knew night-active moths only as far as he found them during the day at lamps.

When uncle Cyril had to move to the Franciscans, where Bastiaan was visiting with him almost daily, he became aware of their large garden, situated in the centre of the city and hidden away by the buildings along the Čopova and Nazorjeva streets. The garden had no open water and it seems it served in the first place as breeding grounds for flowers used in church decoration. At that time all flower-beds were richly rimmed by phlox and garden tobacco, along the walls there were dozens of large pots with blooming oleanders and in the middle of the garden liliac bushes formed a shady place with a table and bank. There they were often sitting in evening twilight, equipped with field glasses and watching the sphingid moths on the flowers. There were plenty of them, referable to at least five species. Bastiaan was not supposed to catch them as, during many days, uncle, using a light meter, was systematically recording the time and illumination intensity at which each species arrived, the period it stayed, and noting down their particular behaviour. Once they noticed the appearance of a *Daphne nerii*, which identity uncle recognized right in the air, but apparently he did not trust himself and ordered Bastiaan to catch it. During that summer this moth appeared several times and, following uncle's suggestion, he had tried to find its caterpillar in the oleanders and was assisted in this by the old Franciscan sacristan, Tone, but they failed.

Not until the first dragonflies emerged from a glass jar and from a larger aquarium, cast in a single piece and which originally served as the housing of an accumulator, Bastiaan perceived the connection between such a larva and the dragonfly. Later uncle gave numerous useful hints on dragonfly home-breeding, such as, e.g., that larvae survive a short transport better if kept in a jar with moist vegetation rather than in water; that animals of different sizes should be kept separate, as otherwise the larger larvae often attack, damage or even eat the smaller ones; that *Calopteryx* should be bred in jars containing *Fontinalis* moss since otherwise they lack the required oxygen; that some species prefer a stony substrate for emergence rather than plants and which should be improvised in aquarium by pieces of tile standing up out of water; that some species are easier to be collected standing in water, turned towards the river bank, using a sieve rather than an aquatic net. These and many other of his directions were very useful when Bastiaan had later specialized in dragonflies.

Bastiaan's entomology "school" with uncle Cyril lasted until his death in 1952. When he died, Bastiaan "inherited" everything what was left of his fishing, aquaristic, entomological and other biological equipment, a splendid Reichart microscope with numerous accessory instruments, implements and a collection of aniline dyes, some minor laboratory equipment, a minor telescope, a microtome anything like two shelf meters of biological books, other biological literature and (Slovenian) natural history publications.

Notably, upon the death of the celebrated lepidopterologist, Julius Bučar (1857-1919), uncle Cyril had purchased his entomological inventory, which included also his handwritten field diary. When Bastiaan departed to The Netherlands (1962), he deposited the document with Dr Jan Carnelutti. Mid-1990s Bastiaan approached him in this matter, but unfortunately he was unable to find it. Consequently, Bastiaan had requested his successor in the institute of Biology of the Academy of Sciences, Tatjana Čelik, to try to trace it, but so far she was not successful either. Uncle Cyril's diary possibly contained also some information on dragonflies and on some other Lower Carniolan insects. Bastiaan, though disappointed with the aparrent loss, did not lose his heart entirely and kept patiently pursuing it, just as John Ruskin, the leading English art critic of the Victorian era as well as a prominent social thinker and philanthropist, has once said, "*Patience is the finest and worthiest part of fortitude, and the rarest too . . . Patience lies at the root of all pleasures, as well as of all powers. Hope herself ceases to be happiness when Impatience companions her.*"

●●●

4

Period of the Early Friendship with Mathias Gogala (1943-1953)

"Who finds a friend finds a treasure"
— (an Italian adventure movie from Hollywood by the same title directed in 1981 by Sergio Corbucci)

In no lesser extent than by the "school" of uncle Cyril, Bastiaan's "entomological profile" was shaped by the friendship and comradeship with Mathias Gogala, called Matjaž (pronounced "Matyazh"), the later Professor of Invertebrate Physiology and member of the Slovenian Academy of Sciences and Arts. They met in 1943/44, as schoolmates in the First primary school form at "Vadnica" which was not an exclusive school for the children of "privileged classes", as insisted by the communist authorities after the war. It was rather the exercise establishment for the students of Teachers' College, located in the same building. Therefore, their teachers were exceptionally qualified and everyone was specialized for giving the instruction in a certain form. Most of his schoolmates were living in the city centre and their parents practiced in a variety of professions; some children came also from working class families. In the first form their instructor was Rudolf Završnik, an elder gentleman, who taught them writing the words in a single stroke and demanded to use in the penholder a pen with a small "head", resulting in the characters of uniform thickness throughout. In this way their script differed from that taught in other schools, where one or two strokes are usual per character and the downward lines are thicker. The specialist for the second form was Viljam Deržaj, who desired to be called as "mister class-master" instead of the usual "mister teacher". He definitely had a much higher education than usually the teachers have, probably in the direction of techniques or physics. In the last but one war year he ventured to keep stressing continuously the importance of patriotism and of mother's tongue. He dictated them some sentences in this spirit, which they had to write down in their school "diary" and provide the text with national ornaments. His instruction programme included a day devoted to beetles. The dictated text Bastiaan wrote in his "diary" and father helped him to illustrate it with the pictures cut out from a duplicate of the above mentioned Schwentner's booklet. On that day Bastiaan brought to school also some living beetles and mister class-master knew all the species. The "Diary" Bastiaan still possesses.

By 1946/47, the authorities started the reorganization of the school, the teachers were successively dismissed and replaced mostly by partisan females, with little or no formal qualification. Their title was "comrade", but the pupils continued calling them "misses". In the fourth form they had a certain Mrs Mara Burja, a very warm person, who treated them like a mother her children. Some of them, certainly Bastiaan, "fell in love" with her. For a school outing she took them to Sveti Urh, a place known for war atrocities, hence one of the compulsory destinations for school excursions. There Bastiaan was of course in the first place catching butterflies and other insects and she kindly feigned a great interest in these. Indeed she knew also the names of some species, while for some others she used appellations different from those Bastiaan knew.

Once a month Bastiaan and fellow pupils were guests in an adjacent classroom, where the instruction was given by a certain Mrs Milica Bartenev, an elder lady and an unparalleled artist in telling fairy tales. He was most impressed by her presentation of the *Adventures of Maya, the bee.* Her animals spoke each with its own voice and the mode of her presentation (without the book) was unique. Even today, whenever he reads the Slovenian edition of Bönsels's book, which was during the German occupation on the list of forbidden reading, Bastiaan hears Mrs Bartenev's voice ringing in his ears. In this way also she strengthened his love for insects.

Bastiaan soon realized that both he and Matjaž had some common interest in nature. At that time he was six years old, but Matjaž did not reach the age of six as yet. Matjaž was in the first place attracted by beetles he was encountering in their small garden somewhere in the outskirts, but where Bastiaan had never visited. He was also an aquarist, whereas his two or three years senior sister, Alenka, called Lenča (pronounced "Lencha"), was more interested in botany. Later she studied piano and became a noted pianist. Lenča and her girl-friends, Mira Avsenik (later a biologist) and Živka ("Zhivka"), often accompanied them on their "biological outings" in the nearby woods.

In the beginning Matjaž and Bastiaan were of course too young to be allowed to go hunting on their own. After school time Bastiaan often accompanied him to his home in Truberjeva street. He was still so small that he had to step on his satchel to reach the doorbell. They regularly visited with each other, inspecting the small beetle collection Matjaž had, discussing on insects and other animals, watching the life in his aquarium and running over the few nature books they had.

Under Matjaž's influence and also under that of the children's book on cockchaffers (*Melolontha melolontha*), by Edo Daržaj, the brother of their teacher in the second form, Bastiaan became more interested in beetles and started to collect them also. Preparing them for the cabinet, they had difficulties with pins. They mostly used taylor's pins, the porcelain heads of which he was removing, but Matjaž did not. Matjaž, however, had a beautiful *Polyphylla fullo* specimen, which species Bastiaan had never seen before. It was pinned on a pin bearing a big yellow head which disturbed Bastiaan aesthetically so much that that specimen remained forever impressed in his memory.

There is seemingly a marked difference in insect and also bird fauna of Ljubljana during Bastiaan's childhood and now, although the butterfly *Iphichlides podalirius* was rare in those days also! In his cabinet Bastiaan had a single specimen caught in Tivoli – in an effort of a whole day and not without the "complications" from the side of a warden. This adventure seemed to him interesting enough to merit a "literary" description. He wrote the story calligraphically in a script of some ten pages and sent it to the daily newspaper, *Slovenski poročevalec.* He never received a response and it was of course not published. In those times, stagbeetles (*Lucanus cervus*) occurred commonly in the woods around Ljubljana. Following uncle Cyril's suggestion, he cut early in spring a moderate wound in a young oaktree. Next summer he visited with Matjaž the tree several times and they were astonished to find up to a dozen of stagbeetles of both sexes and very different sizes per visit, sucking the excreted "pitch". Once was dwelling there also the butterfly *Nymphalis antiopa*, the only representative of this species he had ever seen in Ljubljana.

During the night they were not supposed to stay either in Tivoli or in the garden in Rožna dolina, therefore, their knowledge about night-active insects was limited to those found during the day at some street – or Tivoli lantern. Among these, there was rather common the Japanese silkmoth (*Antherea yamamai*). Brought home, its females often laid eggs from which Bastiaan had several times tried to breed out the large and in wing coloration astonishingly varied moths. All attempts were in vain, since the eggs hibernate and he was unable to delay their development in the spring and to provide in time a sufficient quantity of room-developed, full-green beech or oak leaves the caterpillars require as food. He has never seen a caterpillar of this species in nature. During the summer vacations in Vir, Bastiaan was attracting numerous species of moths by dried apples, soaked in a mixture of beer and honey, suspended between garden trees and at the forest edge. In Vir he also surveyed with Matjaž all kinds of other insects and they collected dragonflies extensively on the Rača and Radomlja rivers, in the Dob area and in the Želodnik wetlands.

In the first classical college form they had as a teacher of natural history (zoology) a certain Mr Bruno Toič. He occasionally conducted Bastiaan and all on various outings and taught them that small insects and aquatic animals are to be conserved in vinegar. This Bastiaan never did, as he was since long familiar with conserving in alcohol and/or formalin. But he had a problem in finding suitable glassware. Therein he was remedied by Dr Franc Derganc, a former schoolmate of his mother, then a doctor in sanatorium "Emona", located close to their home. In this way arose his small collection of dragonfly larvae, each specimen in a separate vial, all vials put together in air-tight glasses. The specimens were labelled, but mostly not identified.

The journey of dragonfly learning was long and tiresome, but Bastiaan had aparrently dictated his destiny and his great success as an odonatologist was the result mainly of his painstaking labour and diligent application. The pristine love for dragonflies bloomed slowly and solidly in the tender heart of Bastiaan, like a marigold does under the rising Sun, so elegantly echoed in a dragonfly-based Malayali song by Dr P.R. Ramachander, an entomologist, a translator and children story-teller:

"Oh first tiny dragonfly,
Would you come along with me?"
"If I come with you,
What all things would you give me?"
"I would give you place to play,
I would give a pond to take you bath,
I would give golden seat for you to sit,
I would give a golden plate for you to eat,
I would give a silver water pot, to you to wash your hands,
I would give you a dotted silk to dry your hands
Oh first tiny dragon fly,
Would you come along with me?"

●●●

5

Breakthrough into the Community of Adult Entomologists (1949-1952)

"Who would deduce the dragonfly from the larva,
The iris from the bud,
The lawyer from the infant?
...We are all shape-shifters and magical reinventors.
Life is really a plural noun,
A caravan of selves."
— Diane Ackerman

In the 1940s, Matjaž Gogala and Bastiaan did not maintain any contacts with elder entomologists, though Bastiaan was a regular visitor of the National Museum of Natural History where, in 1949, he met Savo Brelih, then a student of biology and employed as a Museum curator of entomology. He helped Bastiaan with the identifications of some carabid and cerambycid beetles and enabled him to use Museum collections, which were otherwise not open to public, as a reference.

It must have been in 1950 when Bastiaan became acquainted with Rudolph Rakovec, by profession a cashier in the Opera, but otherwise indisputably one of the leading Slovenian butterfly authorities of his time and an extraordinarily broad-minded mentor and maecenas of younger generations. In his apartment, in the Vrtača building, he maintained two rooms dedicated to his passion: in one there was housed his enormous collection, while the other served for the breeding. When he had half grown or older caterpillars of the Japanese silkmoth, their nibble of beech leaves could be heard right on the street. He was in touch with numerous colleagues throughout Europe and was able to provide the young entomologists with any quantity of Czech entomological pins. He was also manufacturing for them potassium cyanide killing jars, whose use was much more economical than that of quickly evaporating ether or chloroform.

On conclusion of the fourth class at the Classical College, the members of their (school) Naturalists' Club set up a public exhibit of their work. Matjaž and Bastiaan participated with parts of their insect collections. The exhibit was visited also by a certain pupil of the eighth class of the Poljanska gymnasium, who introduced himself by his nickname, "Gou". His proper name was Štefan Sušec Michieli, later a famous biologist and at that time already an expert on Lepidoptera and on all related matters. This was the beginning of a friendship that opened to Matjaž and to Bastiaan the admission to the circle of established entomologists, rendering them also the possibility of enlisting in the Entomology Section of the Society of Biologists of Slovenia. In such company their work was becoming more "professional".

"Gou" had a room in Trubarjeva street, with entrance over stairs outside the home, from a little side-lane, leading towards the Ljubljanica river. The room was quite large, but almost completely blocked by his Lepidoptera collection and books. Of course Bastiaan was visiting with him often. Aside of entomology, he was erudite also in various other fields of biology. On his bed-side table there was usually a pile of voluminous, to Bastiaan at that time "indigestible" books, mostly in Serbian Cyrillic alphabet, in the fields like biochemistry, physiology, histology, cytology and similar. He declared reading them for "personal orientation". In university, "Gou" was a few years senior to Bastiaan, but he read the same subjects with the same professors as later Bastiaan did, therefore Bastiaan could ascertain their contents were much beyond the requirements of university examinations. "Gou" also read German and English, while French and Italian biological literature caused him no difficulties either. He always spoke composedly and liked talking on his personal discoveries in insects.

The Entomology Section operated like a "Family of kindled spirits". There were about a dozen of students, including Savo Brelih, Jan Carnelutti, Štefan Michieli, Rajko Rakovec (the son of the above mentioned Rudolf, who was working on his doctoral dissertation), the malacologist Jože Bole and others. Among the older members there was the legendary coleopterologist Alfonz Gspan, but he did not attend to the meetings regularly. Also Egon Pretner from Postojna was rarely seen, but in the gatherings often participated also Lili Kobi (later married Istenič), the subsequent Professor of comparative vertebrate anatomy, who gained fame by her studies on cave salamander and on her theory of proteid transition to subterranean waters. At that time she was working on the preparation of her PhD dissertation related to the physiology of *Perla marginata* larvae.

Matjaž and Bastiaan participated in the meetings regularly. These were taking place in evenings, at least once a month, in the Institute of Zoology, then located in the main University building on Kongresni Square. Every time one of the members prepared a talk on his work, some brought along also various problematical or otherwise intriguing insect specimens and/or some interesting publications. Since senior members attended rarely and the younger ones did not have as yet any own publications, there was no distribution of reprints. Nevertheless, Brelih and Carnelutti often brought small stocks of various older publications from Slovenian and also from foreign periodicals and distributed these among those interested. All presentations were followed by a vivid discussion that often contributed essentially to the clarification of some open problems, indicated by the foregoing talk.

Occasionally and outside the regular program, Bastiaan and Matjaž experienced also some incidental attraction. For example, when Brelih brought in little linen bags living adders and vipers, let them crawling on the table, then quickly lifted the snake by the tip of its tale so that it slowly crept up along itself, but before reaching his hand, he dropped it back in the bag again. Everybody was composedly watching these happenings as something self-evident, but there was none among them who would venture to repeat the performance. Sometime later Matjaž learnt this skill as well. Brelih collected the snakes for Drago Lebez, who extracted their venom for the production of antiserum. Lebez, a biochemist, worked also with poisonous sea-anemones and spiders, but he was not a member of the Section.

Also otherwise, Brelih always had plenty of anecdotes to tell. When he had a talk on his mallophage research, he told on the experience he had in the Zoological Garden with she-bear, "Meta". While he was seating on a tripod chair, collecting her "fleas", she suddenly stood up and thanked him kindly for the courtesy by gently caressing, creating him therewith a problem how to get home in thoroughly tattered clothes.

They always concluded the meeting with a fish-and-chip supper at "Šestica", where after, in summer, they usually went to Tivoli, to inspect the moths and other insects swarming about the lamps. On such occasion they once had an encounter with a policeman. He was sitting on a nearby bank, in the shadow, so they could not notice him. Suddenly he turned up at the light and inquired rather friendly, as to what they were doing. He said, he was watching them during half an hour, but apparently they are not pilfering the bulbs, nor damaging the lamps, what else they were about? In response, Michieli and Carnelutti delivered quite a lecture on moths, whereafter he declared that clearly everyone has to learn lifelong. Of course he knew the moths, but had no idea there are so many kinds of them and that anybody could have an interest in them.

During this period Bastiaan expanded considerably his entomological activities also geographically, in the first place to the mountainous regions of Upper Carniola, in part also to Inner- and Lower Carniola and the Karst, but Styria, the Littoral and White Carniola were too far for cycling. Matjaž's mother invited Bastiaan for a short stay in Bohinj, where they possessed a chalet. Three summer vacations he spent with his own family in Bled. He felt quite at home in the Polhograjski Dolomiti hills and, mostly in father's company, he made a number of protracted tours in the Julian and Kamnik Alps. The Alpine fauna was attracting his special attention. But also closer to Ljubljana, with Matjaž, he shifted his attention to the wider environs. They were dwelling during several vacations in Vir, the village from where the family of his mother's father originates, entomologically exploring the large Domžale district and the regions east and northeast of it.

In winter 1952, "Gou" helped Bastiaan with the manuscript of a brief note on Alpine butterflies, which was to appear in *Proteus* (1953; 15/6:160-161). This was his first entomological paper published. The Editor, Dr Lavo Čermelj, was living close to their home and Bastiaan felt a great tension going to him to read the proofs. When some *Proteus* readers mentioned to him they noticed the article, pride made him almost embarrassed. At the same time, otherwise, the pleasure of having his first ever research manuscript published in a scientific journal had no bounds to express and he, in his mind, felt flying into the 'seventh heaven'!

●●●

6

Decision on Dragonflies (1952)

"Life is a matter of choices,
and every choice you make makes you."
— John C. Maxwell

Matjaž and Bastiaan were the youngest members of the Entomology Section. Most of the others were biology students (one also Forestry), some already graduated and a few were older, established entomologists. In the meetings they soon noticed that "everybody works on butterflies or beetles", therefore, in these orders "there is not much left to explore" and it would be useful to select for their "research" different, less researched insect orders. It must have been late in 1952 when, in Bastiaan's tiny room, they were leafing through the F. Kos' Natural History Museum Guidebook (1933), he decided on dragonflies "which appear similar to butterflies" and Matjaž took Heteroptera, "resembling the beetles". Until this very day they remain true to this choice. Therewith the youthful introduction time to entomology arrived, in a way, to its conclusion and more serious attempts with dragonflies and bugs/cicadas were about to commence. Matjaž found very soon a splendid mentor in Anton Moder, the vicar in Dobrova near Ljubljana and also graduated biologist, at that time the only student of the bug fauna of Slovenia. A little while later Bastiaan also got in touch with Dr Živko Adamović of Belgrade, the sole professional odonatologist in Yugoslavia. A hard task master and trainer, though immensely kindhearted and soft spoken, Adamović was always willing to help, with proverbial inclination, his younger colleagues to enter and persist in the world of science. On numerous occasions he acted as a mentor to M.Sc. and Ph.D. candidates, and assisted in preparation and defence of their work related to Odonata and mosquitoes. Invariably he approached all assignments with ceaseless zest, enthusiasm, feelings of responsibility and consciousness of the importance of the undertaking. He conquered the difficulties inherent to the start of a scientific periodical with perseverance and vigour, which his younger colleagues will find hard to emulate. In the entomological and odonatological world, Dr Adamović was known as a profound intellectual, an expert in languages, and an artist painter – a member of the widely known and distinguished group of the "Four Painters of Obrenovac". He was a prolific correspondent with entomologists and odonatologists in numerous countries of all continents, to whom he rendered every kind of professional assistance whenever required. All these qualities, which inevitably are attributable to all great scientists, were later to become the hallmark of Bastiaan's career himself!

Dr Živko Adamović, was born of a craftsman's family in Obrenovac near Belgrade, on 9th March 1923, and senior enough to young Bastiaan in odonatological research experience to extend a good mentoring hand. After a brief employment as a teacher, in 1945, he joined the Serbian Natural History Museum. In 1952 he graduated from the Faculty of Natural Sciences and Mathematics (Biology) in Belgrade. It was during this period that he published his first scientific papers, which mark the beginnings of odonatology in the post-World War II Yugoslavia. But he had vast knowledge on as diverse insect groups as grasshoppers (Orthoptera), Cicindelidae and Cerambycidae (Coleoptera) and particularly also on Asilidae (Diptera) and Apidae (Hymenoptera). Indeed, robber flies were the object of his special interest. When he joined the Institute for Medical Research, at the University of Belgrade, in 1972, he specialized on the Psychodidae and Culicidae (Diptera). His papers on mosquitoes represent an invaluable contribution to our knowledge of the fauna and ecology of these insects in the territory of the former Yugoslavia. Dr Adamović was, as he himself often stressed, primarily an ecologist by vocation, but he was also a gifted taxonomist and biogeographer, and contributed significantly to applied medical entomology as well. A noted museum curator, he described seven taxa new to science (Fig. 6.1).

Fig. 6.1. During a field trip in Pohorje Mts, Slovenia, Dr Adamović with Professor Dr B. Kiauta and Mrs M. Kiauta. — (Photo Source: Dr J. Ott)

Adamović was a gentleman of Victorian tradition; in later years Bastiaan met him several times in person at various international congresses and symposia. As apparent from his first letter, he initially mistook his family name for the first name and addressed Bastiaan in his reply (in Serbian) as "Dear (lady) colleague". He was not in habit of using the expression, "comrade", that was almost compulsory in the communist times. He was fluent in English and had a command of the other congress languages, but as to Slovenian he declared he does not understand a word, either in speech or in writing. The famous adage "Still waters run deep" is possibly reflecting on his wisdom who, like all great scientists, did not eulogise his potentialities, and rather preferred to underplay them and yet they are prolifically interesting and have profound understanding and complex thoughts. Deep knowledge people are always silent like the sea or ocean is irrevocably silent in the middle.

●●●

7

The First Steps in Odonatology (1953)

"Take up one idea. Make that one idea your life – think of it, dream of it, live on that idea. Let the brain, muscles, nerves, every part of your body be full of that idea, and just leave every other idea alone. This is the way to success."

— Swami Vivekananda

Bastiaan's decision to take dragonflies (Odonata) as objects for his future research was a well measured conclusion, although he was still not fully aware of the knowledge perpetuating in many other parts of the world, nor had he by then known much about the works of or corresponded at all with any of the overseas odonatological greats, some of which had earned them a great name and fame, such as, e.g., J.G. Needham (USA), F.C. Fraser (India/UK), A.E. Gardner (UK), Sajiro Makino (Japan), R.M. Gambles and E.C.G. Pinhey (Africa), A.N. Bartenev (Russia) etc. He, however, had the privilege of having heard about several of them and even read some of their research papers and/or books. Sooner though, as providence will endorse, almost all of these and other great odonatologists worldwide would become his associates, colleagues and/or friends for many decades in the blooming times of his brainchild, the Societas Internationalis Odonatologica (S.I.O.), the first ever global odonatological association. Early in his life Bastiaan had discovered the importance of correspondence with fellow researchers and developing an amicable environment for exchange of material, both literary and dragonfly specimens, to extend knowledge worldwide! The road ahead was apparently difficult but could be certainly treaded fruitfully with grit determination, focused attention and persistent hardwork which Bastiaan possessed immeasurably.

Bastiaan initially encountered great difficulty in finding all the relevant research publications he needed. Aside of some notes and minor papers, scattered in a few books and in various foreign periodicals, some of which were not available in Slovenia, the literature in Slovenian dragonflies was meagre. It included, for example, the works of Valvasor (1685, discovered only recently), Scopoli (1763), Lippich (1843), Schweighofer (1895, 1905), Klapalek (1900), Puschnig (1905-1935) and Strobl (1906). He never paused even for a moment and continued to work hard to access literature in all possible forms as much as possible. At that young adolescent age of fifteen he had already understood the far-fetching meaning of the famous maxim: "*It's not about never falling down—it's about getting up every time you fall down*".

The Austrian, Dr Roman Puschnig, was helping Bastiaan with reprints and in various other ways, therefore he very much liked to visit him at his home in the nearby Klagenfurt, but in those days the Yugoslav frontiers were hermetically closed to ordinary citizens and Bastiaan was unable to get the passport. Also Triest, Gorizia, the western valleys of the Julian Alps and all other ethnically Slovenian territories across the border were inaccessible. But when the Regime introduced also a 16 km border zone, a permit was required even for large parts of the Upper Carniola province, though it was provided to the members of the Alpine Society automatically, by a stamp in the membership card. The so called "liberation" of 1945 restricted the citizen's freedom in every possible way, during several decades.

To overcome the initial serious constraint of lack of adequate literature, confounding his endeavours to comprehend the science of odonatology on a global basis, first Bastiaan embarked on a review of the history of odonatological efforts in Slovenia and on inventorization of dragonfly material in the collections of the National Museum of Natural History. In October 1955, Bastiaan presented a report in the Entomology Section of Society of Biologists of Slovenia, titled "*A sketch of historical entomological efforts in Slovenia, with an inventory of the older dragonfly collections*". The preserved handwritten manuscript has 24 pages, but it is missing some information discovered not until decades later. Bastiaan did never prepare it for publication, despite its great historical value. *This work is being appended herein this book, for the benefit of, first, the international odonatological community, secondly, the Slovenian odonatologists' comprehension of their own taxa and history, and lastly, the ultimate security of preserving an invaluable hand-written document, to arouse interest of expert calligraphers to quest deciphering what I call "a phenomenon" of the world's one of the greatest odonatological geniuses ever who all along his life strived hard to preserve the sanctity of his brainchild, the S.I.O. (the International Odonatological Society), as a democratically synchronized "family of friends" rather than a hierarchically constituted association!*(see Appendix 7). This thesis of referring a scientific confrerie a "family of friends" by Bastiaan is altogether a new thought within the scientific arena anywhere in the world. It deserves a thorough interpretation in the book elsewhere with a deeper insight into the phenomenon to comprehend its message – *hook, line, and sinker*!

At the same time, Bastiaan commenced exploring systematically the Slovenian fauna, initially in the first place that of the Ljubljana environs. This was, though a rather small-looking research activity, actually surrounded with very many challenges: first, no thorough and recent catalogue was available already for Odonata either for Slovenia or Ljubljana; secondly, the information was scattered in both Slovenian and other languages, and; thirdly, and most challengingly, there was no definitive volume(s) of taxonomic keys available in a set to facilitate identification of the dragonfly fauna collected by Bastiaan and to bring about this he had to depend on various different literature! Nevertheless, Bastiaan was resolved to encounter any constraint head-on, which reminds me of the ever invigorating lines by the Nobel Laureate and the Founder of Shanti Niketan, Rabindranath Tagore (1905):

"If they answer not to your call walk alone

If they shut doors and do not hold up the light
when the night is troubled with the storm,
with the thunder flame of pain
ignite your own heart
and let it burn alone."

To accomplish this daunting task Bastiaan reviewed essentially all the existing literature in Slovenian as well as other languages, updated the Slovenian, and Ljubljana, faunal checklists by his own field work, and a note on the preliminary results he then published in *Proteus* (1954; 16/8: 220-222). Since he did not have modern literature at that time, the taxonomic nomenclature adopted therein, however, is obsolete now and no longer in vogue, but his brief note had already cut the ice!

During 1953-1961 Bastiaan kept a still preserved field diary, where he noted all the species encountered, with dates, locality data and notes on such environmental circumstances as he considered to be useful. In 1958 he deposited his specimen collection of 16 cabinets with the Museum. It is very well eulogized that the best part of wisdom is dissemination of knowledge, since this is the only earning that grows with every investment.

●●●

8

Bastiaan's First Encounter with Professor Jovan Hadži (1953)

"Learn from yesterday, live for today, hope for tomorrow. The important thing is not to stop questioning."

— Albert Einstein

By mid-1950s Bastiaan had irrevocably decided to do research on dragonflies (Odonata) and almost accessed all possible literature prevalent in Slovenia. The best part of his learning science was, however, still to come! It was a rather dim late November day, when Bastiaan was heading with his father for a walk to the Rožnik hill. In the avenue leading to Tivoli Castle they noticed an elder gentleman watching carefully something on the rind of a chest-nut tree. Father greeted him respectfully and Bastiaan bowed courtly. The gentleman looked up, said, "Oh, good day" and stepped up to father. Since father did not introduce Bastiaan and the gentleman did not address him, he withdrew a little and was waiting. They talked for a few minutes. Bastiaan did not listen, but he did notice the gentleman's Slovenian was somewhat unusual. When they parted, father told Bastiaan, "This was Professor Hadži." Bastiaan was shocked and regretted he did not look at him better.

Prof. Hadži was a zoologist of Serbian origin. He had a much chequered career in early life but soon got settled in Ljubljana for his entire future. He was a rare scholar of both plant and animal life. He could also speak many languages, including Slovene although his accent was slightly quaint to local dialect. There is a Jovan Hadži Institute of Biology in Ljubljana, established in his honour as one of the greatest Slovenian zoologists ever. For his contributions to zoology, Hadži received the Prešeren Award in 1956. In 1969, he received the honorary doctorate by the University of Ljubljana. Several invertebrate species were named after him by other zoologists, such as *Astagobius hadzii*, *Cyclopina hadzii*, *Isohypsibius hadzii*, and *Niphargus hadzii*.

Prof. Hadži was born in a Serbian family in Temišvar (today Timişoara, Romania) in what was then Austria-Hungary, on 22nd November, 1884. He graduated in Zagreb. In 1920, he moved to Ljubljana where he became the head of the Zoological Institute at the University of Ljubljana. Between 1951 and 1972, Hadži was the head of the Biological Institute at the Slovenian Academy of Sciences and Arts (SASA). In 1938, he became a full member of SASA.

Dr Hadži was perhaps most widely known for his unique theories of animal evolution. He devised a system of classification, known as Hadži's system of animal classification, in which he divided the animal kingdom into six phyla: Protozoa, Parazoa, "I" (animals with no segments), "Oligomeria" (animals with few segments), "Polymeria" (animals with many segments) and Chordata. His invertebrate phylogeny and classification were widely, though not as yet, generally accepted by zoologists and, due to its simplicity, his system was widely used in science education in the former Yugoslavia. His principal work was his very well documented "*Turbellarian theory of Cnidaria*". Later, when Bastiaan was a Professor at the State University of Utrecht, he was using Hadži's phylogenetic system, was teaching his *Turbellarian theory of Cnidaria*, and scooped greatly from his other monumental works in this field.

Hadži's faunistical work focused on the invertebrate fauna of caves and mountains where he described more than a hundred new species and genera. He was also an active cave explorer and acted as a President of the Slovenian Society for Cave Exploration (Društvo za raziskovanje jam Slovenije) between 1927 and 1945. Later, when Bastiaan came in touch with Prof. Hadži, these two traits of his had an important bearing on the young Bastiaan who himself is an ardent mountaineer and has a special eye for the cave fauna.

Bastiaan's providential rendezvous with Prof. Hadži in the Rožnik hill near Tivoli did spark down his memory lane and soon he realized that he already knew something about the professor: first, from father's recollections of his student time at Biology, and secondly, from mother, who was acquainted with his consort, but most information came from conversations in the Entomology Section meetings. Bastiaan even had a chance to take a glance at his study in the Institute of Zoology, an elegant drawing-room, full of books, instruments and conserved animals. He became Bastiaan's "idol" already in the secondary school times: "the man who knows about animals everything". But Bastiaan spoke to professor for the first time in person only shortly prior to matriculation. Incidentally Bastiaan mentioned to professor on their first encounter in the chest-nut avenue near Tivoli. He said, "I remember", and smiled. This brought a great feeling of joy and relief in young Bastiaan's mind.

As a young man of 17, in 1954, Bastiaan acquired a beautifully bound second-hand copy of his *Turbellarian theory of Cnidarians*, which he kept reading and re-reading continually, though he did not comprehend everything. For the unfamiliar animal groups and for their larvae he was consulting father's copy of Claus Grobben's "*Lehrbuch der Zoologie*", which long found a place in his library. As said above, many years later, when Bastiaan was at the University of Utrecht lecturing on invertebrate zoology, he was using Hadži's phylogenetic system.

When Bastiaan left for the Netherlands (1962), he remained in touch with professor Hadži up to his death in 1972. In 1964, when he informed him about his embarking on the preparation of his doctoral dissertation on dragonfly cytology, he expressed the opinion that it is likely that Bastiaan would be going to get involved with holokinetic chromosomes and with post-reductional type of meiosis and suggested to read publications on this subject produced by the Finn, Tarvo Oksala. Naturally, he was regularly sending him the reprints of his papers the receipt of which professor always acknowledged with some kind of comments. One of his last letters Bastiaan received is in response to his dissertation (1969). The book is in English, but he noticed also the few lines in Slovenian, where he tendered gratitude to his greatest teacher in Zoology. Professor Hadži warmly congratulated and kindly assessed the work. His appreciation meant to Bastiaan more than the "*cum laude*" mark, rendered by the University of Utrecht.

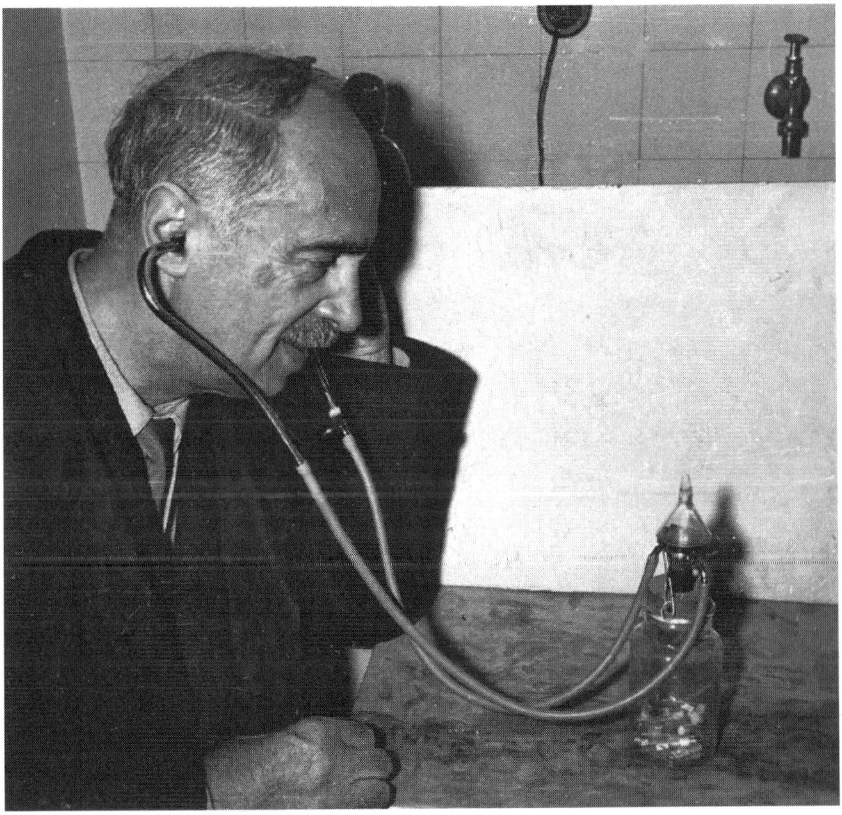

Fig. 8.1. Professor Dr Jovan Hadži, the doyen of Slovenian zoologists, listening to a song of a heteropteran species, in the Institute of Biology of the Slovenian Academy of Sciences and Arts. 1959. – [Photograph Source by Matjaž Gogala, Ljubljana]

Professor Hadži's photograph (Fig. 8.1), showing him with a stethoscope, listening to a stridulating bug and taken by Matjaž Gogala, has a place on Bastiaan's desk ever since half a century! By now Bastiaan was fully prepared to take on the long and meandering odonatological peregrination – a journey which would bring happiness in the life of all the world's odonatologists in manifold ways!

"For all your days prepare
And meet them ever alike
When you are anvil, bear –
When you are the hammer, strike"

●●●

SECTION II: RESEARCH AND ADMINISTRATION MANAGEMENT

9

Beginning of the Professional Career in Odonate Cytogenetics and Cytotaxonomy (1957)

Great teachers emanate out of knowledge, passion and compassion.

— A.P.J. Abdul Kalam

Rocket Scientist, 'Bharat Ratna' and President of India

Bastiaan grew up in an aesthetic environment of abundant scientific literature, field explorations and free discussions, mostly guided by his parents but also other family members, including grandparents, and, above all, teachers, friends and colleagues at both school and college levels. At home, his parents, especially mother, saw to it that he mastered many languages, took interest in free debates and worked on all his matters by himself. He thus developed a very strong taste for book reading, almost voracious, at a very early age. For the forming of his character during childhood, the most important book was perhaps "*Il cuore*" [=*"The heart"*] of Edmondo De Amicis, which was read to him by his mother and (almost daily) by his grandfather. The interest for nature, animals and particularly insects was triggering in him by numerous animal children books, such as, e.g., those of Waldemar Bönsels ("*Adventures of Maya, the bee*"), Edo Deržaj ("*The May-bug*"), Marija Jezernik ("*Godrnjavček" – "The little Growler*", i.e., on the life of a brown bear cub) and he was continuously reading and re-reading Alfred Brehm's "*The life of animals*" and a similar monumental work of Fran Erjavec. Unlike most of the teenager boys, he was not interested in detectives and the like, whereas he liked to read history books and history novels. As a teenager he was swallowing every accessible book and journal article on mountains, karst caves and also on the philosophy of alpinism. Henrik Tuma's book on the "*Meaning and the development of alpinism*" certainly played an important role in the forming of his adult view of the world and conception of life. In his young years he required and received much help from his neighbourhood, therefore he is lifelong inclined to render every kind of help and assistance he can to young people who are asking for it or otherwise he feels they need it.

Aside of his parents and the family, quite a few persons had a profound influence of his choice for the study of nature. In High School, it was Pavel Kunaver, the teacher of Geography, who opened to him the mountain and the cave world. At University, Professor Dr Jovan Hadži triggered and much enhanced his interest in invertebrate systematics and particularly in invertebrate phylogeny. The odonatology he has learnt while going, but his principal teacher and mentor in this field was Dr M.A. Lieftinck.

The value of 'self-help' he learnt early in life and never ever looked for any outside help in any of his undertakings, be it personal or professional. Father, Milan Kiauta, consistently encouraged him with field explorations involving nature resource management and entomofauna biodiversity which rendered Bastiaan enough opportunities to develop an ever perpetuating love for mountains and forests, and lakes, all of which ecosystems boasted of varied and copious dragonfly fauna. His mountaineering-oriented exercises resulted into his becoming a towering 6.5 ft tall personality, influensive enough to cast a feeling of brimming confidence.

Although, Bastiaan was never an addict to athletics and team sports, he developed himself into a passionate alpinist lifelong and mountain wall climber, and a (tour) skier. Even as an octogenarian, he remains a passionate long-distance swimmer. In his younger years he participated sporadically in regional ski matches, but otherwise he was never the type of a competitive sportsman. Cycling and (horseback) riding were also much exercised, though rather for the sake of work requirements than as a passion for its own sake. Horses remain his passion to this very day, and his wife and himself are running an asylum for old, disabled or discarded horses and other animals! For such noble cause of charity only, Mother Teresa has possibly sermonised the following:

"*Every time you smile at someone,*
it is an action of love,
a gift to that person,
a beautiful thing."

Bibliophily and story-telling by bedside in the night by the mother, in particular, but also by the father occasionally, during long walks through forests, were two important happenings that laid the foundation of love for books in Bastiaan's life. He remains a thorough reader of all kinds of scientific works, especially rare ones, he comes across, but post-retirement he has special interest towards those having historical accounts. Book reading formed a formidable character during his childhood (*see above*).

Bastiaan particularly developed a special bond with Dr Lieftinck in whom he saw, besides a teacher and mentor, also a fatherlike figure. Combining, therefore, cytogenetics (including cytology, cytotaxonomy and cytophylogeny) with odonatology for the rest of his life work, least to mention being the core subject of his doctoral dissertation both at PhD and DSc levels, Bastiaan was now ready to move out, to fly unstopped, and to experience the world on his own, and to write his destiny by the sheer dint of hardwork and application.

'tis not in mortals to command success,
'll do more, deserve it.

Bastiaan Kiauta's professional career commenced in 1959, when he was appointed a biospeleologist and (as a university qualified geologist) a stratigrapher of the Quaternary period (in cave strata) at the Institute of Karst Research of the Slovenian Academy of Sciences and Arts, in Postojna, Slovenia. In this capacity he met Professor Dr James Kesar of the University of Oregon (USA) who visited the Institute in connection with his cytology studies on the cave salamander (*Proteus anguinus*) and introduced Bastiaan into the principles and techniques of chromosome studies, which was decisive for his later career. During 1961 Bastiaan was involved in the geological mapping of Slovenia, with the State Institute of Geology, Ljubljana. Additionally, he was associated with the Department of Nature Conservation of the State Institute for the Conservation of Monuments, in Ljubljana, where he was co-founder and first Editor of the journal, *Nature Conservation* (= *Varstvo narave*), the foremost Slovenian periodical in this field. His Director there was the famous botanist, Dr Angela Piskernik.

Bastiaan was interested in entomology since his childhood, as outlined in his autobiographic recollections (KIAUTA, 2015b). With dragonflies he met first (1942) in the collection and under guidance of his grand-uncle Professor Dr Cyril Ažman, whereas the friendship and collaboration with his school-mate, Matjaž Gogala, the later Professor of Invertebrate Physiology and Director of the Slovenian Museum of Natural History (Fig. 9.1), also had a great impact on his work (see GOGALA, 2013).

Fig. 9.1. Professor Dr Matjaž Gogala: field work on the cicade bioacoustics –(Photo Source: Dr Tomi Trilar; 2017)

As often is the case, Bastiaan started with simple faunistics around his native city and elsewhere in Slovenia (KIAUTA, 1954a, 1954b, 1961b). Although quite well acquainted with the identification and biology of dragonflies, Bastiaan soon felt that acquiring a deep knowledge on identification and taxonomy of Odonata would be absolutely essential if he had to comprehend their phylogeny and evolution as well as distribution on global level – subjects which have been attracting attention of the scientists since times immemorial and still continue to challenge their wits. Therefore, he set out on a long study tour to Britain to work at the dragonfly identification in the British Museum (Natural History). There, in 1959, he met Dr D.E. Kimmins, who was a great taxonomist genius. Bastiaan was interested in the odonate material collected by British odonatologists, who served in military service during World War I in SE Europe and Near East, i.e., in the Otoman Empire of that time and whose material is housed in the BM(NH). It was for this purpose he first visited the Museum. Dr Kimmins had a very tumultuous and chequered career at BM(NH), most of the time engaged on temporary and *ad hoc* basis, without however allowing it to distract him from his immaculate and thorough attention to curating the museum assets of entomological collections. The life and work of Dr Kimmins had immense bearing on Bastiaan's own systematic disposition of all his material, literature or biological samples, all along his more than six decades of odonatological peregrination. Bastiaan learnt from Dr Kimmins more about his life and works which were all stimulative, motivative and inspirational. Kimmins' life was a perfect example of upright character, never accepting defeat under any circumstance and making a great teacher.

Dr Douglas Eric Kimmins was born on 4th August, 1905. For the whole of his professional career he was connected with the British Museum (Natural History), where he was a part-time "unofficial worker" from 1925 until 1948. He then became a fully established member of the Museum Staff, retiring in 1970. After this, he was re-employed in a temporary capacity for a further two years, subediting the Museum Bulletins. He was primarily concerned with that heterogeneous group of orders embraced by the Linnaean name "Neuroptera", and affectionately known to many entomologists as the "Kimminsioidea". Although his contributions to Odonatology alone were sufficient to assure him an honourable place in the history of Entomology, he worked as fully on the other orders, was the author of handbooks on the Ephemeroptera and on the Plecoptera published by the Royal Entomological Society of London, and wrote extensively on the Neuroptera and Trichoptera.

Prior to the outbreak of the 1939-1945 War, he, along with the late Dr W.E. China, was responsible for evacuating the Museum's Entomological collections to the Headquarters of the Freshwater Biological Association at Wray Castle, where he continued to work on them for some time. He helped to teach in some of the Field Courses held there, and many a student owes him a debt of gratitude. He was a great walker, and his former colleagues at Wray Castle remember with pleasure walking and collecting with him round the Lake District. He later joined the Royal

Air Force, where he did valuable work in the interpretation of aerial photographs. In due course, back in civilian life, he supervised the return of the collections from Wray Castle to South Kensington.

Another valuable service Dr Kimmins did to the British Museum (Natural History) was to arrange for the purchase of the McLachlan Collection, when he heard from the late Colonel F.C. Fraser that this collection was about to be put on the market by the executors of McLachlan's nephew. As McLachlan was an intimate friend of de Selys, and they freely shared and exchanged their material, most of the specimens described in the latter's classic monographs which are not to be found in the Brussels Museum are in the McLachlan Collection, so now readily available to those using the British Museum. Obviously, Bastiaan was fortunate enough to go through all these during his sojourn at the BM(NH), courtesy Dr Kimmins.

Dr Kimmins was a meticulous worker, and very skilled both as photographer and draughtsman. Many papers by other workers, especially in the pre-war part of his career, were beautified by figures from his pen. He was also very much at home among the intricacies of the International Code of Zoological Nomenclature, and was always willing to guide younger workers in their application. He was extremely helpful and had an immense knowledge in odonatology. He and Bastiaan had long hours of discussion whereupon Kimmins will reveal many dark spots in the odonate taxonomy and as to how an evolving discipline like cytotaxonomy can help resolve these issues. Bastiaan was of course also greatly impressed by the enormous BM(NH) collection and by the system it was managed. So he spent at least a month in the Museum, had countless talks with Kimmins, who outlined him numerous pending problems, guided him through the material and literature and Bastiaan was inclined to consider him his first teacher in this field.

Upon his retirement from the Museum, Kimmins apparently "lost interest in active research, had problems with the diminishing capacity of his memory" and in 1975 he informed Bastiaan sadly that he was not going to participate in the Third International Symposium of Odonatology in Lancaster, "since there will be many workers whom he was supposed to know and he felt embarrassing, if the progressive failure of his memory would create unpleasant situations". In his retirement he did mainly concentrate on nature and insect photography. They stayed in regular touch until Kimmins' death. As an homage to a teacher and one of the most respected insect taxonomists, Bastiaan Kiauta, as the Executive Editor of *Odonatologica*, brought out a commemorative volume on his 70th birthday, and essayed by Dr R.M. Gambles (Vol. 4, No. 2, pp. 61-64).

In May 1962, Bastiaan moved to The Netherlands, where he first worked as a hydroentomologist with The Netherlands State Institute of Basic Nature Conservation Research (RIVON), in Bilthoven and Zeist. In 1964 he received a grant from the Netherlands Organization of Pure Scientific Research (ZWO) and moved to the Institute of Genetics of the State University of Utrecht (Fig. 9.2), where he completed his studies with the PhD dissertation (1969f, titled "*Studies on karyotypic evolution in Odonata*") (Figs. 9.3 and 9.4). Dr Kiauta completed his PhD research by following the time-tested rigorous methods that included invariably (i) preparing the "Author Index Cards", with full details of research publication, including also the address of the author wherever possible, (ii) collecting reprints or xeroxed/photostated copies of all the scientific research publications in journals, (iii) reading critically all research publications and relevant books by making personal notes, (iv) making correspondence with contemporary odonatologists and/or cytogeneticists, and exchanging views on issues baffling enough, (v) searching *inter alia* a mentor who could guide externally to not only accomplish the task of dissertation but also pave the way for future, and, more importantly, (vi) visiting many entomological collections/museums and discussing on the subject with accessible seniors even if engaged in different disciplines.

Fig. 9.2. The State University Utrecht, The Netherlands: the ancient central building, where all traditional formal performances are taking place, such as, e.g., the introductory public lectures of newly appointed professors and the public defence of doctoral dissertations.

The core thesis of the subject of Bastiaan's PhD dissertation looked simple but it entailed gigantic efforts within stipulated time period as allowed by the university to complete the work. After giving lectures and attending other assigned duties at the university, he would generally rush to the field to capture dragonflies for his cytotaxonomic study in the laboratory. He would return to his laboratory late in the evening with his sample of dragonfly catch, and settle down immediately to prepare the chromosome slides. To make the slides, he had to spend 3-4 hrs to process the specimens (mostly males, as a rule) through dissection, removal of testes and transfer to the microscopic slides ready with a drop of lacto-acetic orcein. After following the protocol the chromosome slides were prepared by squash method and the vivisected plesiotype was kept for reference. He would leave his laboratory only past midnight. Next day, he would be at his university department at 8 am sharp. Almost always, seven days-a-week, this ritual went on till he had completed all his PhD thesis related works. The result of his perseverance was also as outstanding since he was able to propound in his thesis not only several new theories alluding toward evolution of new chromosome numbers and sex-determining mechanisms, chorogeographic significance of the *m* (micro)-chromosome but also throwing new light on the phylogeny of the Order Odonata. Obviously, therefore, his PhD thesis, bejewelled with so many novelties, was recommended highly for publication in international journals, for benefit of scientists engaged in similar research across the world. This brief PhD research period at the University of Utrecht brought Dr Kiauta in contact with the leading cytogeneticist, Professor Janny van Brink who, as a senior colleague, not only rendered full support in his ensuing odonate cytotaxonomic research projects, but also emerged as his strongest associate in the life of SIO, through all her golden as well as thick and thin times!

Thereafter, in 1969, Dr Kiauta entered the scientific staff of the Institute of Genetics of the State University of Utrecht, where Professor J.M. van Brink was Head of the Department. There he, through sheer dint of hard work and application, developed many important projects, some of them overseas too, e.g., the Khumbu Himal Research Expeditions of the Netherlands Centre for Alpine Biological Research (NCABR), and he made many new friends, some of whom participated in his scientific research missions. One of these is Dr H.J. Dumont, who named in his honour a new rotifer species, discovered in Nepal, *Scaridium bostjani* Deams & Dumont, 1974.

At the University of Utrecht, Dr Kiauta promptly passed through the usual professional service grades up to Full Professor of Invertebrate Cytogenetics and Cytophylogeny.

Fig. 9.3. B. Kiauta (1969f): cover and title page of his doctoral dissertation. (Source: B. Kiauta archive)

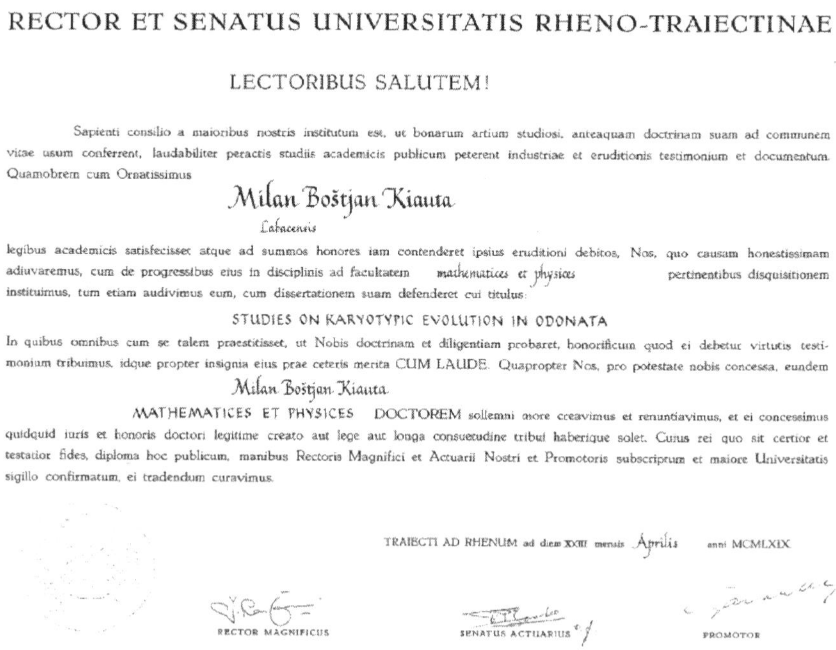

RECTOR ET SENATUS UNIVERSITATIS RHENO-TRAIECTINAE

LECTORIBUS SALUTEM!

Sapienti consilio a maioribus nostris institutum est, ut bonarum artium studiosi, anteaquam doctrinam suam ad communem vitae usum conferrent, laudabiliter peractis studiis academicis publicum peterent industriae et eruditionis testimonium et documentum. Quamobrem cum Ornatissimus

Milan Boštjan Kiauta

Labacensis

legibus academicis satisfecisset atque ad summos honores iam contenderet ipsius eruditioni debitos, Nos, quo causam honestissimam adiuvaremus, cum de progressibus eius in disciplinis ad facultatem mathematices et physices pertinentibus disquisitionem instituimus, tum etiam audivimus eum, cum dissertationem suam defenderet cui titulus:

STUDIES ON KARYOTYPIC EVOLUTION IN ODONATA

In quibus omnibus cum se talem praestitisset, ut Nobis doctrinam et diligentiam probaret, honorificum quod ei debetur virtutis testimonium tribuimus, idque propter insignia eius prae ceteris merita CUM LAUDE. Quapropter Nos, pro potestate nobis concessa, eundem

Milan Boštjan Kiauta

MATHEMATICES ET PHYSICES DOCTOREM sollemni more creavimus et renuntiavimus, et ei concessimus quidquid iuris et honoris doctori legitime creato aut lege aut longa consuetudine tribui haberique solet. Cuius rei quo sit certior et testatior fides, diploma hoc publicum, manibus Rectoris Magnifici et Actuarii Nostri et Promotoris subscriptum et maiore Universitatis sigillo confirmatum, ei tradendum curavimus.

TRAIECTI AD RHENUM ad diem XXIII mensis Aprilis anni MCMLXIX

RECTOR MAGNIFICUS SENATUS ACTUARIUS PROMOTOR

Fig. 9.4. B. Kiauta: doctor's diploma of the University of Utrecht (original size: 47 × 60 cm, in two colours); dated 23 April 1969. – [The official Latin name of the University of Utrecht is Universitas Rheno-Traiectina]. – (Source: B. Kiauta archive)

At the University, Dr Kiauta taught various subjects, such as General insect phylogeny and evolution, General insect cytotaxonomy and karyotype evolution, Insect speciation under extreme environmental conditions as prevailing in high mountains and caves, Population biology, Population genetics and Invertebrate systematics and phylogeny.

Through his diverse research explorations he had carved a special niche, among all the insect orders, in Odonata particularly, though he had developed good knowledge on many others such as, e.g., Trichoptera (1968b, 1969a, 1971a, 1979), Collembola (1970c), Coleoptera (1971b, 1973a), Orthoptera (1973b), Neuroptera (1977a) and Ephemeroptera (1977b). With two of his M.Sc. students, L.J.M. Butot (of the RIVON Institute) and G.A. van Griethuysen, he worked intensively on molluscan cytology. They produced some dozens of publications, including a book on complete molluscan cytological bibliography (1970a) and series of papers on the Rissoacea and Succineoidea, the latter taxon redefined and elevated to the order rank by BUTOT & KIAUTA (1967b).

Kiauta published several minor papers on the Slovenian cave fauna and on the history of its exploration (e.g. 1960b, 1960c, 1960e, 1961a, 1961c, 1961d, 1962c, 1962d). The Thompson & Kiauta review (1994b) of odonate occurrence in the hypogean environments has received much attention worldwide. The reference, listed under (1995b) is his sole other note on the Odonata in caves.

In his young years, Kiauta also produced several faunistic papers, notably on Homoptera (1962g) and Heteroptera (1963c), but also on bats (Chiroptera) (1960e, 1961c, 1962d).

With such a background that includes also various other research projects, e.g., on the fauna and ecology in the saline aquatic environment, the cytology of Turbellaria and Cnidaria, etc., he was well qualified lecturing also the Invertebrate systematics and phylogeny (for senior students), which course was given cyclically and its completion required 6 hours of weekly lectures, during 3 subsequent years and included the systematics of Protozoa as well.

He emeritated from University in 2002.

In 1963 he came in contact with another extraordinary Odonata taxonomy genius, Dr M.A. Lieftinck. At that time Bastiaan was working with the Netherlands State Institute of Nature Conservation Research (RIVON), where the Netherlands dragonfly fauna was one of his main tasks. For this purpose he visited Leiden Museum (RMNH) in order to talk with Dr Lieftinck, whom he had never met before. Incidentally, at the time of his visit Lieftinck was absent, being on a field trip in SE Asia. Later Bastiaan visited with him at the Museum and at his home countless of times and he was also a regular guest at their home in Bilthoven. During these conversations Bastiaan learned from him enormously, in all fields of odonatology. When his wife died and he stayed alone in Rhenen, Bastiaan's wife, Marianne, got accustomed to his kitchen and she was serving the incidental guests whenever they were there. Dr Lieftinck was an institution himself under whose tutelage several great odonatologists gained unfathomable knowledge, including Bastiaan, who always held him in the highest reverence. Bastiaan's own life imbibed most of Lieftinck's inimitable character of displaying inexhaustible energy in conducting field work and absolute precision in detailing each and every feature while describing a new taxon!

Maurits Anne Lieftinck was born in Amsterdam on the 18th of February, 1904. He was second son in the family of tobacco-broker Gerard Lieftinck and his wife, Mrs Elisabet Lieftinck nee Esser, a family enlivened by three younger sisters. Maurits attended the Amsterdam Lyceum (1916-1922) and studied biology at the University of Amsterdam (1923-1929). During his undergraduate years he was an enthusiastic votary of various athletic sports, including boxing. Meanwhile his vivid interest in nature, especially entomology, increased constantly as profoundly evidenced by the fact that, already at the young age of 15, in 1919, he became a member of the Netherlands Entomological Society, and in 1921 he was among the founder-members of the Netherlands Youth Federation for Nature Study (NJN). He was also promptly chosen as editor of the NJN's still existing periodical *Amoeba*, from its first issue (January 1922) until 1926.

Fig. 9.5. Dr M.A. Lieftinck on a field trip in the 1970s. – (Source: B. Kiauta archive)

Dr Lieftinck liked field work and he made numerous field trips to the lakes of Naarden and Ankeveen, not far from Amsterdam, to the brooks and peat-bog areas in the eastern and southeastern parts of The Netherlands, and in the remote districts of Brabant and Limburg, in search of rare dragonfly species such as *Aeshna subarctica* (Fig. 9.5). Based on his extensive odonatan explorations in nature he published his *magnum opus,* "Odonata neerlandica" (1925-1926), which to date is considered a base document for the Dutch dragonflies. On 21st August, 1929, when Dr Lieftinck moved out of his country to take up a position at the Buitenzorg Museum, Java, now Indonesia, he converted the opportunity in his favour to embark on odonatological subjects which were so far remaining unattended and had a great bearing to explain odonate phylogeny and evolution. Soon on 28th February, 1939, he was appointed Head of the Buitenzorg Museum, a position which he held till March 1954.

Following a nerve-breaking wartime tumultuous experience, during a recuperation leave to Holland, he visited musea and other institutions in Belgium, Italy, Sweden, France and England, examining type material of Indo-Australian Odonata and solitary bees. Back in Buitenzorg, now named Bogor, many duties were awaiting him again. He continued to serve as Editor of *Treubia* (1934-1959, honorary editor since 1959), secretary of the Royal Netherlands East Indies Society of Natural Sciences (Natuurkundige Vereeniging), and editor of the semi-popular monthly *De Tropische Natuur* (1935-1942). On 24th April, 1948 Dr Lieftinck became a Corresponding Member of the Royal Netherlands Academy of Sciences, Amsterdam; and on 5th July, 1950 received the same membership of the Swiss Nature Society (Naturforschende Gesellschaft) in Basel.

Dr Lieftinck never sat idly and, accompanied often with many scientists from abroad interested in various aspects of tropical insect life, he was always back-packed for field work, despite hot dampness, sand flies, and fiercely sucking leeches, traversing almost impenetrable forest tracks en route to the site of collection, not without the pleasure of life, on one such occasion, in having discovered the scarlet little libellulid dragonfly, *Lyriothemis salva.* As he had become more and more absorbed in working out life-history studies of dragonflies, he whole-heartedly welcomed that occasion to explore the interior of the "Vogelkop" of West New Guinea — *terra incognita* as to Odonata! — as well as to visit some islands to the west of it, in particular the island of Misool. Shortly afterwards, Lieftinck also spent some time in the south Moluccan islands Ambon and Saparua, studying their odonate faunas.

In 1949 he was a delegate of the Netherlands-Indies' Government to the 7th Pacific Science Congress in New Zealand, and on his way to Auckland, spent a few happy days in the national park near Sydney, collecting and observing several of the endemic Odonata occurring there. In September-October 1950, he went to Malaya and Sarawak (in Borneo), to advise and reorganize the Sarawak Museum insect collections in Kuching, and to have discussions with the Director of the Raffles Museum in Singapore. In the same year he was sent by the Indonesian government to south Borneo, to accompany a German botanist professor, on which occasion important dragonfly discoveries were made by him.

By the end of 1983 the number of Lieftinck's publications exceeded the 300 mark, many of them comprising several tens of printed pages. Interestingly, the first 40 of these appeared during his student years in Amsterdam, with his "*Odonata neerlandica*" being the most important one. His work was suddenly brought to a standstill by the Japanese invasion, early in 1942. A few manuscripts, already in "last-minute" proof print, however, appeared soon afterwards in *Treubia*, during the very first months of enemy occupation. It contains his article on the Anisoptera of the 3rdArchbold New Guinea Expedition. In that publication two subspecies of the libellulid genus *Rhyothemis* were described.

The numerous publications produced, besides his editorial work and many other time-consuming duties, is but a sublimation of what Dr Lieftinck really has achieved. The fine camera lucida drawings illustrating his articles were made all by himself. Not only the number, also the quality of his works is astonishing. Lieftinck's descriptions, for the most part written in perfect English, occasionally in German or French, give evidence of his painstaking observations on the specimens under consideration. He redescribed and figured many older types and often completed species descriptions by providing information and figures of previously unknown allotypes. In this way hundreds of pages were published by him describing the odonate fauna of the Sunda Islands and New Guinea. A very important handlist of Malaysian Odonata summarizes his efforts during a stay of 25 years in the tropical Sundaic area. In his descriptions he always had an open eye for ethological and phylogenetic aspects.

Lieftinck was a perfect polymath, an authority in many insect groups. He studied many other aspects of insect life in Indonesia. He wrote a very important paper on the curiously specialised members of *Leptopanorpa* (Mecoptera), which he observed in the Javanese mountains. Next to dragonflies solitary bees, like Anthophoridae and Carpenter-bees, still had his vivid interest, as is apparent from his prominent taxonomic publications on these groups. Small details like the oothecae of little mantids also fascinated him, as well as various stream-dwelling insects (e.g. Ephemeroptera, gyrinid and other aquatic beetles), cave-dwelling Hemiptera, etc. He often communicated his observations at meetings of entomological societies.

During his POW-camp time, when any field work was virtually impossible, Lieftinck nevertheless had some adventurous experiences. To the delight of his fellow captives, he once happened to come across a fecundated specimen of the common libellulid *Potamarcha congener*, whose eggs he assembled because its early stages had remained unknown to him. As all eggs taken from this female hatched, the offspring was carefully nursed, first fed on minute pond-water animals, followed by mosquito larvae, etc. Several of these dragonfly larvae successfully moulted, but only few survived in their inconvenient confine: only a single full-grown specimen reached maturity. Shortly afterwards, on a railway transport to another camp (at Tjimahi), this unique specimen was taken along with him and happened to keep alive in a moistened tin; immediately upon arrival it was released into a water-containing vessel and after a few days it finally transformed.

Sometimes walking an extra mile may produce exceptionally beneficial results, as most meaningfully resonated by these two short quotes: *He who plants a tree, plants a hope* and *an ant on the move does more than a dozing ox*. Dr Lieftinck always walked that extra mile in his quest for odonatological knowledge. His principal subject of lifelong investigation was and remained the Odonata. An excellent work is formed by his series of reports on the odonate fauna of New Guinea, dealing with over 600 species. As a sequel to his work on New Guinea, Dr Lieftinck studied the odonate faunas of several archipelagoes in the Pacific and that of northern Australia. The zoogeographical relationships of these with those of the Asian and Australian mainland were also considered. Of numerous Pacific odonate taxa he described and figured the larvae as well. He left Indonesia on March 10, 1954, but still remained actively associated with the "Micronesia Project" sponsored by the B.P. Bishop Museum (Honolulu), the United States National Museum, Washington, and the institutes of the Australian Pacific Mandated Territories (from 1952 on). He also served on the Editorial Board of *Nova Guinea*, published in Leiden (1955-1966).

On 1st July, 1954, Dr Lieftinck was appointed Curator at the Museum of Natural History in Leiden, under the direction of Professor Dr H. Boschma. During subsequent years he continued research on Indo-Australian and New Guinean Odonata, and on several other insect orders. In addition, he was intrigued by the late A. Bilek's discovery of *Coenagrion hylas* in Europe (Bavaria) in 1964, and produced special articles on this and the other European species, *Macromia splendens*, including its larva (1965). His sound contributions to the knowledge of the African fauna are laid down in some papers on Zygoptera of Madagascar (1963, 1965) and on Anisoptera of parts of Kantanga and Zambia (1966), while his survey of the odonate fauna of Morocco (1966) still remains the key-stone of north African odonatology. Upon the retirement of Professor Boschma, Dr Lieftinck was invited to accept certain University positions and main responsibilities at the zoological institutes in Amsterdam and Leiden. These offers were declined as he preferred to continue his own research, unhampered by administration duties, inevitably involved with such positions.

His exceptional scientific productivity, his activities as Editor of scientific periodicals and as President of several scientific societies, and, not the least, his cooperation with foreign scientists and participation in international scientific expeditions earned Dr Lieftinck many high awards. At the VIth International Symposium of Odonatology, held in Chur, Switzerland, in 1981, he was elected Honorary President of the Societas Internationalis Odonatologica. When in March 1969 he retired from his office at the Leiden Museum, Lieftinck was appointed Officer in the Order of Oranje-Nassau. Relieved from his formal duties in Leiden, he did not give up his scientific activities. On the contrary; more than

20 substantial papers published after his formal retirement bear witness to his never diminishing working power and productivity. He also continued identifying thousands of newly acquired dragonfly specimens from tropical Asia, the Pacific area, and other faunal regions.

After the unexpected grave loss of his wife, in August 1980, Dr Lieftinck was trying hard to conquer sorrow and distress, fighting against disillusion, by concentrating on the study of dragonfly life and the many other gifts of nature. At that time, an annotated catalogue of the Odonata of Taiwan, prepared jointly with the Chinese colleagues in Taiwan, Drs J.C. Lien and T.C. Maa, was scheduled to appear early in 1984, while other faunal surveys and various revisions were in the course of preparation.

Dr Lieftinck was a true taxonomist in every sense. He was always very precise, accurate and thorough. I was one of those luckiest Indian odonatologists who have made prolonged correspondence with him, mostly seeking help with the identification of specimens. During the course of my doctoral research period (1974-1978) when I wanted confirmation of the species used in my thesis (TYAGI, 1978), it was the kindness of Dr Bastiaan Kiauta, who generously put me in touch with Dr Lieftinck, a relationship that lasted till his death in 1985. One of the greatest attributes of his taxonomic genius was to bring 'an ocean of information within a miniscule of space'! Such was the nature of his correspondence, wherein he would never allow unutilized even the slightest space in a post-card or an aerogramme –all very neatly, clearly and without an over-writing!

Fig. 9.6. During a lunch, at the First European Symposium of Odonatology(1971): from left to right: Dr D.C. Geijskes, Dr M.A. Lieftinck and Dr B. Kiauta. (Source: M. Kiauta)

With Dr Lieftinck, Bastiaan made also a prolonged odonatological tour to France, during which he again learned much. At the meetings of the Netherlands Entomological Society, Drs Lieftinck, Geijskes and Fischer (a trichopterologist) and Bastiaan always had an own table (Fig. 9.6). Occasionally, Lieftinck was also a humorist. On one such occasion, during a meeting on the Netherlands Entomological Society, he was unexpectedly projecting some slides, showing Bastiaan stripped to the underwear, standing in water, collecting *Macromia splendens* larvae. Needless to say, Bastiaan felt somewhat embarrassed, but the slides triggered much laughing of the public.

Dr Lieftinck was a very good friend, a splendid teacher and often treated Bastiaan like a father. For his 80th birthday, Dr Kiauta and his wife organized a celebration at his home and invited also some other odonatologists (e.g. Corbet, Geijskes, Rudolph) and his brother and sister. All preparations were done secretly. They all met at some remote place and then suddenly arrived at his home, without him knowing anything in advance – as he was rather "shy" and therefore reluctant in the presence of "strange" people. They presented him then the tribute issue of *Odonatologica*, devoted to his birthday. It was a very memorable day and he felt very happy. In the evening he invited all to a special festive dinner in a nearby restaurant, which he arranged straight at the spot by telephone.

Dr Kiauta, all along his life, has held both Dr Kimmins and Dr Lieftinck in very high reverence and respected them always as his true teachers in odonatological learning. This reminds me of what Kabir, the famous 15th Century Indian poet and saint, had said in teacher's glory:

गुरु गोविन्द दोऊ खड़े काको लागूं पायं।
बलिहारी गुरु आपने जिन गोविन्द दियो बताय।।

Transcription–

"Guru Govind dou khade, kake lagoon paay,
Balihari Guru aapno, Govind diyo batay"

Translation–

"If both, Teacher and God were to appear at the door, whose feet will I worship first? It has to be the Guru's feet first, because without him, how would I have recognized God?"

●●●

10

New Insights in Odonate Cytotaxonomy and Cytophylogeny

"The voyage of discovery is not in seeking new landscapes but in having new eyes."

— Marcel Proust

By 1950s some research work on odonate cytology had been already done in some countries, notably Japan, India, The Netherlands, Germany, France, USA, Brazil, Poland, UK, Australia, New Zealand etc. Most of these research works depended on the functionalities of various cellular organs and but distantly attempted to unravel the riddles inherently associated with the evolution of karyotype, the sex-determining mechanisms, the significance of *m*-chromosome etc. Professor Kiauta developed a crystal clear plan of research work on cytotaxonomy and cytophylogeny for the order Odonata at the Department of Cytogenetics & Cytotaxonomy in the Institute of Genetics (University of Utrecht), where Prof. Dr J.M. van Brink, a world famous geneticist in a variety of animal groups such as calyptrate Diptera, reptiles and birds, was Head of the Department. The first odonate chromosome note he published was in collaboration with Dr van Brink (1964d).

Though individual perseverance is the central strength to any great research activity, it is the team work with a masterly touch of leadership which ensures success. In Professor J.M. van Brink, Dr Kiauta found a highly dependable senior colleague with whom he teamed up to accomplish many important projects. She served for many years in the Institute of Genetics, Utrecht University, The Netherlands and edited international journals, *Genetica* and *Genen en Phaenen*, together with Professor Kiauta, was also the charter member and the Treasurer General of the Societas Internationalis Odonatologica (SIO), in addition to being Associate Editor of *Odonatologica*. Dr van Brink was a great dipterologist-cytogeneticist and worked in many parts of the world, including India. She was a highly determined scientist, an astute administrator and a very kind-hearted person. Her deeply-rooted sense of commitment to the various responsibilities undertaken was phenomenal. Extremely meticulous in executing all her undertakings, she was well known about her timely fulfilment of all her tasks. One instance that remains firmly ingrained in my mind keeps guiding me going committedly in my own scientific pursuits. It was during the 8th International Symposium of Odonatology in Paris, that during a discourse with Prof. Dr Kiauta it struck to me that the SIO National Office in India should come out with a complete bio-bibliography of Col. Dr F.C. Fraser, best known for his odonatological researches in India, particularly the writing of the three famous Odonata volumes in *The Fauna of British India* series! I discussed the prospects of its accomplishment with Dr Kiauta, and set out to complete the task with their help and guidance. Dr Kiauta provided the first draft of the bibliography which I updated and added to it the biographic account of Col. Fraser. To be doubly sure that the bio-bibliographic description was correct I consigned the manuscript to Professor Kiauta, who in turn passed it on to Professor van Brink to seek her critical views. Prof van Brink meticulously reviewed the manuscript for nearly a week and, on that unfortunate fateful night, believing it was safe to keep the document on a nearby study table, she normally retired to sleep. Alas! Then it happened, to her utter horror and disappointment whereupon she found next morning only a myriad bits of brutally torn pieces of the manuscript, by her pet pussy-cat who had incidentally sneaked into her bedroom during the night! For a moment she felt that all her hard work had gone wasted but, true to her character, she bore a lion's heart. She did not mourn over the near extinction of the manuscript, instead laboriously collected all the torn pieces of the manuscript and carefully glued them appropriately page-by-page to 'reconstruct' the manuscript once again, *de novo*!! She passed on Dr Kiauta her comments on the manuscript well in time as scheduled and without a page missed. Such is the order of magnitude of accomplishing a task undertaken by all the greats in science, and that is what makes a great attribution to a teamwork!

Because of her all-round help in the Department, Bastiaan soon was able to set up a working laboratory for himself on odonate cytogenetics and cytotaxonomy. He worked everything by himself, from collecting dragonflies during the daytime to making chromosome slides and microphotography till late night hours. Quite often he would remain in his laboratory past midnight, working on his samples and preparing manuscripts. Yet, without fail, next morning he would be timely available at the Department for lecturing the students – absolutely fit and fresh! Dr Kiauta understands the value of time: "*Time and tide wait for none*", and that in its tiny fold this highly time-tested adage encompasses the meaning of whole life!

Dr Kiauta emerged as a leading authority of cytological investigations on dragonflies, revealing many novelties that had revolutionized our thinking about pathways of evolution of chromosome numbers and sex-determining mechanisms etc. His pioneering investigations on odonate cytotaxonomy and cytophylogeny influenced a large number of young and budding scientists across the world in pursuing their own doctoral research and other projects, more or less directly under his tutelage through correspondence! The journey into redefining the phenomena of evolution of sex-determining mechanisms and chromosome numbers in the order Odonata was not very easy, primarily for three following reasons, viz.,

(i) There was no comprehensive and updated information available on the chromosome numbers for the dragonflies on global basis or, for that matter, insects as a whole,

(ii) Even though some hypotheses with regard to formation of new chromosome numbers were propounded in past, none however was demonstrable considering reasonably wide range in chromosome complements and the higher taxa (families) in the order, and

(iii) Several noted animal cytogeneticists had put forth their theories about development of new modes of sex-determining mechanisms, but all these were neither demonstrable exquisitely for the order Odonata nor were they supported by adequate examples to help in arriving at a reasonable conclusion.

Therefore, keeping in the mind above vagaries in the odonate cytogenetics, Dr Kiauta had clearly cut out his task delving deeply into the subject on global basis and amassing huge information in every possible form, through personal contacts and/or by correspondence. In addition to consulting several thousands of technical papers on insect cytology and cytotaxonomy, Dr Kiauta had meticulously examined the various abstracting services such as, for example, Entomology series of the *Zoological Record*, *Genetics Abstracts* and *Entomology Abstracts* which covered the world periodical literature and provided not only bibliographic references, but also brief English abstracts and addresses of the authors of any paper that have appeared from 1968 onwards on cytology and cytogenetics, and on entomology in general, respectively. He even regularly accessed the American *Biological Abstracts* and the Russian *Informativnyi Zhurnal Entomologiya*, the latter covering a part of Russian local literature that is not covered by any other abstracting service. The texts of abstracts published by these services were prepared in such a way that the abstracts alone often sufficed without consultation of the original source journal.

Two absolutely indispensable books in the field of cytotaxonomy of insects which Dr Kiauta had gone through time and again included, viz., (i) Makino, S., 1956. *A review of the chromosome numbers in animals*. (revised ed.). Hukuryyukan, Tokyo, and (ii) White, M.J.D., 1973. *Animal cytology and evolution*. (3rd ed.). Cambridge University Press. The first of these, though incomplete, he found out, gave a list of all animal species studied cytologically up to 1956 with chromosome numbers, sex determination mechanisms, notes on karyotypes, and abbreviated bibliographic references. The book of White, on the other hand, presents a general treatment of animal cytogenetics, listing more than 3000 full bibliographic references, a considerable portion of which on insects. Unfortunately enough, Dr Kiauta realized, no compilation existed on animal (or insect) cytotaxonomy.

A versatile field entomologist, Dr Kiauta was always full of energy while sampling his specimens for cytotaxonomic studies. Either personally or by correspondence he found specimens from all over the world. Based on his extensive cytogenetic and cytotaxonomic documented investigations on various different types of insects from all over the world, he was able to, for the first time, characterize the 28 hitherto studied insect orders, covering 3 apterygote and 25 pterygote orders, notwithstanding a great amount of variation in information available for different groups. Very little indeed was then known on the cytology of apterygotes other than Collembola and the pterygotes except Ephemeroptera, Isoptera, Neuroptera, Strepsiptera and Mecoptera – to list just a few of them. On the other hand, considerable information was available on Odonata, Orthoptera, Coleoptera, Diptera, Lepidoptera and Heteroptera. For three out of 31 insect orders then known (It is noteworthy here that the 32nd insect order, Mastophasmatoidea, was incorporated only in 2001), viz., Diplura, Grylloblattodea and Zoraptera, cytological observation was completely lacking. There was not a single order in which all major taxa, from the family group onwards, have been sufficiently examined. There were quite a few abstracts and/or summaries of the world information available in the past on range of chromosome numbers, sex-determining mechanisms, type of centromere etc. but Dr Kiauta possibly was the first to inventorize the various cytogenetic and cytotaxonomic characteristics of all the insect orders of which a succinct account is given below based on KIAUTA (1974).

Table 10.1. Cytological account of insect orders (based on KIAUTA, 1974, 1975a)

S.No.	Order	Range of chromosome numbers	Sex-determining mechanisms	Centromere
1	Protura	n♂=4-10	?	Localized
2	Collembola	n♂=4-7	XO/XX, male heterogametic	Localized
3	Thysanura	14-18 (♂=27, ♀=28)	$X_1X_2O/X_1X_1X_2X_2$; male heterogametic	Localized
4	Ephemeroptera	n♂=6-9	XO/XX, XY/XX; male heterogametic	Localized
5	Odonata	n♂=3-15	Original XO/XX (male heterogametic); Secondary neo-XY/neo-X neo- neoY	Diffuse
6	Orthoptera	8-68 (2n♂8= ♀=68 in a tetraploid form)	A great majority has XO/XX; recent groups may have the types of XY, neo-XY, and compound sex chromosomes a few cases of thelytoky are known	Monocentric
7	Phasmida	2n♂ bisexual=21-53; highest no. in parthenogenetic species is 88	Males heterogametic, XO, occasionally no-XY	Localized
8	Plecoptera	n♂=5-15, with a mode at n=13	Males heterogametic, in few species XO/XX or XY/XX mode; but in most varies, e.g., X_1X_2 or $X_1X_2X_3$	Localized
9	Embioptera	n♂=10-12, with probable mode at n = 11	Males heterogametic with XO/XX	Localized

10	Dermaptera	n♂=3-30	Males heterogametic with XO/XX, XY/XX, X_1X_2Y/ X_1 X_1X_2, $X_1X_2Y/X_1X1X_2X_2$, (and probably in male only) $X_1X_2X_3Y$	Diffuse
11	Blattodea	n♀=8-40, with a mode at 19	XO/XX	Localized
12	Mantodea	n♂=8-20, with a clear mode at 14	XO/XX in majority species; in others X_1X_2Y; rarely X_1X_2Y	Localized (monokinetic)
13	Isoptera	n♂=26(?)	XO(?)	Localized(?)
14	Coleoptera	n♂=10 (4-66)	XO, XY,$Xy_{p,}$ neo-XY, XY	Localized
15	Neuroptera	n♂=5-13	XX/XY	Localized
16	Hymenoptera	n♂=4-30+	N sex chromosomes, hence none	Localized
17	Strepsiptera	n♂/♀=8	male heterogametic XY/XX	?
18	Aphaniptera	n♂=8-11	XX/XY, $X_1X_2Y/X_1X_1X_2X_2$, $X_1Y_1X_2Y_2/X_1X_1X_2X_2$	Localized
19	Diptera	2n♂=4-20 (mode 6,8 and 12)	XY/XX (variation)	?
20	Mecoptera	n♂=20 and 23	XO/XX; in one species X_1X_2Y	Localized
21	Trichoptera	n♂=6-30 (in one species 50)	Female hetrogametic XX/XO	Diffuse
22	Lepidoptera	n=7-253 (31)	Heterogametic female XX/XY (also XX/XYY, XX/XO)	Diffuse
23	Thysanoptera	n=10-53	?	Localized
24	Zoraptera	2n♂=38	XY/XX	?
25	Psocoptera	n=8-15	XO/XX	Diffuse (holokinetic)
26	Mallophaga	n♂=2 (in one species); 2n♂=12, ♀=24 (other species)	?	Diffuse
27	Siphunculata	2n=7-14; n♂=6,5	XY/XX	Holokinetic (diffuse)
28	Homoptera	n=2-20	XO/XX (occasionally neo-XY/XX)	Holokinetic (diffuse)
29	Heteroptera	n=2-26 (modes at 7,12 and 17)	XY/XX	Diffuse

He surmised that of the close to one million known insect species in the world, with several thousands of new species being additionally described annually, the number of cytologically investigated species exceeded the 10,000 mark, but the percentage values varied greatly, with Odonata being by far the best examined group, though even in dragonflies serious gaps existed. It became apparent that it was not the percentage of cytologically examined species of a certain order that can give an impression of current knowledge on its cytology, but rather the representativeness of the coverage of families and other higher taxa, since the cytotaxonomic features are often highly distinct on higher taxonomic levels within one and the same order.

One of the biggest initial questions Dr Kiauta faced was to justify uniform distribution of cytotaxonomic characteristics in a particular insect order, and their comprehensible affinities with those in the allied insect orders in the phylogenetic tree. The classification of insects has passed through many changes and with the increased detailed knowledge an increasing number of orders has come to be recognized. The taxonomic level of some of these, however, is subject to continuous reorganizations and reclassifications. Recently, even the main subdivision like Apterygota and Pterygota has become a matter of reconsideration and in case of the former, there are strong reasons in favour's of the elevation of each of these to an independent class. As in vogue at that time, Dr Kiauta favoured Martynov's scheme of classification and phylogenetic affinities within the class as depicted below in Figure 10.1.

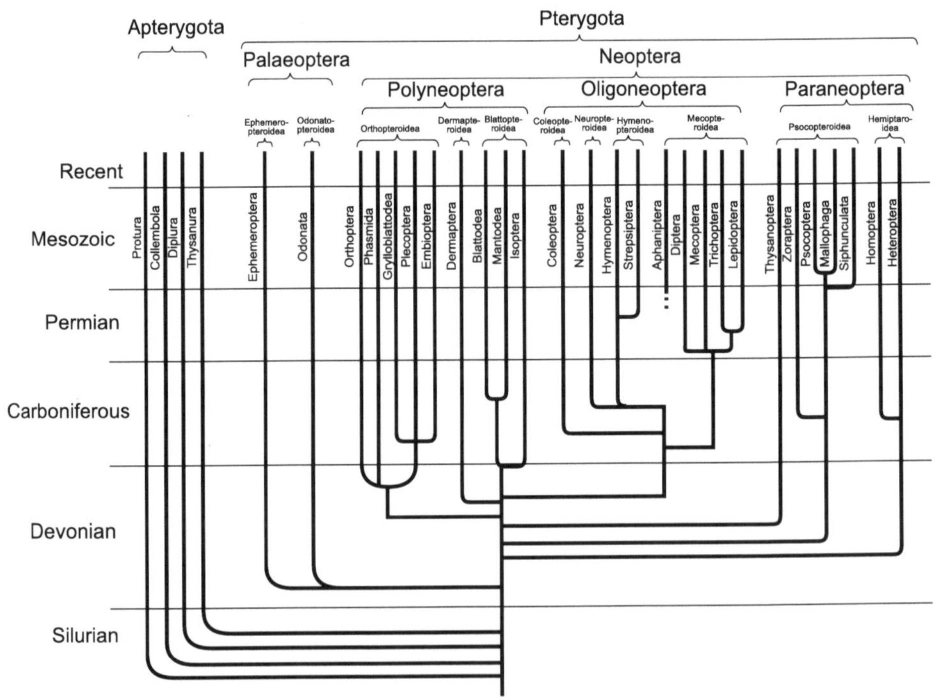

Fig. 10.1. Martynov's scheme of insect classification (From Kiauta, 1974).

Next, Dr Kiauta explained that the insect community called dragonflies (order Odonata, Superorder Odonatopteroidea), in the very first glance, differed in more than one feature from all other insect superorders, including the mayflies (Ephemeropteroidea) –the only other members of the cohort (or pterygote subclass ?) of Palaeoptera which were first to split off from the pterygotan stock during probably the Lower Devonian period. The two superorders are but very remotely allied, and do not at all show any structural affinities with the other pterygotan groups. Both the superorders are characterized by a number of extremely archaic features, have evolved but very little in the course of the past geological epochs, and are, therefore, justly considered as a kind of living fossils.

He drew geological history, genealogy and classification of the superorder Odonatopteroidea in order to place the living dragonflies representing but one order (Odonata) of a once flourishing group, the other order of which, Protodonata, became extinct already in the Upper Permian. He emphasized that in no other insect group was the paleontological evidence as complete as in dragonflies; therefore it was not surprising that the classification of the superorder was nearly entirely based on those morphological features that were liable to preservation in fossil material, most particularly on wing venation. He further re-emphasized that nothing was known on the direct ancestors and the earliest dragonfly history. In the Lower Carboniferous the two orders, Protodonata and Odonata, were already flourishing. While the former became extinct towards the end of Paleozoic, three out of seven odonate suborders survived to the present time, viz. Zygoptera, Anisozygoptera and Anisoptera. Anisozygoptera were the dominant suborder of the Mesozoic era and are represented in the present day fauna only by two relic species (at present, in fact, there are three *Epiophlebia sinensis* being the additional species), one of which is endemic in the Himalaya (Darjeeling), i.e., *E. laidlawi*. The Zygoptera and Anisoptera, on the other hand, are the dominant groups of our Cenozoic times (Fig. 10.2).

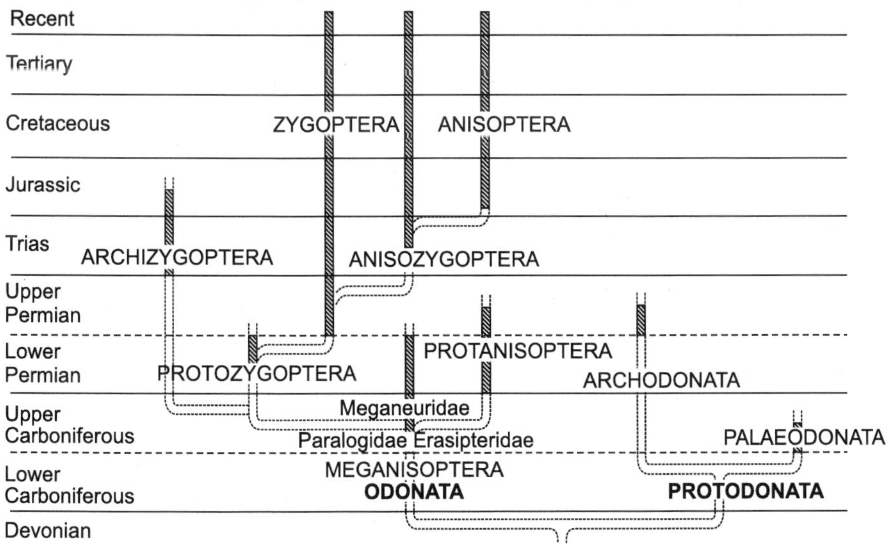

Fig. 10.2. Geological history, genealogy and classification of the superorder Odonatopteroidea. The dotted lines indicate supposed affinities and periods for which no paleontological evidence is available (From Kiauta, 1975b).

Aside from the fact that in several families, too, little material had been examined and that cytological data were lacking for a few phylogenetically very important families, there were few, if any, orders in insects with cytological records as complete as dragonflies (KIAUTA, 1972g). However, he cautioned that more material would be needed to be studied before the cytotaxonomic and, particularly, the cytophylogenetic picture of the order will become clear. From the cytogenetic point of view, on the other hand, it was expected that new material would render few new aspects. Some 6000 known species of the three living suborders were organized into 28 families. He summarized the cytotaxonomic record of 450 species, referable to 22 families of all three suborders. He drew a genealogical tree of the living Odonata, with the names of cytologically studied families framed (Fig. 10.3).

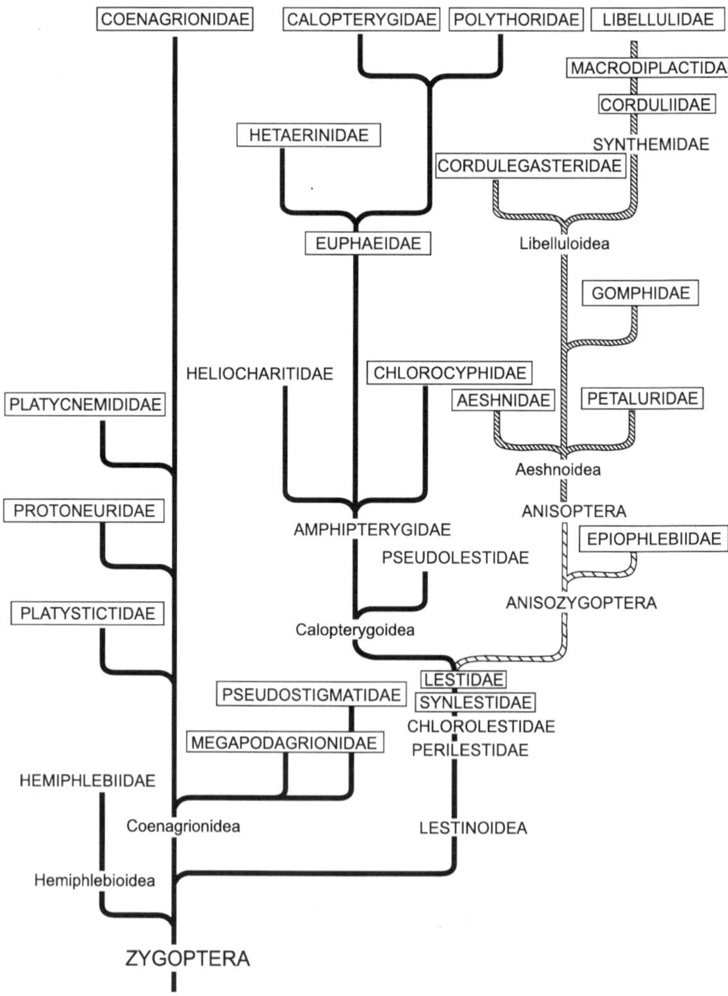

Fig. 10.3. Geanealogy and classification of the living odonate families, with indication of cytologically studied (framed) and unknown (unframed) families. – (From Kiauta 1975a)

At first glance at least two important lacunae in the cytotaxonomic record were brought on record by him:

(a) The superfamily Hemiphlebioidea, with single family *Hemiplebiidae*, containing a monotypic genus *Hemiplebia* (*H. mirabilis* Selys, from Gouldburn River, Victoria, Australia) was considered, on the basis of venational charaters, as a transitional form between the Coanagrionidea on one hand, and the extinct suborder Protozygoptera on the other. It could, therefore, perhaps furnish some evidence on the cytological conditions during the earliest history of Zygoptera.

(b) In view of the peculiar cytotaxonomic situation found in the only member of the primitive *Pseudolestidae* studied so far, i.e., *Hypolestes clara* (Calvert) (n♂=9), it was suggested to gather some information on the karyotypic condition in the *Amphipterygidae*, which family occupied a key position at the base of the higher Calopterygoidea. The more so, since in the *Chlorocyphidae*, one of the direct derivatives of the *Amphipterygidae*, primitive karyotypes seemed to prevail.

Dr Kiauta travelled all over the world a number of times and extensively studied chromosome behaviour in a large number of species belonging to different families. He discovered many novelties in dragonfly cytogenetics, including type of kinetochore the dragonflies have on their chromosomes, role of *m*-chromosome in developing chromosome complements and chorogeography of certain taxa, kinetics of sex-chromosomes etc. etc. Two of his theories were widely accepted, viz., (i) Evolution of new chromosome complements through 'autosomal fusion and fragmentation', and (ii) Development of new sex-determining mechanisms involving sex and autosomal fusions. These theories have offered a new insight in the dragonfly cytology and for next five decades have withstood the test of time and scientific inquiry.

1. Theory of evolution of odonate chromosome numbers

Dragonflies have different sexes; males being heterogametic. The male haploid chromosome numbers in dragonflies range from 3-15 (Fig. 10.4).

Nevertheless, in spite of this rather broad range, there is in general little numerical variation in dragonfly complements. The haploid numbers 12, 13 and 14 are represented in more than 95% of the Odonata examined, hence n = 13, found in more than 60% of the species studied cytologically, is considered as the type number of the order (Table 10.2).

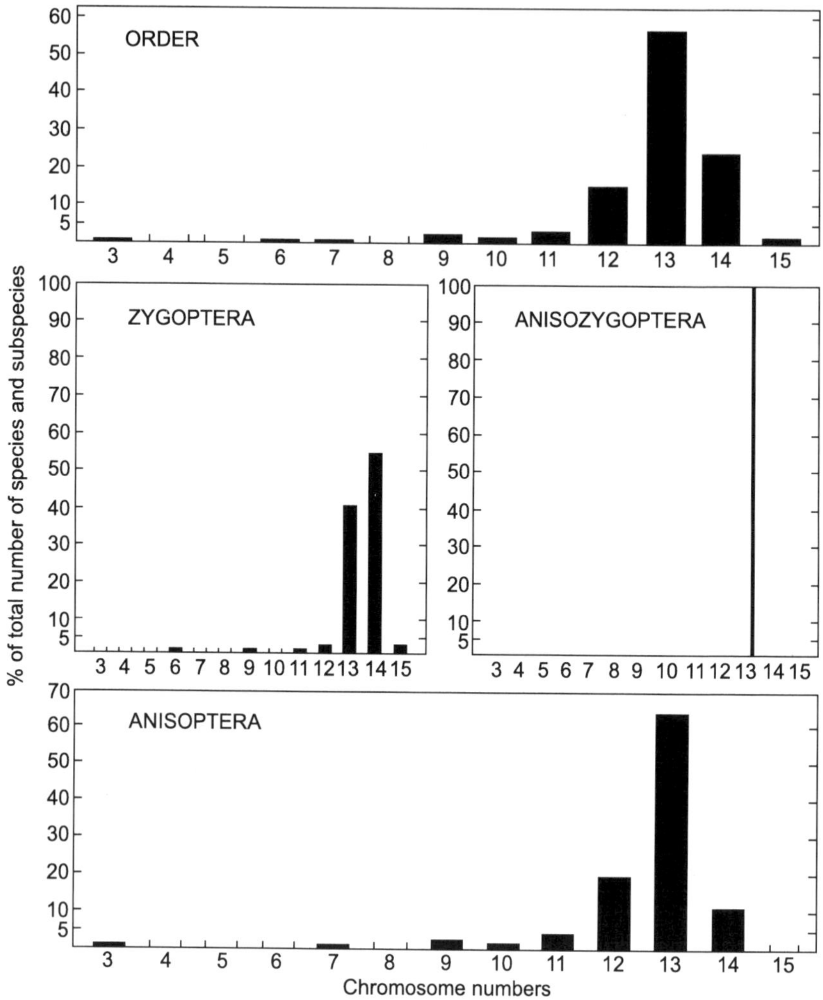

Fig. 10.4. Histogram of haploid chromosome numbers in the order and in the three living suborders (in % of total number of species studied). The width of columns is proportional to the absolute number of cytotaxonomically examined species within the order and suborders. (From KIAUTA, 1975a).

Table 10.2. Family type numbers of some odonate families (From Kiauta, 1975b)

Suborder/*Family*	**Type number**
Zygoptera	
Platystictidae	13
Protoneuridae	14
Platycnemididae	13
Coenagrionidae	14
Synlestidae	13

Megapodagrionidae	13
Lestidae	13
Pseudolestidae	9?
Polythoridae	12
Euphaeidae	13
Chlorocyphidae	12
Hetaerinidae	13
Calopterygidae	13
Anisozygoptera	
Epiophlebiidae	13
Anisoptera	
Gomphidae	12
Petaluridae	9
Aeshnidae	14
Cordulegasteridae	13
Corduliidae	13
Macrodiplactidae	13
Libellulidae	13

Dragonfly Karyotypes can be divided roughly into two groups: (1) the "normal", high-n complements (n = 9 to 14), and (2) the low-n complements (n = 3 to 8). Chromosome size in the high-n species is approximately half that in the low-n karyotypes. In the high-n species a parallel is apparent between specialization and advancement on one hand, and increase of the chromosome number on the other. The low-n species can be found in any systematic group, of any type number. It has been demonstrated in numerous cases that the low-n complement arises by fusion of some or all elements of the original complement, hence nomenclature "primary" and "secondary" complements. Aside from fusions, in several species fragmentation of one or more elements of the original set has taken place, resulting in an increase of the chromosome number. These complements too, are called "secondary" and may occur in single species of any taxonomic group.

Before Dr Kiauta several investigators had discussed the evolution of the odonate chromosome complement. Some expressed the opinion that the chromosome numbers lower than the type number came about by fusion of elements of the ancestral karyotype, others considered the *m*-chromosome as an autosome undergoing a gradual diminution (*cf.* KIAUTA, 1975b). The latter hypothsis became known as the *m*-chromosome theory.

Four main objections were raised against these hypotheses: (1) they either did not consider or misinterpreted the phylogenetic position and structural advancement (specialization) of the taxonomic groups involved, (2) they took as starting point for evolutionary speculations the family type numbers, giving them *eo ipso* the meaning of ancestral numbers, (3) they failed to account for chromosome numbers exceeding the family type number, and (4) the *m*-chromosome theory was built up on an *ad hoc* assumption of gradual diminution and eventual disappearance of one autosome after another, which was in contradiction with the observation that the total chromosome length in, say, n = 14 species is about the same as in n = 13 species.

In support of his hypotheses Dr Kiauta postulated that the lowest of the high-n complements, n = 9, seems actually the most probable approximation to the ancestral chromosome number of the Odonata. On palaeontological grounds one could suggest that this number represents the true ancestral complement of the order. It was found in the most archaic and phylogenetically the most primitive among living dragonflies. It seems most likely that any numerical variation in odonate karyotypes had developed by the occurrence of breaks leading to haploid numbers 10 to 15 and by fusions leading to haploid numbers 3 to 8 in ancestral forms which had a chromosome number n = 9.

The high-n karyotypes are not homogenous by origin either. Fusion of two or few elements gave rise to secondary high-n types (14 → 13 → 12 etc.). In many such cases fusion could be traced because the fused chromosomes are significantly longer than other chromosomes of the set. If only two chromosomes were fused, the sex chromosome might be involved, and a neo-XX/XY sex determination mechanism replaced the usual XX/XO type. The fusion is not always found in all populations of the same species, nor does it always occur in all cells of one individual. Based on this interpretation, he presented a schematic interpretation of the probable evolution of dragonfly chromosome numbers (Fig. 10.5).

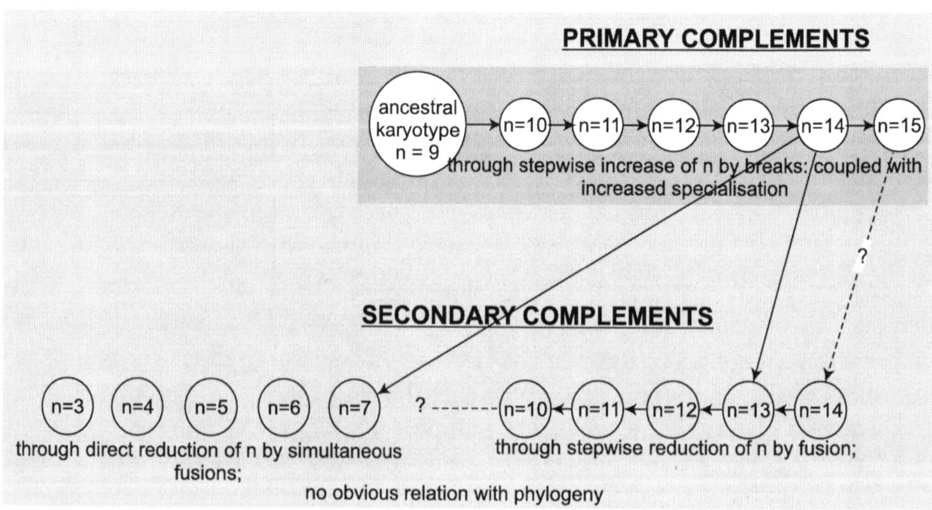

Fig. 10.5. Diagram of probable trends in the evolution of odonate chromosome numbers. (From Kiauta, 1975b).

2. Development of modes of sex determination

A sex-determination system is a biological system that determines the development of sexual characteristics in an organism. The morphological and kinetical features of the odonate sex chromosomes are rather simple-looking, nevertheless, the riddle of trends in the evolution of different modes of sex determination in the order were hardly ever explained thoroughly. The cytological conditions in the family Gomphidae are of particular importance for the understanding of the evolution of sex determination in dragonflies and are discussed in detail. The original mode of sex determination in Odonata is of the XO/XX type, the male being the heterogametic sex, and is observed in all primary complements (*cf.* KIAUTA, 1967a) regardless of the chromosome number, viz., the degree of phylogenetic advancement and specialization achieved by the taxons concerned. The mean length of the sex chromosome and the ratio between the longest autosomal bivalent and X are not characteristic at superspecific taxonomic levels. In the male the sex chromosome is usually positively heteropycnotic at all spermatocyte stages, save metaphase and anaphase, whereas in the female it is nearly without exception isocyclic. The first maturation division is equational for X, the second is reductional. In secondary complements a neo-XY sex determination occurs, in those cases where the original X was involved in a fusion with an autosome. Its occurrence is not related to phylogeny. The neo-XY condition is often reversible: it occurs in some cells (or stages) while not in others of the same individual. If stabilized, it tends to evolve further, as it was demonstrated in Gomphidae, into a secondary XO type. The process has three stages, with an intermediate neo-X neo-neo-Y sex determination and a final numeric reduction of the diploid complement by two elements. It was demonstrated, that the present type number of Gomphidae is of secondary rather than of primary origin.

He innovatively offered a graphical interpretation of the paths of evolution of sex determining mechanisms in dragonflies as depicted in Figure 10.6.

The cyclic three successive stages occurring in the process of stepwise numeral reduction, as far as the sex element is involved, are elaborated here. At the *first stage* one autosome of the primary complement fuses with the original X. The result is a neo-XY karyotypic and numeric reduction of the primary complement by one. The *second stage* in the process is represented by fusion of neo-Y with an autosome, which reduces the chromosome number for another element and gives rise to a neo-X neo-neo-Y sex determining mechanism. The final *third stage* is achieved by translocation by the autosomal part of the neo-X to an autosome. Its result is the reappearance of the XO system, which is not accompanied by a numerical reduction. All three stages were demonstrated in the family Gomphidae, but the relatively common occurrence of neo-XY complements in most advanced families suggest the universality of the process.

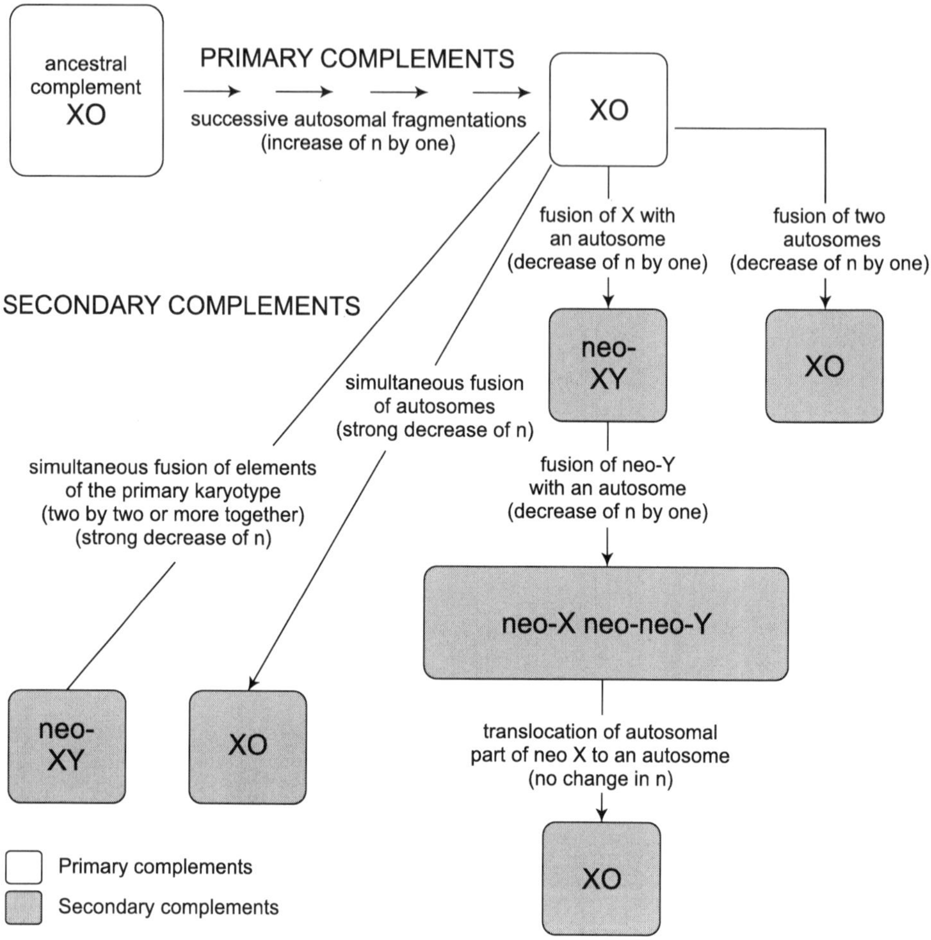

Fig. 10.6. Diagram of the trends in the evolution of sex-determining mechanisms in dragonflies. (For the general evolution of the odonate complement *cf*. Fig. 10.5 above). (From Kiauta, 1975a)

If the reduction of the primary chromosome number is achieved by simultaneous fusion of all elements of the karyotype (two by two or more together) including the original X, a neo-XY system arises. Thus far, this special case could be demonstrated only in a single species (*Hemianax ephippiger*).

From the foregoing arguments Dr Kiauta concluded the following:

1. The XO sex determination is the original mode of sex determination in dragonflies and is the only type found in all primary complements, regardless of the chromosome number, viz., the degree of phylogenetic advancement and specialization achieved by the forms concerned.
2. The secondary complements which originate exclusively by fusion of the elements of the primary karyotypes and do not have any relationship

with the phylogenetic position of the species concerned, both the XO and the neo-XY sex determination occur. The former was either original and inherited from the primary complement if the sex element was not involved in fusion, or it was of a secondary origin and had evolved through the intermediate neo-X neo-neo-Y stage.

3. The neo-XY is characterized by (a) reversibility at the first stage of the stepwise reduction process (it is usually not present in all cells of one individual), (b) by the tendency to evolve through the neo-X neo-neo-Y system into an XO sex determination again, and (c) by its occurrence in forms which have achieved different degrees of advancement and specialization.
4. The neo-XY condition, therefore, could not always be regarded as a character of cytological specialization. Nevertheless, it appears as such in the families Gomphidae, Aeshnidae, and perhaps also in the libellulidan genus *Orthemis*.

By these and many other epoch-making cytotaxonomic and cytophylogenetic investigations in dragonflies, Dr Kiauta has laid foundation of the highest quality of research making and interpretation. Many young researchers (e.g., Dr Mola, Dr Tyagi, Dr Chatterjee, Dr. Sidhu, Dr Walia, to name a few only) followed in his footsteps of doing their own piece of research with the same force of focused attention and dedication as Dr Kiauta did. Many years later, two important publications each by TYAGI (1977) and MOLA et al. (1999), reporting respectively neo-neoX/neo-neo-Y and $X_1X_2Y/X_1X_1X_2X_2$ types of sex-determining mechanisms, actually reaffirmed Dr Kiauta's hypothesis.

Like in any other discipline of life, here too, Dr Kiauta's epochal theories did not smoothly go down to be accepted without resistance or repudiation from some of his contemporarians. Professor M.J.D. White, one of the greatest cytogeneticists during mid 20th century, remarked sceptically in his famous book (*vide supra*) about Dr Kiauta's theory of evolution of karyotypes, and thereby also the evolution of new sex-determining mechanisms in Odonata through increase or decrease of chromosome numbers, popularly referred to as "the theory of chromosomal fragmentation and fusion". Still, a healthy criticism is necessary, and most welcome since it is also a good measure of the strength in one's argument that reminds me of what Winston Churchill, the British Prime Minister and the Nobel Laureate, had had to say:"*You have enemies? Good. It means you've stood up for something, sometime in your life.*"

●●●

11

Research Missions of the Netherlands Centre for Alpine Biological Research (N.C.A.B.R.) to the Nepal Himalaya (1972-1980)

"चाह नहीं मैं सुरबाला के गहनों में गूँथा जाऊँ,
चाह नहीं, प्रेमी-माला में बिंध प्यारी को ललचाऊँ,
चाच नहीं, सम्राटों के शव पर हे हरि, डाला जाऊँ,
चाह नहीं, देवों के सिर पर चढ़ूँ भाग्य पर इठलाऊँ,
मुझे तोड़ लेना वनमाली! उस पथ पर देना तुम फेंक,
मातृभूमि पर शीश चढ़ाने जिस पर जावें वीर अनेक।"

Transliteration:

Chah Nahin Mai Sur Bala Ke Gehnon Mein Guntha Jaaon,
Chah Nahin, Premi Mala Mein Bindh Pyari Ko Lalchaon,
Chah Nahin, Samraton Ke Shav Par He Hari, Dala Jaaon,
Chah Nahin, Devon Ke Sar Par Chadhoon, Bhagya Par Ithlaoon,
Mujhey Tod Lena Banmali, Us Path Par Tum Dena Phaink,
Matra Bhoomi Per Sheesh Chadhaney, Jis Path Jaayen Veer Anek

Translation:

I don't desire to be part of maiden's jewellery,
I don't desire to be in the garland of young lovers luring them with my charm,
I don't desire to be placed on the mortal remains of monarchs,
I don't desire to be placed as the crown of Gods and take pride in my luck,
rather, I would request the gardener to pluck me and strew me on the path,
which would be trodden on by the brave soldiers on their way to sacrifice themselves for my country.

—Makhanlal Chaturvedi

in "Pushp ki abhilasha/Longings of a flower"

Netherlands Centre for Alpine Biological Research is a Netherlands foundation, at that time associated with the University of Utrecht. In Nepal, Dr. Kiauta and team were invited by Professor Dr W. Hellmich, Head of the (German) Association for Comparative High Mountain Research (= *Arbeitsgemeinschaft für vergleichende Hochgebirgsforschung, Munich*), who rendered hospitality in their Nepal Research Centre (Chauni, Kathmandu). The Manager of the Centre, G.B. Kalikote, M.Sc (Geology), was very helpful in many ways, including in assembling the local personnel (Sherpas and porters) and served as a liaison with the Nepalese Authorities. Aside of the high-altitude Sherpas, the missions employed up to 16 porters and some specialized personnel (cooks, kitchen boys etc.).

Since 1970, Dr Kiauta served as Head of the Netherlands Centre for Alpine Biological Research (Utrecht) and did much research also in the European Alps, the Canadian Rockies and elsewhere. In this capacity and in cooperation with the late Professor Dr W. Hellmich, he conducted one international (1974) and four Netherlands (1972, 1973, 1976, 1980) research missions to Nepal Himalaya, with the principle objective of the study of invertebrate communities, their ecology and succession in glacier moraines:

- 1972, September-October. – Pilot Mission to Khumbu Himal (Mt Everest area), with base camp in the Kala Pattar (alt. 5545 m) – Gorak Shep area, on the Khumbu Glacier, SE of Pumori (7145 m). The invertebrate community composition, structure, ecology and succession were studied in the moraines. Also examined were the Pumori SE slopes up to an elevation of ca 6100 m and mountain dragonflies were systematically collected in the Kathmandu Valley (ca 1350 m and up) and along the Lukla (2900 m) – Namche Bazar (3440 m) – Kumjung (3800 m) – Tengpoche (3867 m) trek.
- 1973, April-May. Same area and objectives, extended to Trashi Laptsa (5755 m). The equipment included a tent with a cytogenetics laboratory.
- 1974, August-October, with international participation, including Dr H.J. Dumont and some of his collaborators. Same area and objectives, but much time was spent also in the Kathmandu Valley. From Tau Daha lake, Dr Dumont described a new rotifer species, *Scaridium bostjani*. At the Tribhuvan University of Kathmandu, Dr Kiauta has given a few-days seminar for students and Staff (lectures-cum-laboratory practice) on insect cytotaxonomy, with special reference to the fauna of the Indian Subcontinent. For this purpose, two booklets were published by the Netherlands Centre for Alpine Biological Research, in cooperation with the Nepal Research Centre (KIAUTA, 1974, 1975a).
- 1976, April-May. Helambu Himal, Gosaikund. Mostly invertebrate collecting, cytotaxonomy research and odonate collecting at lower elevations (550-4400 m).
- 1980, July-August. Arun Valley, mostly elevations up to 2500 m, the highest point worked at was Popti La (4300 m), close to the Tibetan border. During the monsoon, the fauna was rich, but at times the work conditions inconvenient. The mission was extended to Singapore, the Philippines, Japan and Taiwan, all devoted to dragonfly cytology.

Dr Kiauta's findings concluded that in nival biota, the odonates are replaced as principal predators by some representatives of various lower arthropod groups (Chilopoda, Chelicerata) and by some insects, pertaining to the orders of Dermaptera and Diptera, and to the coleopteran families of Carabidae and Staphylinidae though, as migrants, dragonflies do occur even at elevations exceeding the 6000 m mark (*cf.* KIAUTA, 1974). Bastiaan's work was largely centred at higher altitudes (up to ca 6100 m), at lower elevations (up to ca 4000 m), however, the odonate fauna and odonate cytology also received much attention (e.g. KIAUTA, 1975a). He even dealt with the dragonfly folk names in Nepal. Yet, perhaps more importantly, his wife Marianne (then M.A.J.E. Kiauta-Brink) has deeply explored about some Tibetan expressions for "dragonfly" with special reference to the biological features and demonology (KIAUTA-BRINK, 1976,1977).

At that time the Nepalese odonate fauna was still insufficiently known, though more than 120 species had been brought on record from the Nepalese territory. To him, as a cytogeneticist and cytotaxonomist no less than 42 of these were studied cytologically, which made dragonflies by far the best studied insect order also from the Nepalese point of view. In the past quite a few such expeditions were undertaken, e.g., Botanical Expedition of Tokyo University, 1963; Rikkyo University Himalayan Expedition, 1964 (ASAHINA, 1964, 1965), Yugoslav 1969 Himalaya Expedition (KIAUTA, 1972a). Prof. S. Asahina (Japan) and St. Quentin (Austria) were among the few experts on the Nepalese fauna. Asahina undertook a special expedition to Nepal in quest of *Epiophlebia laidlawi/superstes* in the 1980s. Dr Kiauta tracked many remote and odonatologically rich areas during his sojourn in Nepal under the NCABR such as, e.g., Western Nepal (Humla and western Terai), Kathmandu valley, the area between Kathmandu valley (including Godavari area famous for classical dragonfly localities) and Khumbu Himal, Eastern Nepal: Arun valley, Langtang valley, and Kathmandu to Chinese border.

In the process of executing these expeditions Dr. Kiauta interacted with a large number of odonatologically interested and actively publishing workers, zoologists and other colleagues at the Tribhuvan University and Entomology Department of the Nepalese Ministry of Agriculture, Nepal Research Centre of the German Oriental Society (Deutsche Morgenländische Gesellschaft), Kathmandu Natural History Museum, Nepal-Holland Friendship Society and Nepal-German Manuscript Preservation Project, most prominent of which were; C. Smith, L. Shrestha, Prof. Dr T.C. Majupuria, S.S. Singh, and Prof. Dr. R. Uprety, Dr P.N. Rana, Dr K.C. Sharma, G.B. Kalikote, R.N. Sharma, Dr W. Voigt, Dr F. Kollmannsperger and Dr M. Witzel. He even helped at that time the sole and very active native odonatologist Mahendra Mahato, whom upon his request during the Ninth S.I.O. International Symposium of Odonatology held in Madurai in 1988, found him a position as a research scholar under Prof. Dan Johnson of Zoology Department, Tennessee University, United States of America, where Mahato moved in 1988, made his PhD and settled to reside there permanent basis.

Dr Kiauta's research on dragonfly cytotaxonomy is of particular interest in Nepalese reference since it set the formation of a major project under which a couple of *magnum opuses* on odonate cytotaxonomy could be prepared following a great legacy in the valley state. The Langtang Khola, visited during 1976, arises not far over the Tibetan border on the shoulder of the Sisha Pangma and flows West under the slopes of Langtang Lining (7245 m) into the Bhote Kosi, coming from far inside Tibet. The floor of the upper valley, beyond Ghora Tabela, lies at elevations between 3200 and 4500 m. It is one of the most frequently visited valleys of the Nepalese central Himalaya. Lying only a 10-day trek north of Kathmandu, the Langtang Valley was one of the first geographically explored regions in the Nepal Himalaya in the early days of Nepalese mountaineering (Fig. 11.1).

Fig. 11.1. Langtang Valley with the village of Langtang (3430 m) and its walled potato fields. In the foreground marshy meadows with small pools. Typical site of *Indolestes cyanea* (Selys) and its highest locality in the Langtang Valley (early May). (Source: M. Kiauta)

As early as 1949 it was visited by H.W. Tilman, accompanied by the botanist O. Polunin and the geologist J.S. Scott. As sirdar and cook of the party served Tenzing Norgay of the later Everest fame, at that time already one of the most experienced Sherpas. To his vivid account of the expedition TILMAN (1952) appended an essay on the natural history of the valley by the hand of Polunin. This, and an earlier paper by POLUNIN (1950), became the basis for all forthcoming biological, and particularly ecological work in the area.

At present, the valley is classified under the Nepal Government's Remote Areas Development Program as a National Park (surface 1224 km^2) and for this reason now receiving considerable attention from foreign biologists. Although Tilman already collected some insect material in the region, so far no systematical collecting has been carried out and the only few records on the odonate fauna of this vast area are those published by ASAHINA (1974).

The greenest and most pristine part of the Kathmandu valley lies at its southeastern edge, around Godavari (ca. 1600 m). Nestling at the foot of forested Phulchoki, the highest peak of the valley rim, are the pleasant National Botanical

Garden, the temple at Naudhara and the shrine of Bishanku Narayan, hidden in a gorgeously rural side-valley. The Godavari area is one of the classical dragonfly localities in the Kathmandu Valley, being very rich in various types of running and stagnant waters. It is the type locality of the platycnemidide *Coeliccia dierli* St. Quentin, 1970. The list of species recorded from this area includes, among others, also *Calicnemia nipalica* (Kimmins), *C. pulverulans* (Sel.), *Coeliccia renifera* (Sel.), *Ceriagrion azureum* (Sel.), *C. cerinomelas* Lieft., *Enallagma parvum* Sel., *Ischnura aurora* (Brau.), *I. forcipata* Mort., *I. mildredae* Fraser, *I. rufostigma annandalei* Sel., *Anisogomphus occipitalis* (Sel.), *Onychogomphus schmidti* Fraser, *Chlorogomphus olympicus* Fraser, *C. selysii* Fraser, *Anax nigrofasciatus nigrolineatus* Fraser, *Gynacantha incisura* Fraser, *Acisoma panorpoides* Ramb., *Crocothemis servilia* (Drury), *Diplacodes trivialis* (Ramb.), *Neurothemis fulvia* (Drury), *N. intermedia intermedia* (Ramb.), *Orthetrum devium* Needh., *O. glaucum* (Brau.), *O. japonicum internum* McL., *O. luzonicum* Brau., *O. pruinosum neglectum* (Ramb.), *O. sabina* (Drury), *O. triangulare triangulare* (Sel.), *Palpopleura sexmaculata* (Fabr.), *Pantala flavescens* (Fabr.), *Trithemis aurora* (Burm.) and *T. festiva* (Ramb.).

Dr and Mrs Kiauta visited the Godavari area several times and encountered some more additional odonates (Fig. 11.2). Among the species found breeding in a characteristic rivulet in Godavari, near the Royal Botanic Gardens, were *Rhinocypha quadrimaculata* Sel., *R. unimaculata* Sel., *Bayadera indica* (Sel.), *Neurobasis chinensis* (L.), *Onychogomphus bistrigatus* (Hag.), *Temnogomphus bivittatus* (Sel.), *Macromia moorei* Sel., and *Anotogaster nipalensis* Sel.

Fig. 11.2. Dragonfly collecting by Marianne Kiauta at a characteristic rivulet in Godavari, near the Royal Botanic Gardens. (Photo: B. Kiauta).

During the 1976 mission (April-May) of the Netherlands Centre for Alpine Biological Research and the University of Utrecht, among other projects, a systematic survey was made of the odonate fauna of the region as demarcated by the line Trisuli Bazar – Ramche – Dhunche – Syabru – Ghora Tabela – Kyangjin Gompa – Gosaikund. For his cytological preparation Dr Kiauta used a part of the material, collected along the localities shown in the trekking map "Gosaikund – Helambu – Langtang", at elevations measured on the spot with the help of a Thommen altimeter. Only imaginal males were used for the chromosome preparations (Fig. 11.3). They refered to the following taxa: Zygoptera: *Drepanosticta* sp. (Platystictidae), *Indolestes cyanea* (Selys) (Lestidae), *Anisopleura comes* Selys (Euphaeidae); Anisoptera: *Neallogaster* sp. near *hermionae* (Fraser) (Cordulegastridae), and *Orthetrum japonicum internum* McLachlan (Libellulidae). Of these, only the last named had been previously studied cytologically (KIAUTA, 1975a). *Drepanosticta* sp. (not *D. carmichaeli* [Laidlaw, 1915], recorded from Nepal earlier by St. QUENTIN (1970) and *Neallogaster* sp. near *hermionae* are new to the Nepalese fauna. The slides were fixed in the field in acetic carmine, temporarily sealed off, and transported in this stage to Kathmandu, where they were further processed (Feulgen) in the field laboratory set up in the premises of the Nepal Research Center. They were photographed in Utrecht, and stored, along with the specimens, in the collections of the Netherlands Centre for Alpine Biological Research.

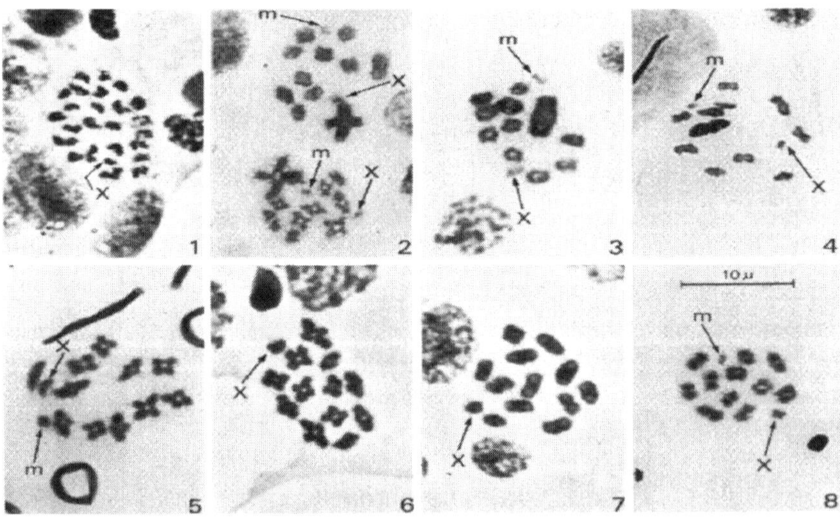

Figs. 11.3 (1-8). Male germ cell chromosomes of five dragonfly species from the Langtang Valley, central Nepal Himalaya (Feulgen squash, 1500 X): (1) *Drepanosticta* sp., spermatogonial metaphase (Manigaon 1340 m); — (2-4) *Indolestes cyanea* (Selys) (note the extremely large autosomal bivalent/dyad); (2) late diakinesis / early metaphase I (note the delayed chiasma terminalisation and chromosome contraction of the large bivalent (Langtang Khola, 1900 m), (3) metaphase I (Langtang Khola, 1900 m), (4) metaphase II (Langtang Village, 3430 m);— (5) *Anisopleura comes* Selys, late diakinesis (Ramche, 1530 m); — (6-7 *Neallogaster* sp. aff. *hermionae* (Fraser), late diakinesis and metaphase I,respectively (Langtang Khola, 2150 m); — (8) *Orthetrum japonicum internum* McLachlan, early metaphase I (Syabru, 2000 m).

This study revealed the male germ cell complements of *Drepanosticta* sp. (Platystictidae), *Indolestes cyanea* (Sel.) (Lestidae), *Anisopleura comes* Sel. (Euphaeidae), *Neallogaster* sp. *aff. hermionae* (Fraser) (Cordulegasteridae), and *Orthetrum japonicum internum* McLachl. (Libellulidae). In all of these the chromosome number was n = 5-13, while a pair of *m*-elements was lacking in *Drepanosticta* sp. and in *Neallogaster*. Save for *O. j. internum* none of the species had been previously studied cytologically. *Indolestes cyanea* was characterized by an extremely large autosome pair (bivalent). The karyotype of *A. comes* was similar to that of *Bayadera indica* (Sel.), but differed considerably from the complement of the more "primitive" member of the family, *Epallage fatime* (Charp.). *Neallogaster* sp. was the first cordulegastride species lacking definitely the *m*-chromosomes. The latter and *Drepanosticta* sp. were new to the Nepalese fauna.

Dr Kiauta, an established dragonfly taxonomist by then, trained first in taxonomy under Dr D.E. Kimmins at the British Museum (Nat. Hist.), London, United Kingdom during 1959 and thereafter in biogeography and evolution under Dr M.A. Lieftinck, Rhenen, The Netherlands, also rendered identification service to many colleagues involved in odonate collections in Nepal. Thus, following the 1972 post-monsoon mission to Humla and the western Terai, Professor H. Franz (Hochschule fur Bodenkultur, Vienna, Austria) collected also some odonate material and kindly passed it on to Dr Kiauta for identification. The collection consisted of 32 adult specimens, referable to 12 species of nine genera, arising out of Rara lake north of Jumla (alt. 2900 m app.) and Amlekhganj in Terai. Since hardly anything was known on odonate fauna of Terai and no collections were ever made in the western part of it and in the Humal region, it seemed to him worthwhile to identify them and briefly describe them, more so also because there were two new records for the Nepalese fauna. He identified the following species: *Ceriagrion cerinomelas, Ischnura elegans, I. aurora, Agriocnemis pygmaea* (all Zygoptera), *Orthetrum glaucum, O. sabina, Acisoma p. panorpoides, Crocothemis servilia, Diplacodes nebulosa, D. trivialis, Sympetrum haematoneura* and *Trithemis pallidinervis* (all Anisoptera). He subsequently deposited all specimens in the collections of the NCABR, Utrecht.

It is noteworthy to mention here that during his assignment in Nepal under the NCABR, Dr Kiauta prepared, among many significant odonatological publications, two most invaluable books in the form of lectures delivered at the Tribhuvan University, Kathmandu (KIAUTA,1974, 1975a) which, even after four decades, continue to serve as indispensable reference monographs on the subject in the whole of Indian Subcontinent (Fig. 11.4).

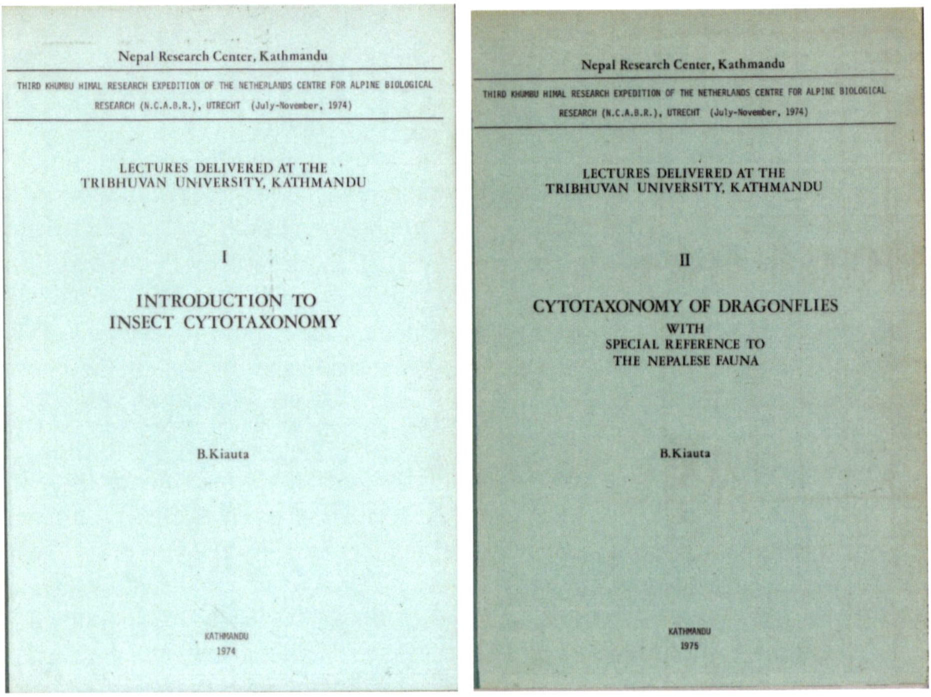

Fig. 11.4. B. Kiauta (1974, 1975 a): covers of the two booklets, published by the Nepal Research Center (Kathmandu) for the Tribhuvan University of Kathmandu, Nepal. Vol. 1 was published in two editions and it was widely used at various universities in Europe and, particularly, in southeastern and eastern Asia, South Africa and at the McGill University of Montreal, Canada. – (Source: B. Kiauta archive)

The first of these books is truly global in treatment since it has the purpose and scope of presenting a very generalized outline of the main cytotaxonomic and cytogenetic features met with in insects, even if it was intended as an introduction to insect cytotaxonomy for biology students of the Tribhuvan University of Kathmandu, with emphasis on bringing on record the information pertaining to the following cytotaxonomic features:

(i) General taxonomic classification and geographic dispersal, number of known and of cytotaxonomically investigated taxa, with a brief reference, where necessary, to the Indian (and Nepalese) species; mentioning the occurrence in the Himalaya;

(ii) Indication of tissues and ontogenetic stages in which divisions can be followed;

(iii) Type of centromere, form of chromosomes, and range of chromosome numbers;

(iv) Type of sex-determining mechanisms, and some features of sex chromosome morphology;

(v) Notes on chromosome behaviour;

(vi) Notes on taxonomic significance of cytogenetic features – where possible and considered useful;

(vii) Selected references, divided into general references for the order and those on the cytology of Indian (and Nepalese) material.

Dr Kiauta soon realized that the Nepalese students of cytogenetics did not have access to most of the world literature in vogue, and needed to be guided into the intricacies of odonate cytogenetics and cytotaxonomy. He, therefore, redefined and simplified the terminology used in describing science of cytotaxonomy which, according to him, is the study of the relationships of organisms by a combination of cytology and taxonomy. He further defined and simplified three main constituents of cytotaxonomy:

1. Taxonomy: It deals with the study of the characters of organisms, their identification and the analysis of variation. Information obtained in this way determines the position of organisms in the system of classification.
2. Classification: It deals with the arrangement of individuals into groups and groups into a system, and finally,
3. Systematics: It combines cytology, taxonomy, classification and many other aspects of the study of organisms. At its upper level it is concerned with phylogenetic affinities and evolutionary trends.

Considering the scope and objective of the science of cytotaxonomy, Dr Kiauta ventured to redefine cytotaxonomy which combines information on the kinds of organisms at all three levels of organization:

(1) On the basic (taxonomy) level it collects information on the cytological situation within the given material and, after due evaluation of cytological chatacters and their variation, it can be used for purposes of identification,

(2) In classifying the species into a system, their cytological characters are also taken into account, and

(3) The system built on the basis of combined characters, reflects, to some extent, also the cytophylogenetic trends.

He cautioned that under the term "cytotaxonomic character" are considered all those morphological, structural, chemical and kinetic features of the chromosomes that are subject to variation on any taxonomic level. He re-emphasized that knowledge on the variation of cytotaxonomic characters, combined with understanding of the mechanisms governing the process of heredity, opens up an enormous field of practical applications – not only those within the scope of "modern systematics", but particularly also those needed in the improvement of agriculture, animal husbandry, and in economic and medical entomology.

In insect cytotaxonomy he included a large number of variable morphological and kinetic features of chromosomes and chromosome sets (karyotypes) which can be used in cytotaxonomic classification if their variability does not extend below the level of a population and/or above the level of the class. Some of these are enlisted below:

(i) *Type of centromere:* Most insect chromosomes possess a special region which plays an essential role in the development of the spindle at mitosis and in the separation of the daughter chromosomes at anaphase. It is generally called centromere or kinetochore, and the chromsomes possessing it are called monocentric or monokinetic chromosomes. Chromosomes are called telo-, acro-, or metacentrics, if the centromere lies at the end, near the end or in the middle of the element, respectively.

If such a special region on the chromosome is lacking, and the element shows centromeric activity along its entire length, the chromosomes are called holocentric or holokinetic and the chromosomes are considered diffuse – in contrast to the above described localized centromere.

It is nearly certain that the condition of the centromere is uniform within an order. Thus, diffuse centromeres (holokinetic chromosomes) are peculiar at least to Odonata, Dermaptera, Trichoptera, Lepidoptera, Mallophaga, Siphunculata, Homoptera and Heteroptera. All other insects possess monokinetic chromosomes, save for a few orders where the nature of the centromere is still uncertain.

(ii) *Chromosome number:* Chromosome numbers are of utmost importance in insect cytotaxonomy and cytogenetics, since they determine the recombination index, which represents one of the fundamental properties of a species genetic system. The recombination index is the sum of the haploid chromosome number and the chiasma frequency and represents the mean number of "blocks" of genes segregating independently at meiosis. Thus, low chromosome numbers with high chiasma frequency are equivalent to high numbers with few chiasmata, as far as the extent of meiotic recombination is concerned.

The range of chromosome numbers, from n=2 to n=253, is greater in insects than in any other class, and both the lowest and highest numbers known in the animal kingdom occur in insects. Within single orders the range is, of course, different. Thus, Collembola, Odonata and Diptera have a small range, Lepidoptera an enormous range.

The most common haploid number of a group is called the type number; the highest taxonomic unity for which it can be determined is the order, and the lowest is a geographic population. While some orders, like Odonata and Lepidoptera possess clear-cut type numbers, others have several modes at different numbers, e.g., Coleoptera, Heteroptera and Orthoptera. The family type number can usually be determined if the sample of species studied is sufficiently large.

In orders possessing holokinetic chromosomes, higher chromosome numbers are usually in more advanced taxa, while in those with monokinetic elements the situation is just the opposite, but there are very many exceptions to this rule.

Usually the chromosome number is uniform within a species, but often variations occur even within one and the same individual.

(iii) *Karyotype morphology:* This term implies to the total appearance of the chromosome set of a given organism, including the number, size and shape of chromosomes. It can be characteristic on all taxonomic levels below the order, though it is most important if varying on population level.

(iv) *Polyploidy:* Polyploidy occurs within single taxa of some orders, but is completely lacking in many other orders. It may be peculiar on the population level also. Very interesting cases, related to the geological history of the territory inhabitated, were brought on record in Lepidoptera.

The known cases of polyploidy in animals can be divided into three groups: (a) polyploidy connected with parthenogenesis, (b) in hermaphroditic organisms, and (c) in species where fertilization is the rule. In insects only the cases under (a) and (c) are found.

(v) *Sex chromosomes and sex determining mechanisms:* The modes of sex determination can be uniform within an order, but often there is a great diversity both in modes as well as in number, types and behaviour of sex chromosomes.

Male heterogamety is general in insects, but in a few orders female heterogamety either prevails or is exclusive (Lepidoptera, Trichoptera).

Variation in sex determining mechanisms may occur on all levels, including the population. It is nearly always combined with variation in chromosome number.

In various groups, but notably in Coleoptera, morphology of sex chromosomes may be characteristic of higher taxa.

(vi) *Parthenogenesis:* Various types of parthenogenesis may be peculiar to certain taxa or even populations, but may also be found generally in a certain group. In a number of orders parthenogenesis never occurs, in others it is exceptional (e.g., Odonata, Trichoptera), while in a few it represents a very common feature.

Haplodiploidy (arrhenotoky) is a type of parthenogenesis in which males arise from unfertilized eggs, while females develop from fertilized ones. It is general in Hymenoptera, lacking in a few secondary cases. Occasionally it occurs also in Coleoptera, Thysanoptera and Homoptera.

In the process of thelytokous parthenogenesis females produce exclusively females without fertilization. Thelytoky was found in representatives of

Orthoptera, Phasmida, Embioptera, Blattodea, Mantodea, Coleoptera, Diptera, Homoptera, and secondarily also in a few Hymenoptera.

Cyclical parthenogenesis may involve an alternation of thelytokous and bisexual generations, or thelytokous and hermaphroditic ones, and was found in a few cases in Coleoptera, Hymenoptera, Diptera and Homoptera.

(vii) *Chiasmata:* The formation, morphology and, particularly, the number of chiasmata are often peculiar on all taxonomic levels. In connection with the recombination index as described above, the chiasmata morphology is of great importance.

At least a single chiasma occurs per bivalent in most orders, but usually there are differences between the sexes. The dragonflies have one chiasma in spermatogenesis and two in oogenesis. There are no chiasmata in oogenesis of Lepidoptera and Trichoptera, and they are lacking in spermatogenesis of one family of Mecoptera and in that of the higher Diptera.

The mean chiasma frequency may vary in different populations of the same species, but its determination is rather difficult and published data often based on too few observations, therefore one should be extremely cautious in drawing conclusions on this kind of literature records.

Chiasma formation in sex chromosomes is often quite different from that in autosomes of the same species. Several orders are highly aberrant in this respect, notably Neuroptera.

About books, in general, it is so wisely said that whenever you read a good book, somewhere in the world a door opens to allow in more light – the light of wisdom. The above referred two books too are the testamount of Dr Kiauta's eternal wisdom – to benefit all in the world. There is no friend as loyal as a book which reminds me of the famous Latin proverb "*Verba volant, scripta manent*" meaning while spoken words fly away, written words remain forever.

●●●

SECTION III: GLOBAL LEADERSHIP AND STATESMANSHIP

12

Origin of Societas Internationalis Odonatologica (S.I.O.)—The 'Family of Friends'

"ubi concordia, ibi victoria."

'*Where there is unity, there is victory*' aptly says the Latin maxim on the power of togetherness and application of joint wisdom. In the 1960s, there was in the Netherlands much interest in dragonflies. Several dozens of people were working on various local faunal surveys, on ecology of some species or on populations of dragonfly assemblages of certain biota, many were breeding larvae incidentally or systematically, some were gathering information on adult behaviour and/or on dragonfly migratory flights, a few workers constructed identification keys for the adults and larvae that were published by the Netherlands Youth Federation of Nature Friends (NJN) and were widely used. The dragonfly interested people came from all walks of life: many were high school or university students, some were officers of the State Forestry Service (Staatsbosbeheer), some were teachers or general naturalists of any kind of profession, few were professional biologists, but three of them were widely acknowledged professional odonatologists of high international standing: Dr J. Belle, Dr D.C. Geijskes and Dr M.A. Lieftinck. Aside of the latter three workers, most of the others were not very prolific writers, though some of them published a few minor notes or produced reports of limited circulation. Consequently, most of these remained lonely workers and did not cultivate any mutual contacts. Due to his position with RIVON, an institution of the State Forestry Department, Dr Kiauta had the privilege of the possibility to get easily in touch with almost all of the dragonfly workers in the Netherlands.

Such was the situation, when Dr Kiauta decided to convene a meeting of all dragonfly interested people in the Netherlands. The meeting took place at the Institute of Genetics of the University of Utrecht, on 7th March 1970. The reception list was signed by 23 domestic and 5 foreign participants, from Belgium, Czechoslovakia and Germany. The enthusiasm of participants triggered the venue of a second meeting (20th November, 1970), the foundation of a rather primitively produced newsletter (*Contactbrief Nederlandse Libellenonderzoekers*, wherein its inaugural issue Dr D.C. Geijskes, Dr B. Kiauta and Dr M.A. Lieftinck had categorically emphasized, "*The morning session will be devoted to discussions on organizational topics, particularly the central registration of the European odonate fauna, the setting up of a central odonatological bibliography and on international coordination and cooperation in odonatological research.*"), and the desire to organize similar gatherings at regular intervals. Consequently, the third meeting was thus convened, upon Dr Kiauta's request, by Dr H.J. Dumont, at the University of Ghent, Belgium, during 22-23 October 1971. It was called The First European Symposium of Odonatology, with 39 participants in person and 17 registered "in absentia", from 17 countries in all. At that meeting were born the Societas Internationalis Odonatologica (SIO) and its journal *Odonatologica* (see KIAUTA, 2013c). Dr Dumont was elected Secretary of the Society and Dr Kiauta the Editor and Manager of the journal, entrusted also with the calls for membership and subscription. Dr Janny van Brink was elected the Treasurer.

The International Odonatological Society (Societas Internationalis Odonatologica, S.I.O.) was thus founded by a melange of both the young and seasoned odonatologists drawn from various parts of the world. This was the only worldwide society in this field, having by the late 1970s, over 600 individual members in nearly 65 countries. The Society has a very unique 'philosophy' underlying its constitution and which holds that it is a 'large family of friends' rather than a hierarchically organized society. This is possibly best exemplified, at the initiative of Dr Kiauta, by the united action of the entire dragonfly fraternity after Japan was devastatingly struck by an earthquake in early 1995! The entire SIO family stood in every possible manner with our Japanese colleagues (Fig. 12.1).

The membership in the Society is open to all bonafide odonatologists, professionals and amateurs alike.

The organization of any society is governed by its Constitution. Drafting Constitution and By-Laws for an emerging society at the international level is always a great challenge, and the responsibility of drafting it was entrusted with Dr Kiauta who not only had experience of certain other associations, but also could understand the legal aspects of various articles. It was a Herculian task which he completed with much aplomb from all his colleagues then willing to join the SIO.

The International Odonatological Society is a non-profit society and the sole monetary resources are its membership for whom alone the Society invests all her funds, in a number of constructive ways. The Society's general affairs are taken care of by the Executive Committee. The Council is elected from the floor by the membership, while the Executive Committee is appointed by, and is responsible to, the Council. The SIO is defined to be a non-political society which is devoted to develop the science of odonatology through adherence to certain well-defined aims and objectives. The Society's general aims are the promotion of odonatology and the conservation of dragonflies and their habitats throughout the world. The message 'SIO is a family of friends' touched the hearts of many workers across the globe. Freedom of expression of scientific thoughts without an incapacitating feeling of fear of being either a less knowledgeable, budding or junior odonatologist, and that all are equal, was the hallmark of all SIO activities such as, e.g., biennial symposia, *Odonatologica* (and other publications) etc. Dr Kiauta and other senior members of the society saw to it that researchers from all over the world, particularly developing and under-developed nations, got opportunity to participate in these and many other research activities. Younger scientists were always assisted, e.g., by way of assisting financially to attend its biennial symposia, waiving off reprints costs, supply of bibliographic material, and guiding to prepare manuscripts. The atmosphere in 'the family of friends' was invariably that of a well-knit, happy and interactive family. With such a unique philosophy the SIO became soon the cynosure of the global odonatological fraternity.

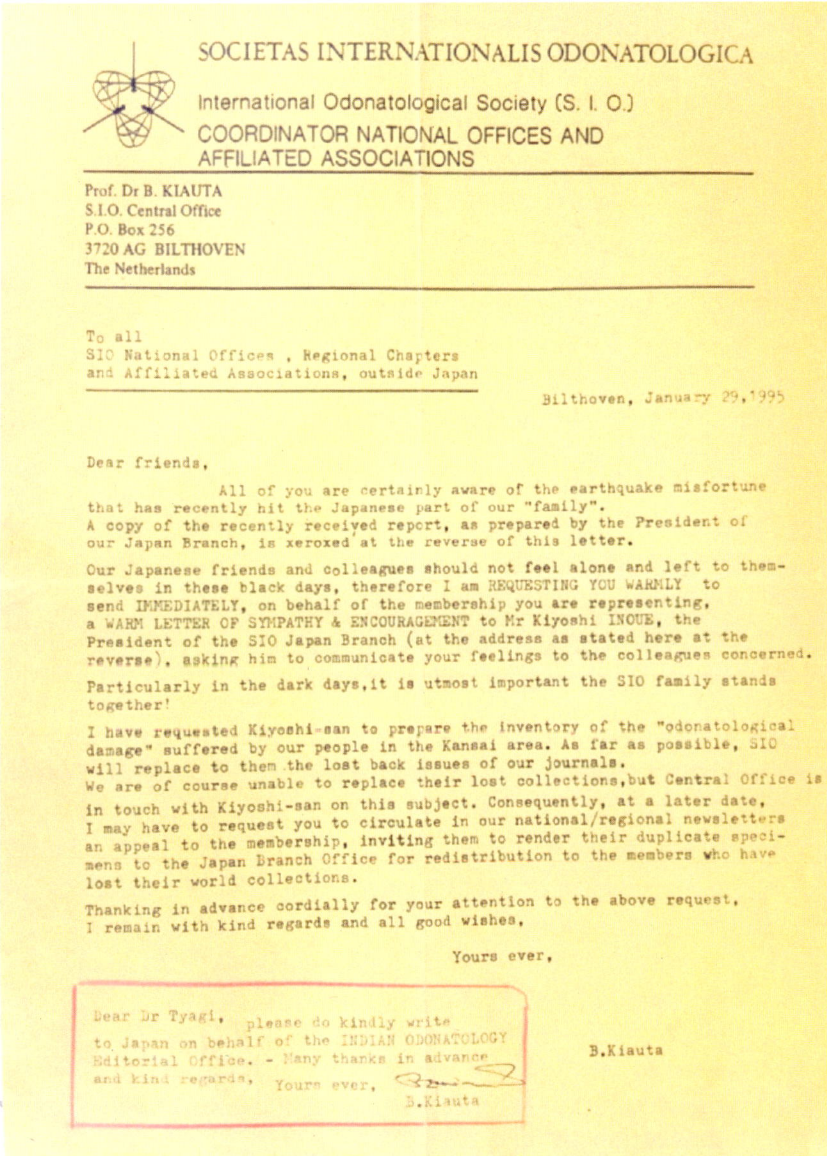

SOCIETAS INTERNATIONALIS ODONATOLOGICA

International Odonatological Society (S. I. O.)
COORDINATOR NATIONAL OFFICES AND
AFFILIATED ASSOCIATIONS

Prof. Dr B. KIAUTA
S.I.O. Central Office
P.O. Box 256
3720 AG BILTHOVEN
The Netherlands

To all
SIO National Offices , Regional Chapters
and Affiliated Associations, outside Japan

Bilthoven, January 29,1995

Dear friends,

All of you are certainly aware of the earthquake misfortune that has recently hit the Japanese part of our "family". A copy of the recently received report, as prepared by the President of our Japan Branch, is xeroxed at the reverse of this letter.

Our Japanese friends and colleagues should not feel alone and left to themselves in these black days, therefore I am REQUESTING YOU WARMLY to send IMMEDIATELY, on behalf of the membership you are representing, a WARM LETTER OF SYMPATHY & ENCOURAGEMENT to Mr Kiyoshi INOUE, the President of the SIO Japan Branch (at the address as stated here at the reverse), asking him to communicate your feelings to the colleagues concerned.

Particularly in the dark days,it is utmost important the SIO family stands together!

I have requested Kiyoshi-san to prepare the inventory of the "odonatological damage" suffered by our people in the Kansai area. As far as possible, SIO will replace to them the lost back issues of our journals. We are of course unable to replace their lost collections,but Central Office is in touch with Kiyoshi-san on this subject. Consequently, at a later date, I may have to request you to circulate in our national/regional newsletters an appeal to the membership, inviting them to render their duplicate specimens to the Japan Branch Office for redistribution to the members who have lost their world collections.

Thanking in advance cordially for your attention to the above request, I remain with kind regards and all good wishes,

Yours ever,

B.Kiauta

Dear Dr Tyagi, please do kindly write to Japan on behalf of the INDIAN ODONATOLOGY Editorial Office. - Many thanks in advance and kind regards, Yours ever,
B.Kiauta

Fig. 12.1. Dr Kiauta's letter dated 29th January, 1995 to all the National Offices, Regional Chapters and Affiliated Associations, outside Japan, regarding extention of all possible help to our Japanese counterparts in their testing time!

The following are some of the services the Society is rendering to the membership, (a) abstracting service of the whole world odonatological literature (published as a special section in each issue of *Odonatologica*), (b) publication of original papers in the Society's periodicals, (c) advice on the supply of odonatological literature, (d) taxonomic identification of material, (e) liaison with specialists who could give advice on any special odonatological topic, (f) circulation of important communications by members to the memberships, (g) information on names and addresses of odonatologists interested in any particular topic of research, (h) supply of full bibliographies on any subject, and (i) financial support, to young workers and in special cases only.

SIO's various research activities are narrated in the following pages.

However, the first and foremost task before the newly founded SIO was the daunting activity of preparation of its Constitution and By-Laws. Many names were proposed and after deftly short-listing of names of the experts in the field the seniors in the society unanimously agreed to bestow with Dr Kiauta the responsibility. In preparing the first draft he was open to advices and suggestions from one and all, and, after several tedious exercises, finally came out almost single-handedly with the society's marvelous and acceptable write-up of its Constitution & By-Laws. The Constitution and By-Laws of the Society, as documentedly existed till December, 2013, are briefly described in the following pages without probably omitting any essential feature - just to exemplify the colossal amount of planning, labour and execution needed to be put into this kind of Himalayan commitment in a specified time period, by Dr Kiauta!

CONSTITUTION OF THE INTERNATIONAL ODONATOLOGICAL SOCIETY (SOCIETAS INTERNATIONALIS ODONATOLOGICA, S.I.O.)

Article I

Objectives: The Societas Internationalis Odonatologica (S.I.O.) is a non-profit society interested in:

(1) Promoting odonatology through:

- (a) The encouragement of understanding, cooperation and friendship among odonatologists throughout the world.
- (b) The encouragement of international collaboration in odonatological research activities and services.
- (c) The planning and organization of international symposia and regional meetings of odonatologists.
- (d) The dissemination of information of interest to odonatologists and others by means of publications.
- (e) The establishment and maintenance of an international research centre, which would house the collections of the Society.

(f) The encouragement of the development of Odonatology in countries where it is not strongly represented.

(g) The maintenance of liaison with other odonatological societies or institutions.

(h) The representation of odonatology in other international organizations.

(2) Conservation of dragonflies and their natural habitats.

Article II

Membership

(1) Four types of membership are recognized:

(a) *Ordinary Membership*, open to all persons interested in odonatology, on application to the Secretary General,

(b) *Honorary Membership,* on election,

(c) *Associate Membership,* applicable to members of affiliated Associations who are not Ordinary or Honorary members, and

(d) *Sustaining Membership.*

(2) Ordinary, Honorary and Sustaining members have full voting privileges in the Society.

(3) Ordinary and Associate members pay fees.

Article III

Administration

The administration of the financial and other affairs of the Society will be under the direction of a Council.

Article IV

The Council and the Executive Committee

(1) There will be a Council consisting of elected *ex-officio* members as determined in the By-laws.

(2) There will be an Executive Committee consisting of those members of the Council who are deemed in the By-laws to be the Officers of the Society.

Article V

Business Meetings

At every International Symposium of Odonatology a properly advertised Business Meeting will be convened by Secretary General.

Article VI

Publications

(1) The Society will publish a journal Odonatologica and Odonatological Abstracts.

(2) The journal will be supervised by the Executive Editor in consultation with the Editorial Board.

(3) Other publications may be produced from time to time with the approval of the Council.

Article VII

Finance and Property

(1) Sources of revenue of the Society are:
   (a) Membership fees.
   (b) Subscription fees to the journal *Odonatologica.*
   (c) Special charges for pages, illustrations etc., for the journal.
   (d) The sale of other publications.
   (e) Miscellaneous other sources.

(2) All funds collected are the property of the Society, together with all material purchased with revenues mentioned above.

(3) The accounts of the Society will be audited by two members of the Society, approved by a regularly convened Business Meeting. The Auditors will hold office during the entire period between two consecutive Business Meetings of the Society.

Article VIII

The By-Laws

Other matters affecting the conduct of the Society are determined by the By-laws.

Article IX

Modification of the Constitution

Proposed amendments to the Constitution must be submitted in writing to the Secretary General six months before the meeting at which they are to be discussed and must be circulated to the Membership well in advance of such meeting. The amendment must be approved by a two-thirds majority of members present before going into force.

The Constitution was approved by a two-thirds majority of members present at a formally convened meeting of the Society, held on 21st September, 1973, during the Second International Symposium of Odonatology in Karlsruhe, German Federal Republic. Subsequent modifications were effected at the Plenary Business Meeting of the Society held on 18th August, 1981, during the Sixth International Symposium of Odonatology in Chur, Switzerland.

BY-LAWS OF THE INTERNATIONAL ODONATOLOGICAL SOCIETY

Article 1

Objectives

The S.I.O. aims at promoting odonatology. It is a non-profit Society without political or philosophical convictions. The seat of the Society is at the Treasurer's address.

Article 2

The Council

(a) There will be a Council, consisting of a President, a President-Elect, a Past-President, a Secretary-General, a Second Secretary, a Treasurer, the Executive Editor and four ordinary members. In addition, the Council is empowered to co-opt members as appropriate and to appoint specialized working groups, including standing committees.

(b) The President, President-Elect, Secretary General, Second Secretary and the four ordinary members are elected by the membership. The Secretary General, the Second Secretary and the four ordinary members are elected for a period of two years and are eligible for immediate re-election. The President-Elect is elected for a period of two years and will automatically become President for a further period of two years. The retiring President is not immediately eligible for re-election as President-Elect.

(c) Council may produce a list of nominations for the elected posts. These will normally be made available, together with a request for any further nominations, to the membership at least five months before the biennial Plenary Business Meeting. Nominations must be received by the Secretary General within eight weeks of the mailing date. Each nomination must be supported in writing by two voting members and include the written consent of the nominee. A voting slip will be circulated promptly after the closing date for the receipt of nominations. To be valid all votes must be received by the Secretary General within six weeks of the mailing date. The results of the election will be announced at the Plenary Business Meeting and the membership duly informed. The new Council will take office immediately following the Plenary Business Meeting.

(d) The Treasurer, the Executive Editor of *Odonatologica* and the National Representatives are appointed by the Council and are the *ex-officio* members of the Council.

(e) The President-Elect, Secretary General, Treasurer, and the Executive Editor of *Odonatologica* are the Officers of the Society. An officer may hold more than one office, except that the President may not also be the current President-Elect, Secretary General, or Treasurer, and the Executive Editor may not also be the current Treasurer.

(f) The Council is responsible for the general administration of the Society, including the appointment of the Associate Editors and Editorial Board of *Odonatologica* and the S.I.O. Librarian, and the nomination of the Society's representative on the Odonata Specialist Group of the Species Survival Commission of the I.U.C.N. Any member of the Council may act as a representative of the Society at international meetings of other societies.

(g) Decisions of the Council are taken by simple majority vote.

(h) Correspondence of the Society, implicating the Society's responsibility in financial matters, may be signed only by the Officers and the National Representatives.

(i) The members of the Council are not rewarded financially for services rendered to the Society.

Article 3

The Executive Committee

(a) There will be an Executive Committee, consisting of the Society.

(b) The Executive Committee is empowered by The Council to make decisions on its behalf and reports to the Council. The Executive Committee will normally consult the Council on matters which it considers to be controversial.

Article 4

Membership

(a) Ordinary membership is open to all persons interested in the field of odonatology, upon written application to the Secretary General or a National Representative. Ordinary members receive the journal *Odonatologica* and have full voting privileges in the Society.

(b) Distinguished odonatologists may be elected as Honorary Members of the Society. Candidates may be proposed by any member of the Society. A two-thirds majority of the business meeting is required for election. Honorary members receive the journal free of charge and have full voting privileges.

(c) Associate membership is defined in Article 7(b)

(d) Sustaining membership.

Article 5

Fees and Cancellations

The fees (in Dutch Guilders at present) of Ordinary and Associate Members are recommended by the Treasurer and subject to approval by the Council. Fees should be paid one year in advance by December 1st of the current year. Cancellations, likewise, should be brought to the notice of the Treasurer or a National Representative before December 1st. Otherwise the membership shall stand automatically and the fee for the next year become due to the Society.

Article 6

National Representatives

(a) In countries where circumstances make it appropriate the Council may appoint a National Representative.

(b) The structure and operation of National Office will be decided by the Council.

Article 7

Affiliated Associations

(a) Odonatological Societies and other odonatological Associations may become affiliated to S.I.O. The conditions for affiliation are:

  (i) that the Association should produce a circular, newsletter or other serial publication.

  (ii) that the Association's publication should include the S.I.O. emblem in its heading, and keep members of the Association informed of S.I.O. activities, *e.g.*, symposia and publications.

(b) that, if an Affiliated Association forms a committee, the structure of the committee should be such as to include one member of S.I.O. as an *ex-officio* member of that committee.

  (i) All members will be encouraged to join any Affiliated Association formed in their country.

(c) Associate members shall pay a subscription to S.I.O. (which will be in addition to a national subscription). This will be fixed as in Article 5.

(d) Associate members shall be eligible to:

  (i) attend the International Symposia of the Society on payment of an appropriate conference fee.

  (ii) Purchase the publications of SIO at the same price as ordinary and Honorary members.

(e) The organization of each Affiliated Association shall be entirely within the hands of the members of that Association.

(f) All Ordinary and Honorry members of SIO will be encouraged to join any Affiliated Association formed in their country.

(g) The Editor of the National Publication may reproduce any Odonatological Abstracts and summaries of papers in *Odonatologica* which are of national interest, in the Affiliated Association's publication.

(h) Any expenses incurred by the Affiliated Association shal be the responsibility of that Association, not of SIO.

(i) If an Affiliate Association decides that a National Subscription is necessary, Ordinary members of SIO who are also members of that Association, will have the National Subscription added to their SIO subscription. The appropriate amount will be refunded to the Association either directly by National Representative or, if there is no National Representative, by the Treasurer of SIO. Honorary members of SIO should be considered for the same status by their Association.

Article 8

The Honorary President

(a) The title *Honorary President* of S.I.O. may be awarded to a person who has made distinguished contributions to Odonatology.
(b) The title is personal and persists for life.
(c) There will be only one Honorary President at any one time.
(d) The nomination procedure will be decided by Council and made known to the membership at the appropriate.

Article 9

Publications of the Society

The S.I.O. publishes the journal *Odonatologica* at regular intervals. Other publications may be also released. In all cases, the volume of publications shall depend on financial possibilities. The prices of all publications are to be fixed by the Treasurer.

Article 10

Symposia

The venue of the next Symposium shall be arranged at the biennial Plenary Business Meeting.

Article 11

Finances

All funds collected by the Society must be spent for the benefit of the Society and its activities. A detailed account of all revenues and expenses shall be prepared by the Treasurer on an biennial basis. This account shall be audited by two members, elected at the biennial Plenary Business Meeting. The Auditors should reside in the same country as the Treasurer. The Auditors should reside within a reasonable distance of the Treasurer.

Article 12

Modification of the By-Laws

Any change in the By-laws requires a two-thirds majority at a quorate Plenary Business Meeting of S.I.O. or in a referendum of all voting members (*i.e.*, Ordinary, Honorary, and Sustaining members). A quorum shall be defined as 10% of the membership eligible to vote, namely Ordinary, Honorary, and Sustaining Members.

THE S.I.O. NATIONAL / REGIONAL OFFICES

The growth of SIO was spectacular and some 600 members and institutional subscribers, from anything like 60 countries of the globe, were registered within a few years. The Society set up 11 National or Regional Offices, all publishing their local newsletters, viz. in Canada (newsletter *Walkeria*), Germany (*Hagenia*), India (*Fraseria, Occasional Publications of the SIO National Office in India* and some other incidental publications, including the periodical *Indian Odonatology*), Italy (*Lindenia*), Japan (*Sioja*), Slovenia (transformed into the national 'Slovenian

Dragonfly Society', with the periodicals *Exuviae* and *Erjavecia*), Spain and Portugal (*Navasia*), Thailand and SE Asia (*Malangpo*), United Kingdom (*Kimminsia*), United States (*Selysia*) and the USSR, now Russia (*Belyshevia*). Dr Kiauta served as an informal general overseer of the National Offices.

GENERAL FEATURES, STRUCTURE AND OPERATION OF THE S.I.O. NATIONAL OFFICES

Preamble:

(i) If the Office covers a larger geographic region rather than a single country, the term "National" is replaced by "Regional". For the sake of convenience only the former is used in the below text.

(ii) The institution of the SIO National Offices is one of the principal means towards the achievements of the Society's objectives as outlined in Art. 1/a, b, e of the Constitution.

(iii) The basic tasks are outlined in Art. 5 of the By-Laws. Further details were specified in *Selysia* 1980, 9(2):17, but some are here partly modified.

Article 1

Location

(a) By appointing a National Representative the Society may open a National Office in any country or region where this is considered opportune.

(b) The opening of a National Office is considered particularly opportune in the following cases:

  (i) The local membership is too high to be dealt with and/or served adequately by central Society's executive.

  (ii) The actual membership in the region is low, but there are many active workers that are not easily reached by the central Society's executive.

  (iii) Due to prevailing local currency regulations and/or language and communication difficulties, the odonatologists are unable to fulfil the membership dues and/or the Society is unable to communicate with them directly.

  (iv) The establishment of a National Office should be considered in all regions where this could contribute towards the increase of the local S.I.O. membership, and/or where this is likely to enable the Society to effectively promote odonatological research and substantially support the local odonatologists.

Article 2

Organization

(a) A National Office is considered an organ of S.I.O. offices *not* a national society. Its function is comparable to that of a "legation", therefore the National Offices (= Representatives) are appointed by and are responsible to S.I.O. rather than to the local membership. It is in the interest of the Society, that persons that do not have the confidence of the local membership, are not appointed to or continued to the Office.

(b) The business of the National Office is conducted by the National Representative, who is appointed by and is responsible for all the activities of the Office to S.I.O. As in the case of all S.I.O. offices, his/her appointment is for two years, and can be extended for number of terms if so desired by both parties.

(c) The internal organization and technical mode of operation of the Office are left entirely to the discretion of the National Representative.

(d) If, during the duration of his/her term the National Representative becomes unable to continue in the Office, the S.I.O. will take the appropriate measures, whenever possible in cooperation with the retiring National Representative and/or in consultation with the local membership.

(e) If, for the adequate performance of the task of the Office, the National Representative considers useful, opportune or necessary to enlist the help of one or more odonatologists, he/she may appoint any reasonable number of assistants. These assistants must be members of S.I.O. and are responsible for their activities to the National Representative rather than the S.I.O. directly. As soon as the appointment has taken place, the Executive has to be notified of the names, addresses, and tasks of the assistants, as well as of the period for which the appointment has been made. Any changes in the status of the personnel of a National Office have to be communicated to the Executive.

(f) A National Representative has to be Subject of the country in which the Office operates (or of one of the countries of the region, if the Office covers more than one country). The assistants have to be residents of the country (region) for which they are responsible.

Article 3

Name and Emblem

(a) The name of the S.I.O. National Offices is standardized in a Latin and in an English version as follows:

– "Societas Internationalis Odonatologica [country] Office", and

– "[country] Office of the International Odonatological Society (S.I.O.)

In non-English texts the English name can be given in the respective language translation, but the abbreviation, S.I.O., should always be added.

(b) When appropriate, the National Office may use the S.I.O. emblem, with one dragonfly abdomen in vertical position (as in *Odonatologica* and *Notulae*).

The emblem must be shown on the Newsletter and on any other publications of the Office.

Article 4

Tasks

(a) The scope of the tasks of a National Office depends largely on the technical possibilities available, hence it can be different for different Offices.

The tasks of each Office are defined in agreement and upon consultation with the National Representative. They are subject to modifications, by the same procedure, and have to be laid down in the "Regulations of Responsibilities and Tasks" for each Office.

(b) The following are among the principal tasks of a National Office:

(i) Administration of the local membership.

(ii) Collecting of the local membership fees and other Society's dues, and their transfer to the S.I.O. Treasurer.

(iii) Mailing of SIO publications to the local membership.

(iv) Editing and production of a local newsletter or similar periodical circulars. These have to show the S.I.O. emblem, and have to be numbered consecutively in order to enable their bibliographic citation.

(v) If a part of the regional SIO fees and dues is scheduled to be used for financial support of local odonatological research, the National Representative is requested to advise on the appropriate use of the funds. The final decision is reserved to the SIO Executive in all cases.

(vi) Organization of local odonatological meetings.

(vii) Liaison with local odonatologists and assistance in all matters concerning the local odonatological research.

(viii) Assistance to no-resident odonatologists visiting the country (region) for odonatological purposes.

(ix) Representation of SIO in regional meetings and elsewhere, where appropriate.

(x) Assistance in all matters concerning odonate conservancy, including regular cooperation with the SIO Representative on the IUCN Survival Service Commission.

(xi) Other tasks and representation responsibilities as defined in the "Regulations".

(c) During the initial period of operation of a National Office, a part of the above mentioned tasks can be delegated by the National Representative to the SIO Executive, particularly those involving extra financial costs, e.g., mailing of SIO publications, production (but not editing) of the local newsletter, etc.

This has to be specified in the "Regulations", and is considered as a temporary measure only.

Article 5

Finance

(a) The National Representative is responsible for all financial matters to the SIO Treasurer, and keeps him/her informed on the balance and position of all assets and liabilities of the office.

(b) If a part of the assets is needed for operation of the Office, this can only be done in consultation with the Treasurer.

(c) If the Society is furnishing grants to sponsor local odonatological research, the National Representative has the right to suggest the appropriate use.

(d) It is understood that the National Representative is able to use some of his/her professional facilities (postage, Xerox etc.) also for the purpose of the SIO Office.

Article 6

Position of the national Representative

(a) The official position of the National Representative is regulated by Art. 5 of the By-Laws.

(b) If the odonatologists of the country where the National Office operates have set up a national odonatological association, and have applied to SIO for its status as "affiliated national association', the National Representative has to be adopted into the Executive of the Association if the latter is to be granted the affiliated status. However, to avoid the possibility of a "conflict of interest", he/she cannot be a President, Vice President, Chairman or Treasurer of such an association.

Article 7

Final Provision

(a) The (first and original) version was prepared by B. Kiauta for the 1980-1981 Executive Board; it was checked, on behalf of the Executive Board, by P.S. Corbet, R. Rudolph and M.J. Westfall, and it was sanctioned by the Secretary General, K.J. Tennessen.

(b) If and when the need arises, modified versions may be issued by the Executive Boards then in function, The document becomes operative by the sanction of the Secretary General, which automatically makes all earlier versions void.

PLENARY BUSINESS MEETING AND THE MEMBERHIP OF COMMITTEE FOR 1985-1987

The Plenary Business Meeting is the highest authority of the Society. It is convened in the framework of every international symposium. During the course of each symposium various officers are elected to represent its several departments, and for the period 1985-1987 the election of various different positions was conducted during the 8th International Symposium of Odonatology held in Paris, which was attended by me too. The following officers were elected and appointed for the Council, the Executive Committee and several other departments:

The Council:

President	: R.M. Gambles
President – Elect	: F. Schaller
Past President & Director, IORI	: M.J. Westfall
Secretary	: K.J. Tennessen
Second Secretary	: G. Pritchard
Treasurer	: J.M. van Brink
Executive Editor	: B. Kiauta

Members elected under Article 2b of the By-Laws:

J. Legrand	M.J. Parr
M. May	J.G. Pilon

Members elected under Article 2d (National Representatives):

K. Deacon	R. Rudolph
K. Inoue	(Bro) A. Pinratana
P.J. Mill	B.K. Tyagi
C. Utzeri	

Members co-opted under Article 2a:

P.S. Corbet	A.B.M. Machado

The Society had an Honorary President, elected for life: Dr M.A. Lieftinck (Rhenen, The Netherlands). Unfortunately, he passed away on 13th April, 1985, and the post is vacant since then. The late Dr Lieftinck, one of the great odonatologists ever, had not only played a significant role in founding the Society but had always been a great inspiration for the ever growing Society. He was an internationally reputed odonatologist, specializing in odonate systematics and biology. Dr Kiauta had always banked on him for multiple taxonomic issues and, a few months before his passing away, reverencedly brought out a special issue of *Odonatologica* (Vol. 13, No. 1), highlighting his life and odonatological works including research publications, to commemorate his 80th birthday!

NATIONAL OFFICES AND AFFILIATED NATIONAL ASSOCIATIONS

The Society has National Offices in Canada, the Federal Republic of Germany, France, India, Italy, Japan, Thailand, the United Kingdom, and in the United States of America. In addition, it has Affiliated National Associations, in Germany, The Netherlands and the United Kingdom. The Head of a National Office is called the

National Representative in-Charge he/she directly represents the Society in his/her country. An Office governing membership in many countries in its region is called the Regional Office.

INTERNATIONAL ODONATA RESEARCH INSTITUTE

The SIO established the International Odonata Research Institute (IORI) in the Florida State Collection of Arthropods (FSCA) in Gainesville, Florida in 1985 with Prof. Dr Minter J. Westfall, Jr. as Director and Dr
Sydney W. Dunkle as General Manager of the Institute. The IORI is associated with the Centre for Arthropod Systematics and the central office of the institute is located in the state-operated Doyle Conner Building of the Florida Department of Agriculture on N.W. 34th Street.

There is a Policy Board of IORI approved by the SIO Council which includes Dr M.J. Westfall, Dr Sydney Dunkle, Dr George Bick, Dr Allan Davies, Prof. Philip S. Corbet, Prof. Dr B. Kiauta, Dr Jean Legrand, Prof. Angelo Machado and Dr Michael May. The Policy Board is appointed by and responsible to the SIO Council and its task is general policy formulation.

Following are some of the main tasks and activities of the IORI:

(a) The collection provides for the continuing maintenance of all systematic specimens, a substantial part of which is donated by retiring or deceased SIO members. It also provides a repository for voucher specimens submitted to the Editors of *Odonatologica* by the authors of papers published therein.

(b) The rapid and authoritative Identification Service operates in accordance with internal regulations.

(c) The Library aspires to be a complete collection of the world odonatological literature, and is continuously updated through acquisitions of current titles through the usual channels. Material used for the preparation of the "Odonatological Abstracts" section of *Odonatologica* constitutes part of the acquisitions.

(d) The Library Xerox Service operates in accordance with internal regulations.

(e) The Archives provide for the continuing storage of, and easy access to, unpublished material concerning odonatology and odonatologists. The SIO Archives are an integral part of the IORI Archives. Photographic and tape recorded materials are also systematically collected.

(f) The computerized Data Retrieval System is being developed with the aim of covering the world literature and the world fauna. The cooperation in the area with other institutions is subject to agreements between the IORI and these bodies.

(g) Laboratory and Research facilities should be adequate to the requirements of the scientific staff and guest research workers. Cooperation in this area with other institutions is subject to agreements between the IORI and these bodies.

(h) The basic publication of the IORI is its *Annual Report*. Other incidental and serial publications may be released if and when required. The coordination of the publication programme with that of the general SIO publication programme is subject to agreement between the SIO Executive Editor and the IORI Director.

(i) Coordination of research planning on an international scale is among the principal objectives of the IORI.

(j) It is endeavoured to concentrate the SIO Administration and some SIO Editorial Offices, as well as other activities in conjunction with SIO operation in the IORI. Each case is considered separately by the SIO Executive and the IORI Director.

The Institute now has the beginning of a series of cabinets and cases for the storage of the SIO Odonata collection. It is the goal to bring together a study collection, which with the existing large collection on permanent loan from the FSCA, will contain most of the species of Odonata of the world.

Library and active materials will also be housed for use of visiting workers. Any donations of each material by any worker throughout the world will be appreciated. Also the odonatologists can donate specimens of even the common species in their countries which may be lacking in the collections now in Gainesville. The specimens will be transferred to 3x5 inch cellophane envelops, with data on backing cards, including the name of the donor, and kept separate from the FSCA material. Specimens of larvae and exuviae are also desired. They will usually be stored in vials of alcohol.

ANTIQUARIAN DEPARTMENT

The Antiquarian Department is one of the SIO service institutions. It has been set early in 1984, and has the following main objectives:

(1) Circulation and recirculation of odonatological literature throughout the world.

(2) Financial aid to the production and other departments of SIO.

(3) Enabling the odonatologists in the non-convertible areas to pay the literature, provided the SIO has there a National or a Regional Office.

[Since there is a SIO National Office in India covering membership in many countries, i.e., in addition to India, also Bangladesh and Nepal, the payments by local workers are to be made to the SIO India Office and the process help buyers of the SIO literature in many ways. The buyer concerned gets about 66% discount on the actual cost of the material, pays in the local currency without undergoing the burden of currency-transfer procedures and the overseas mailing, generally by air, of the remittances].

The Antiquarian Department is an invaluable treasure of the odonatological literature. It houses literature of very great value since many of the publications remain either unseen or undiscovered and, in many cases, unapproachable by many workers. So far 2000 works of different types of literature (monographs, reprints, books, mimeographs, hand-written/unpublished documents, letter correspondence etc. etc., just to mention a few) are catalogued by the Antiquarian Department. The list is available from the SIO Central Office and the National Offices in various countries on request.

JOURNALS

The Society is publishing several periodicals and other serial works, viz., the quarterly *Odonatologica* (since 1972), the semi-annual *Notulae odonatologicae* (since 1978), the irregular *Soc. Int. Odonatol. Ser. Repr. Class. Odonatol. Monogr.* (since 1974), *Soc. Int. Odonatol. Rapid Comm.* (since 1980), and the *Rep. Odon. Specialist Group Int. Un. Conserv. Nat.* (since 1982). At intervals various other publications are also issued, incl. the volumes of *Abstracts of papers of the International Odonatological Symposia*.

The members are receiving *Odonatologica, Notulae odonatologicae* and semiannual newsletter, *Selysia.* Those having a membership in the German or in the Netherlands National Association are receiving also semiannually *Libellula* and the *Contactblad,* respectively. The various national newsletters are also supplied to members outside the jurisdiction of the respective national office on request.

INTERNATIONAL SYMPOSIA

An International Symposium of Odonatology is sponsored by SIO once every two years. The first eighteen were held in Ghent (1st,1971), Karlsruhe (2nd 1973), Lancaster (3rd, 1975), Gainesville/FL, USA (4th, 1977), Montreal, Canada (5th, 1979), Chur, Switzerland (6th, 1981), Calgary, Canada (7th, 1983), Paris, France (8th, 1985) [The famous fossil of *Meganeura monii* in the Paris Museum, was one of the “highlights” of the Symposium in Paris], Madurai, India (9th, 1988), Johnson City/TN (10th, 1989), Trevi, Italy (11th, 1991), Osaka, Japan (12th, 1993), Essen, Germany (13th, 1995), Maribor, Slovenia (14th, 1997), Novosibirsk, Russia (15th, 2001), Schwerin, Germany (16th, 2003), Hong Kong, China (17th, 2006), and Nagpur, India (18th, 2008).

ASSOCIATION WITH INTERNATIONAL UNION FOR CONSERVATION OF NATURE & NATURAL RESOURCES (IUCN)

The prime conservation institution of IUCN have a special chapter on dragonfly conservation, and the SIO is intimately associated with it. The SIO has a permanent representative on the Odonata Specialist Group of Species Survival Commission (IUCN) and it has been the goal of the SIO to conserve rare and endangered dragonflies all over the world by –

(i) Making in-depth evaluations of all such species and their habitats which are essential for the survival of these rare dragonflies;

(ii) Making reserves in appropriate places in countries where endangered species can best thrive, and

(iii) Influencing upon the SSC (IUCN) to bring the endangered species in the *Red Data Book* for international attention.

As a result of continuous efforts of SIO, a number of species are now protected all over the world. The list of endangered species placed in the *Red Data Book* is available from the SIO Central Office.

For convenience in administration in conserved dragonfly fauna and their habitats throughout the world, SIO/IUCN have deputed a number of "Representatives" in different zones/regions of the world. The SIO also holds a regular meeting of the IUCN in the framework of its biennial symposium.

In the framework of activities of the SIO National Office in India, TYAGI (1988a) prepared a compendium of the '*Activities of the SIO and its National Offices*'– a treasure of memories, of course! Most material in this chapter has been drawn from this compendium.

"*Without memory,*
there is no culture.
Without memory,
there would be no civilization,
no society, no future."

●●●

13

The S.I.O. National/Regional Offices Worldwide

"Science knows no country, because knowledge belongs to humanity, and is the torch which illuminates the world."

— Louis Pasteur

The science of odonatology was emerging at a great pace during the late 19th and early 20th century, covering all continents, even though most of research work done appeared scattered and uninteracted with others. Some countries made a mark in odonatological research and produced their own journals but none of these had a worldwide circulation: Japan's *Tombo* was restricted mostly to the Japanese odonatological fraternity, *Selysia*, published in USA circulated to many countries, but greatly restricted to a few countries due to obscurity of many odonatologists particularly in developing and underdeveloped countries. However, after the Societas Internationalis Odonatologica was founded in 1971, odonatology became a truly global science, embracing close to 600 members from over 65 countries in all continents, and making the SIO by far the best entomological association in the world with the largest number of members in a single insect order! The credit for such a grand achievement went directly to Dr Kiauta who laid a successful plan for spreading the message "SIO being a family of friends" and developing *Odonatologica*, the key journal of the SIO, as the most celebrated periodical among all the insect order-specific scientific journals across the globe. Dr Kiauta envisioned a true worldwide growth of the International Society of Odonatology (SIO) by developing an infallible and truly democratic mechanism to disseminate the odonatological knowledge throughout the world. He designed a scheme of establishing a network of national and/or regional offices throughout the world which would carry on the objectives of the Society. He placed his idea before the SIO Council during the 2nd International Symposium of Odonatology, held in Karlsruhe (Germany) in 1973, with the first list of countries where the SIO National/Regional Offices could be likely opened. All the Council members supported the view and saw immense benefits in executing decentralization of certain administrative responsibilities such as, e.g., increase in membership, collecting fees in local currency, distribution of SIO periodicals to local members etc. Over a period of more than 40 years the SIO had been able to establish eleven National and/or Regional Offices, each of which characteristically had a newsletter as the spokesperson of their diverse activities. Below mentioned are the eleven National and/or Regional Offices of the SIO in different countries (name of the respective National Representative in-Charge is given in parenthesis). Each National Office is listed along with its newsletter and the year of its inaugural publication:

1. Canada, 1995 (Mrs N.L. HOUSE); *Walkeria*, 1985
2. Germany, 1991 (Dr M. SCHORR); *Hagenia*, 1991
3. India, 1981 (Dr B.K. TYAGI); *Fraseria,* 1981
4. Italy, 1984 (Dr C. UTZERI); *Lindenia*, 1984
5. Japan, 1989 (Mr K. INOUE); *SIOJA*, 1981
6. Russia, 2001 (Prof. Dr A.Yu. HARITONOV); *Belyshevia*, 2001
7. Slovenia, 1994 (Dr A. PIRNAT); *Exuviae* and *Erjavecia*, 1994
8. Spain & Portugal, 1992 (Dr M. FERRERAS ROMERO); *Navasia*, 1992
9. Thailand, 1985 (Dr A. PINRATANA); *Malangpo*, 1985
10. United Kingdom, 1990 (Mrs J. SILSBY); *Kimminsia*, 1990
11. United States, 1974 (Prof. Dr M.J. WESTFALL); *Selysia*, 1963

Erecting peripheral national or regional offices in different countries across the world was perhaps one of the greatest challenges Dr Kiauta had encountered. This was largely because of the differences in understanding of the science of odonatology in these countries, implying that the number of active odonatologists greatly varied among the developed and developing nations, with advanced nations being far forward in boosting their quantum of dragonfly researchers. In some countries there were obviously no odonatologists at all! Dr Kiauta desired that science of odonatology must become popular in all countries. Therefore, he developed a far-sighted plan whereby in countries where odonatology developed strongly he resolved to establish there a national or regional office, depending upon its size of coverage, and those countries where science of odonatology was only recently evolved he used his indomitable personality to influence the native odonatologists to initiate publication of some kind of a dragonfly newsletter, periodical or journal so as to first popularize the science in their midst! This *modus operandi* worked marvellously as he opened up, one after another, not only nearly a dozen national offices but also impacted several countries' Affiliated Associations and National Sections to bring out as many as 60 periodicals devoted to dragonflies which is, in a way, so to say, some kind of a world record! The SIO has even stimulated the foundation of the German and British dragonfly societies and that of the South Asian Council of Odonatology in India. Various local groups and organizations also have their roots in SIO, for example the Slovenian Dragonfly Society, whereas the Netherlands Dragonfly Society and its periodicals also sprang up historically from the 1970 meetings of the Netherlands dragonfly workers, convened by Dr Kiauta.

The first National office of SIO was opened up in United States of America where a large number of serious dragonfly researchers existed already and many were members of the society. In USA, however, already an odonatological association was continuing, and opening an office of the SIO there seemed to be a risky proposition. Yet, Dr Kiauta's uncanny and skilful style ensured a smooth establishment of the SIO National Office in United States of America. Two events particularly favoured this decision; (i) Eminent odonatologist and a close colleague and friend of Dr Kiauta, Professor Minter J. Westfall, who as a much revered professor of zoology was already famous for his book, "*A manual of the dragonflies of North America (Anisoptera)*", as a co-author with Prof. J. G. Needham, and had a great influence on young and budding odonatologists (some of which famously led the group at a later stage), such as, e.g., Drs Michael L. May and Sydney Dunkle, took upon himself the responsibility as the National Representative in-Charge, and, (ii) the universally distributed dragonfly newsletter, *Selysia*, was already in place. *Selysia* was given such a majestic name in commemoration of the founder of odonatology, Michel Edmond de Selys Longchamps (1813-1900), in short Selys (Fig. 13.1). This also set the trend of naming a country's newsletter under its SIO National Office, in most cases, after the name of odonatologist famous in that country or region (e.g., *Fraseria* in India, *Kimminsia* in UK *etc. etc.*). Professor Westfall remained to Dr Kiauta a highly trusted and dependable senior colleague till his passing away in 2003, at the mature age of 87. With Minter's passing, a chapter in the history of odonatology was closed. He would be revered along with the likes of other great 20th century odonatologists before him, such as Calvert, Walker, Williamson, Montgomery and Lieftinck, to name just a few. He touched the lives and careers of nearly all members of SIO but his reach went to all corners of the globe as he corresponded with and assisted nearly the entire world dragonfly community; in short, he was a cornerstone and anchor for the science of odonatology, the SIO and *Selysia*.

Opening a National Office in Canada was a great adventure considering the ever growing number of odonatologists in the vast north American country, and the SIO envisioned great prospects of disseminating odonatological knowledge particularly among young nature lovers studying dragonflies. The reason of establishing a National Office in Canada was based on the fact that Canada, exorbitantly rich in dragonfly fauna, has been a major centre for odonatological research for well over a century, with the odonatologist *par excellence* Edmund Walker's 50 years of important systematic and distribution works (Royal Ontario Museum and University of Toronto) and those of his important amateur collaborator, Francis Whitehouse. The latter was a banker who primarily collected and wrote on Odonata from Alberta and British Columbia, in Canada's West. Walker (1887-1969) produced many important works, especially monographs on *Aeshna* and *Somatochlora*, and two volumes (1953, 1958) of his monumental '*The Odonata of Canada and Alaska*', the third volume of which was completed by Corbet (WALKER & CORBET, 1975). Corbet, who worked in Canada for several years (late 1960s and early 1970s) during his long odonatological career, along with Adrien Robert (Université de Montréal), who studied the Quebec fauna, acted as a sort of transition between the studies of the first half of the 1900s and the second half, when odonatology expanded enormously in Canada.

SELYSIA

THE NEWSLETTER OF THE SOCIETAS INTERNATIONALIS ODONATOLOGICA
AND THE U.S. NATIONAL OFFICE

Vol. 22, No. 1 Johnson City, Tennessee March 1, 1993

REQUEST FOR DAMSELFLY ECOLOGY/ BEHAVIOR PAPERS

Alex Córdoba-Aguilar and Fabricio Capistrán-Hernández
Instituto de Ecología, Apartado Postal 63
Xalapa, Ver. 91000, MÉXICO

We are Mexican students working on ecology and behavior of adult damselflies. We want to interpret our work in the context of a basic framework on natural history of the order, with reference to other researchers. But we lack several references on important ecological parameters (maturation time, longevity, survivorship, sex ratios, etc.) because in our position it is difficult to get them. We are mainly interested in papers published before the 1980's on African, Asiatic, and European species. We can pay for photocopies, but will be happy to receive free contributions. Any papers you can send will produce beneficial results in our work, and we acknowledge in advance all kinds of help.

Two dragonflies
mating in the sunlight-
water over stones

tombo

PRECOPULATORY BEHAVIOR: CORRESPONDENCE REQUESTED

Milen Marinov
Macedonia Str. 112, BG-7500, Silistra, BULGARIA

I am a student of Biology at *Sofia University*. I have been working for a year now at the *Natural History Museum* at the Biosphere reserve "Srebarna." This job gives me the possibility to study dragonflies much more thoroughly. My interest in these insects dates back to my first year of study at the University. It is very exciting and interesting to observe their behavior (and especially their precopulatory behavior) and I would be very pleased to correspond with somebody having similar interests and working over the same questions and problems.

SHORT NOTICES

An encouraging increase in volume of material submitted to *SELYSIA* has forced me to initiate a new format for short items written by me, often based on information sent by someone else. These "short notices" will be dispersed throughout the newsletter to optimize use of space. They can be identified by distinctive borders, and lack of headings with authors' addresses.

DOES *CORDULEGASTER* EXIST IN NORTH AMERICA?

The 10 November 1992 issue of *Opuscula zoologica fluminensia* [96:1-18] contains a paper (in German) by H. Lohmann [Biologisches Institut I (Zoologie), Albert-Ludwigs-Universität, Albertstrasse 1a, D(W)-7800 Freiburg i. Br., GERMANY], "Revision der Corduligastridae. 1. Entwurf einer neuen Klassifizierung der Familie (Odonata: Anisoptera)." This paper will interest many odonatologists, perhaps especially those in North America, because it proposes to change genus names for many of our species. Minter Westfall and Sid Dunkle are beginning to study the paper; but an English translation would help them (and others) a lot if there is someone willing to do the work!

Bastiaan Kiauta has provided the following annotation of Needham and Westfall (1955) *A Manual of the Dragonflies of North America*, page 79, to reflect Lohmann's proposed changes:

former *Cordulegaster* species	new genera	comment
sayi	*Archegaster*	gen. n. (type sp. of genus!)
diadema	*Lauragaster*	gen. n. (type sp. of genus!)
erroneus	*Kalyptogaster*	gen. n. (type sp. of genus!)
fasciatus		merely a subspecies of *obliquus*
obliquus	*Taeniogaster*	gen. n. (type sp. of genus!)
diastatops	*Zoraena* Kirby,	1890 (type sp. of genus!)
maculatus	*Pangaegaster*	gen. n. (type sp. of genus!)
dorsalis	*Lauragaster*	gen. n.

A reprint, or subscription to odonate issues of *Opusc. zool. flumin.*, is available from S.I.O. with reference to respective catalog numbers 1209 or 1223.

1

Fig. 13.1. Facsimile of front page of *Selysia* Vol. 22, No. 1, published March 1, 1993.

One of the main trends in Odonata studies in the past 20 years has been the increased interest in conservation of species and the resultant burgeoning of collections, surveys and reports on species at risk. These activities have involved many amateurs and students and, along with the great interest in Odonata watching and digital photography by the general public, have immensely increased the knowledge of the fauna. Some excellent Centres in Canada (with leading professors) for dragonfly study included: University of Calgary, Calgary, AB (Gordon Pritchard), Université de Montréal, Montreal, QC (Jean-Guy Pilon), Carleton University, Ottawa, ON (Tom Sherratt), Acadia University, Wolfeville, NS (Tom Herman), University of Toronto, Toronto, ON (Rob Baker), and Bishop's University, Sherbrooke, QC (Don Hilton). Other academic institutions (significant graduate student research) existed: Royal British Columbia Museum, Victoria, BC (Rob Cannings), Spencer Entomological Collection, University of BC, Vancouver, BC (Syd Cannings, Rex Kenner), Provincial government environment ministries (Conservation data Centres, Natural heritage programs) charged with wildlife conservation efforts, especially BC (Leah Ramsay, Syd Cannings), Ontario (Colin Jones). Besides there were innumerable amateurs and professional biologists working, collecting and publishing in odonatology out of personal interest, e.g., John Acorn (Edmonton), Paul Catling (Ottawa), Gord Hutchings (Victoria), Dave Halstead (Saskatoon), Paul Brunelle (Halifax), Raymond Hutchinson (Ottawa/Gatineau).

The SIO Canadian office was formed in 1985 after discussions at the SIO Symposium in Calgary (1983) and Ken Deacon (Lakehead University, Thunder Bay, ON) accepted the leadership of the office, to collect SIO dues, liaise with the Central Office in The Netherlands, edit the newsletter *Walkeria* (named after Edmund Walker; two issues per year), and dispense other information as needed (Fig. 13.2). Deacon published two issues of *Walkeria* in 1985 and collected dues in 1986 but resigned owing to pressures of his university work. The job was transferred to Syd Cannings at the University of British Columbia in Vancouver. Syd's brother, Rob (Royal BC Museum, Victoria) helped edit *Walkeria*, which was produced out of Vancouver from 1987 to 1990. When Syd moved to Victoria, BC, Dr Nancy House, Ottawa, ON, took over and Rob continued as SIO manager and *Walkeria* editor.

Two SIO international symposia were organized in Canada; Montreal (1979) and Calgary (1983). The Montreal meeting was organized by Jean-Guy Pilon, Université de Montréal, with the able help of Raymond Hutchinson, an avid amateur odonatologist. The Calgary symposium was led by Gordon Pritchard, University of Calgary; the symposium field trips were planned and run by Rob and Syd Cannings.

WALKERIA

Newsletter of the Canadian National Office of the International Odonatological Society,
Bulletin du Bureau national canadien de la Société odonatologique internationale,
SOCIETAS INTERNATIONALIS ODONATOLOGICA (S.I.O.)

Vol. 2, No. 1	pp. 1 - 4	1 June 1987	Vancouver, B.C.

Editorial Address/Adresse du Redacteur: S. CANNINGS, Department of Zoology, University of British Columbia, Vancouver, B.C. V6T 2A9

THE REBIRTH OF WALKERIA

After a one-year hiatus, Walkeria has returned! This is the third issue, technically Volume 2, Number 1.

As Ken Deacon wrote in the first issue, this newsletter is designed to bring Canadian odonate enthusiasts together and to stimulate an increased interest in dragonfly biology. We will publish (in French or English) research interests, articles, thesis abstracts, requests for information or specimen exchange, and even articles or books for sale. Walkeria is not intended as a medium for publication of scientific or technical papers, but general articles (for example, those in this issue) would be appreciated.

Ken Deacon has also stepped down as the manager of the Canadian office of SIO, and I have assumed those duties. From now on, all membership fees and address changes should be sent from now on to the editorial address above. Please pay in Canadian funds and make cheques out to me personally. To minimize bank charges I will send the fees to the Netherlands in one cheque.

The Editor,

Syd Cannings

NINTH INTERNATIONAL SYMPOSIUM OF ODONATOLOGY

The Ninth International Symposium of Odonatology will be held during the third week of January 1988 in the School of Biological Sciences, Madurai Kamaraj University, Madurai, India.

Madurai, in southernmost India, has a population of 900,000 and is called the Temple City. Winter temperatures average around 24°C with an average annual rainfall of 80 mm.

Madurai Kamaraj University is located 15 km outside of the city, and boasts beautiful buildings, parks, residential quarters, a health centre, a gymnasium, and three guest houses. The participants in the symposium will be accommodated in local hotels as well as in the University guest houses. Tariffs range from US$10/night for an air-conditioned single room to US$25/night for a king/queen suite.

During the symposium a half-day collecting trip will be arranged, and there will be a post-symposium tour to the Thekkady-Periyar Game Sanctuary and the Kodaikanal hill station.

Please send enquiries to:

Dr. S. Mathavan
Organizing Secretary
9th International Symposium of Odonatology
Madurai Kamaraj University
Madurai 625 021, INDIA

Fig. 13.2. Facsimile of front page of *Walkeria* Vol. 2, No. 1, published June 1, 1887.

The opening of the Japanese National Office was a highly spectacular and scientifically profitable event. True to the traditional love and care for these beautiful creatures – the dragonflies, both the Japanese national office and a large number of local dragonfly enthusiasts and legends (e.g., Dr S. Asahina, Dr K. Inoue etc.) lent SIO Central Office one of the strongest supports, covering not only the membership administration but also developing many new conservation research programmes (Fig. 13.3).

Fig. 13.3. *(a)* Dr Siyoziro Asahina –SIO's Charter Member, Member of Honour, Past President, and doyen of Japanese odonatology, and teacher of most of the Far Eastern dragonfly workers, and; - (b) Dr Kiyoshi Inoue–SIO's Member of Honour and its immediate past President, as well as Member of Honour of the Japanese Society for Odonatology and a charter member of the Kansai Research Group of Odonatology (Osaka), who set up the S.I.O. National Office in Japan and served during four decades as the National Representative there. (Source: B. Kiauta archive)

Since, however, the Japanese members had difficulties in remitting the annual dues in Dutch Guilders to the Treasurer in Holland, Dr K. Inoue volunteered in mid-1970s to collect the fees in Japan. He also served as an indispensable liaison between the Japanese odonatologists and the rest of the odonatological community of the world.

The National Office in Japan was sanctioned at the SIO Plenary Business Meeting in Chur, Switzerland, on 18 August 1981 and the first two issues of its newsletter appeared on the 1st of December 1981, named *Sioja*, a word combined from "SIO" and "Japan". The objective of the bulletin that looked like a numbered "circular" was to serve as a vehicle for business – and some other minor communications in Japan. It was written in Japanese throughout. No. 2 contained a brief outline and some statistical data on the development of SIO in Japan, covering the 1971-1981 period. It is due to *Sioja* that the Japanese membership increased spectacularly up to and above the 110 mark.Until 2003, there were published 1-3 issues annually.In the subsequent years, the National Office in Japan covered also the membership in Hong Kong and in South Korea. Upon Dr Inoue's retirement, in 2014, he was succeeded in Office by Mr Hidetaka Natsume, serving in the capacity of National/Regional Representative until this very day.

SLOVENSKO ODONATOLOŠKO DRUŠTVO

ERJAVECIA

številka 22 BILTEN 31.10.2007

izhaja enkrat na leto ISSN 1408-8185 uredil: Matjaž Bedjanič

PROF. DR. BOŠTJAN KIAUTA -
SEDEMDESETLETNIK

Fig. 13.4. Prof. B. Kiauta appearing on the front page of *Erjavecia*, Vol. 22 dated 31th Oct., 2007. (Source: M. Bedjanič)

The seed for SIO National Section in Slovenia was sown on October 23, 1992 when in its first Charter Meeting in Ljubljana eleven members attended, along with Dr Kiauta and his wife, Marianne, from the SIO Central Office. In this meeting, Dr Kiauta outlined the history, objectives, and constitutional set-up of SIO, whereafter the aims, Constitution and By-Laws of the National Section were discussed and accepted. Dr Kiauta always knew that Slovenia is a country where not only a huge legacy in entomological research prospered but also love for dragonflies knew no bounds and several young odonatologists were involved into various significant biogeographic and taxonomic explorations. He painstakingly developed the science of odonatology in a systematic way over nearly six decades, often by personally conducting field surveys and amassing huge data on local taxa. Upon his initiative, therefore, the SIO national Office in Slovenia (later transformed into the Slovene Odonatological Society) was established in 1994 with Dr A. Pirnat as its first National Officer in-Charge. The Office produced its journal, *Exuviae* (edited by Mr I. Geister) and the bulletin *Erjavecia*, the inaugural issue of the former appeared in 1994 and that of the latter in 1995. Since 1996 *Erjavecia* has been edited by Dr M. Bedjanič, a well-known authority on Asian and European dragonflies (Fig. 13.4). From 1995 until the end of 2017, 32 issues of *Erjavecia* have been published.

The Slovene Odonatological Society has been active for more than two decades. It is named after Fran Erjavec (1834–1887), a Slovene writer, naturalist and teacher. In 1864, he introduced the expression 'kačji pastir' for dragonfly in the first Slovene zoological school book. Until 2005 it was published twice a year but since then only one issue is regularly published, usually at the end of October each year. The main aim of *Erjavecia* is to record and publish all available odonatological information from Slovenian geographical area. Around 1,200 pages have been published thus far, including many different articles and reports covering numerous faunistic, ecological, nature conservation, historical, literary and other topics. Each issue of *Erjavecia* also includes many citations of published odonatological works, including 'grey literature', which contain faunistic data for the territory of Slovenia, titled 'Additions to the Slovene odonatological bibliography'. Dr Kiauta has helped Slovene odonatologists to systematically mine all the known published records, information and knowledge on the dragonfly fauna of Slovenia from 1685 to 2016. In this way the Slovene odonatological bibliography currently totals of over 1,100 citations.

Dr Kiauta soon identified a new habitat for the SIO's new national Office, the second in Asia, in the ecologically highly rich country for Odonata, Thailand. The Thai National Office of S.I.O. was established on 1st September, 1985 with Dr Manas Titayavan of the Department of Entomology, Faculty of Agriculture at Chiang Mai University, as the National Representative. His connection with Dr Kiauta and Marianne Kiauta originated from their short visit to Chiang Mai in April 1979, during which Titayavan guided them in the field on the lower slopes of Doi Suthep near the University campus. Titayavan produced two issues of the newsletter *Malangpo*, published on 1st September, 1985 and 1st April, 1986. The issues were entirely in Thai language and contained 5 and 11 mimeographed pages, respectively. Both issues presented details of the organization and publications of SIO., the second one also listed the titles of 52 publications on Thai dragonflies (1904–1985). Although Dr Titayavan was very helpful, he actually was not an odonatologist, therefore, on 12th December, 1986, Dr Kiauta asked Brother Amnuay Pinratana of Saint Gabriel's College in Bangkok, and the formal representative of the 'Thai Committee' of SIO, if he would be willing to take over the editorship of *Malangpo* and eventually become the National Representative. Pinratana agreed 'to take over editing *Malangpo* temporarily'. He began editing *Malangpo* and was formally nominated as the Thai National Representative at the SIO Business Meeting at the 9th International Symposium of Odonatology in Madurai in January 1988.

The third volume of *Malangpo* was published on 30th May, 1987. It was entirely in English, since Pinratana realized that publishing in Thai would make the newsletter inaccessible for all foreign odonatologists, and those very few Thai persons interested, or expected to become interested, in Odonata, could read English. This decision proved to be correct. Had the newsletter continued to be published in Thai, it would have ceased very quickly, as discussed below. The other innovation was that the newsletter was printed by a commercial printing house on a good quality paper, better than that used by any other SIO newsletters. The same printing house produced the newsletter up to volume 10 (November 1993), and thereafter it appeared as computer printouts. In 1987 two volumes were published, but in 1988–2003 (volumes 5-20) only one annual issue was published, in November every year. Two further issues were published in November 2006 (Vol. 21) and June 2008 (Vol. 22). Volumes 3–22 had consecutive pagination and these volumes included a total of 236 pages. The majority of the articles and notes were travel reports and other articles written by over 15 visiting foreign odonatologists, such as Matti Hämäläinen, Allen Davies, John Michalski, Nick Donnelly, Sidney Dunkle, Jill Silsby, Rosser Garrison, René Hoess, Rory Dow, Albert Orr and Oleg Kosterin. Bro. Amnuay Pinratana had either participated or helped in arrangements in visits of most of these people. *Malangpo* became a forum of foreign visitors to publish their travel and dragonfly collecting reports in Thailand and the newsletter elegantly gave news about the activities of odonatologists and odonatology. It is produced as a public service to all of those who are interested in Odonata.

Malangpo also included articles listing the dragonfly species recorded in several national parks or wildlife sanctuaries in Thailand, written by Hämäläinen and Pinratana. A few articles dealt with odonate records from Peninsular Malaysia, Sabah and Sarawak, Laos, Vietnam and Cambodia. Only 8 articles, totalling just 16 pages (published in 1987, 1992 and 2001–2003) were written by Thai nationals other than Pinratana. Among the most notable of these were the two articles by Apisit Thipaksorn and his colleagues (in volumes 18 and 20).

When the Thai National Office was established there were only 3 members; by 1st July, 1991 the membership had increased to 12 members, but in May 1994 it had declined to 9 members. Later the membership again reached about a dozen members, but then fell away again. Over the years the membership total scaled up to 20, although many of them were non-odonatologist friends of Pinratana, some being entomologists working at universities with only a minor research interest in Odonata. Actually, in addition to Pinratana only two or three of the members collected or studied Odonata during the entire period of existence of the National Office, among them was Mr Pisuth Eak-Amnuay in 1980–1990s.

Since the Thai National Office was performing excellently, largely due to field support from Dr Matti Hämäläinen, and odonatological help from the SIO Central Office, Dr Kiauta asked Bro. Amnuay Pinratana to ponder on organizing one of the future SIO international symposia. To provide enough exposure to Bro. Pinratana, Dr Kiauta extended necessary help to him whereupon Bro. Pinratana participated at three International Symposia of Odonatology: Tennessee (USA 1989), Trevi (Italy 1991), and Osaka (Japan 1993). In Trevi the Thai delegation included also Dr Valuli Rojanavongse and Mr Pramote Saiwichian. In Japan there were no fewer than six Thai participants, but among them only two (Bro. Pinratana and Mr Sirichai Divasiri) were active in odonatology. Bro. Pinratana and Mr Pramote also participated in the first WDA (Worldwide Dragonfly Association Symposium in Hamilton, New York in 1999.

Dr Matti Hämäläinen (Helsinki University, Helsinki, Finland), who made a total of 30 visits to Thailand during 1982–2013, most of which lasting about one month, and developed a very close research bonding with Bro. Pinratana, largely because filling the pages necessary to produce an issue of *Malangpo* (minimum of 8) also required much effort on his part, feels himself well placed to write some behind-the-scenes recollections, partly based on Pinratana's letters to him (see Dr Matti's email dated 17th June, 2017 to Dr Tyagi). During these visits he came to know Bro. Amnuay Pinratana well (meeting him for the first time on 6th June 1984, and thereafter in all visits) and see how faithfully and unselfishly he handled his voluntary task as the editor of *Malangpo* and the Thai national representative of SIO. For a busy man with numerous other pursuits, including his other entomological interests and professional work in Saint Gabriel's College, the work for SIO was not always an easy task. More so since getting enough manuscripts for *Malangpo* was always a difficult task for him; especially hard was to get any Thai person to submit to the newsletter. Since the newsletter was printed by a commercial printing office, it was rather tedious for the editor to correct the proofs, because the printers were not fluent in English and made numerous mistakes. *Malangpo #8* took him one full month for proof reading, artwork and the rest. But he still did it most effectively.

As a very kind, humble and affable person it was difficult for Bro. Pinratana to decline help to others and say a definite 'No' to anyone. His personal activity promoted the knowledge of Thai dragonflies in many ways. He helped at least some twenty foreign odonatologists to collect dragonflies in Thailand. During their visits several new species were found and the knowledge of the Thai dragonfly fauna increased tremendously. He also supported a few Thai students in their dragonfly work. He arranged the publication of three books on Thai dragonflies: '*List of the Odonata of Thailand and annotated bibliography*' by PINRATANA, KIAUTA & HÄMÄLÄINEN (1988a), '*A list of the Odonata from Thailand* (Parts I-XXI)' by ASAHINA (1993) and '*Atlas of the dragonflies of Thailand – Distribution maps by provinces*' by HÄMÄLÄINEN & PINRATANA (1999).

(a) (b)

(c) (d) (e)

Fig. 13.5. (a) Bro. Amnuay Pinratana in his collection room at St. Gabriel's College in Bangkok on 16 January 1988. (b). Partial view of the spacious, modern collection room of the 'Insect Museum of Saint Gabriel's College' in the 5th floor of the '90th Anniversary Building of Saint Gabriel's College' built in 2010. Photo taken on 19 February 2011. (c) The pinned specimens of Bro. Amnuay Pinratana's Odonata collection are preserved in these wooden drawers kept in the 'Insect Museum of Saint Gabriel's College'. The specimens in envelopes are kept in a separate cabinet. Photo taken on 25 November 2010, (d) Bro. Amnuay Pinratana collecting dragonflies at Huay Mae Kamin on 13 February 1988 during a joint trip to Kanchanaburi province with Matti Hämäläinen, (e) Three odonatologists resting at Lam Khlong Ngu in Kanchanaburi province on 13 March 2003: Bro. Amnuay Pinratana (right), Dr Albert Orr (middle) and Dr Matti Hämäläinen (left).

Bro. Pinratana's major accomplishment in odonatology was the gathering of an extensive collection of dragonflies, comprising of ca. 260 species in ca. 6000 specimens from different parts of Thailand (Fig. 13.5). The collection still includes a few undescribed species and it must be the most diverse single collection of the over 350 Odonata species so far known from Thailand. His other entomological achievements include publishing several long series of books on Thai butterflies, moths and beetles, and acquiring large and diverse, representative collections of these insects. The collections are now preserved in a museum at Saint Gabriel's College

(Fig. 13.5a,c). As an amateur entomologist, without any training in entomology, his influence in increasing the knowledge of Thai insect fauna surpassed that of any academic entomologist in Thailand, and, in fact, the value of existing insect collections, such as those acquired by Bro. Pinratana is inestimable.

It is remarkable to note that while less than 20 years ago Bro. Pinratana had great difficulties in finding Thai persons interested in dragonflies, interest in these insects has now greatly increased in Thailand, just as in many other countries. To his great surprise, now (June 2017), a Facebook group on Thai dragonflies includes nearly 4700 members, at least some 200–300 of them actively involved in observing and photographing dragonflies. A few persons maintain regular blogs on Thai dragonflies. Obviously, the *Malangpo* perseverance and the SIO National Office's vision have worked wonders!

In the Russian and in the international odonatological community alike, the need of a national odonatological organization in Russia has been strongly felt for some time. Following the discussions at the 14th International Symposium of Odonatology, Maribor, Slovenia, July 1997, the initiative was born to convene a meeting of the Russian workers, where the possibilities of setting up a Russian SIO Section would be explored. The meeting took place in Novosibirsk, on 11th February, 1998. It was attended by thirteen members, viz., Dr S.N. Borisov, Dr A.G. Bugrov, Ms M.V. Dronzikova, Dr V.D. Gulyaev, Prof. Dr A.Yu. Haritonov, Dr Kh.A. Ketenichev, Dr O.E. Kosterin, Mr E.A. Perepelov, Ms O.N. Popova, Ms E.A. Serbina, Dr G.A. Sukhacheva, Ms V.A. Yanybaeva and Dr N.L. Yurlova. They discussed five Agenda items: (1) Discussion on the foundation of the Russian Section of the Societas Internationalis Odonatologica, R.S.S.I.O., (2) Election of the Board of the Section, (3) Publication organ of the Section, (4) Election of the Editorial Board of the latter, and (5) Report by Dr A.Yu. Haritonov & Ms O.N. Popova on the 14th International Symposium of Odonatology.

Taking into account the SIO tradition of national sections, and based on the preliminary opinions of 18 Russian workers, it was considered reasonable to set up the Russian Section in 2001 with Prof. Dr A.Yu. Haritonov, the mentor of Russian odonatology, as its first National Officer in-Charge. The Office produced, in veneration of the doyen of Russian odonatology, Dr Boris Feodorovich Belyshev (1910-1993), the SIO charter member and member of honour (Fig. 13.6), its newsletter, *Belyshevia*, the inaugural issue of which appeared in 2001. The R.S.S.I.O. Board was elected with the *Chairman*: Professor Dr Anatoli Yuryevich Haritonov, Deputy Director of the Institute of Animal Systematics and Ecology of the Siberian Branch of the Russian Academy of Sciences; *Deputy Chairman*: Dr Sergei Nikolaevich, Senior Research Associate at the Institute of Animal Systematics and Ecology of the Siberian Branch of the Russian Academy of Sciences; *Deputy Chairman*: Dr Oleg Engelsovich Kosterin, Research Associate at the Institute of Cytology and Genetics of the Siberian Branch of the Russian Academy of Sciences; *Secretary*: Ms Olga Nikolaeva Popova, post-graduate student at the Institute of Animal Systematics and Ecology of the Siberian Branch of the Russian Academy of Sciences.

Fig. 13.6. The doyen of Russian odonatology, Dr Boris Feodorovich Belyshev (1910-1993), SIO charter member and member of honour (Source: B. Kiauta archive)

The Editorial Board of *Belyshevia* was also elected as, *Editor-in-Chief*: Dr A.Yu. Haritonov; - *Members*: Dr S.N. Borisov, Dr A.Yu. Haritonov, Dr O.E. Kosterin, Ms O.N. Popova, and Dr G.A, Sukhacheva. The Editorial Board was authorized to tackle the membership subscriptions etc.

The Russian National Office of SIO organized the 15th International Symposium of Odonatology in Novosibirsk, in July 2001.

Dr Kiauta is always crystal clear and decisive in his planning and action. He would never mince words and would criticise, if eventually unavoidable, in direct but instructive words with a view to improve in future. At the same time, if he would find something really appreciable he would not take even a second to convey his admiration. So, in respect to the newly formed SIO National Office in India, he openly lauded its progress in his communication dated 27th November, 1986, to Dr R.A. Cannings (Entomology Division, British Columbia Provincial Museum, Victoria, Canada) and to Dr S.G. Cannings (Department of Zoology, University of British Columbia, Vancouver, Canada) as: "*The most active is the Indian National Office. Its* Fraseria *is published extremely punctually, they have started now a new series,* Occasional Publications, *in book print, with very nice cover, so far two issues since*

October. They are also organizing Regional Symposia at 2-year intervals. The last was in October in Dehra Dun (4 countries participated), and the next one is scheduled for 1988. They also have an own Regional Council, and render small grants from the membership fees (occasionally also from the SIO Central funds)". Opening of the SIO National Office in India was a turning point in the nearly one hundred years' history of Indian odonatology; both the science of odonatology and the odonatologists sprang up from virtual oblivion and produced some highly remarkable contributions both at the national and global levels. Today, the office is about 36 years old, and with the founding of an independent yet collaborative new confrerie, the Indian Dragonfly Society, formed in 2017, with its recently concluded first biennial conference on March 21-23 in Dehra Dun, Uttra Khand, India, the future of dragonfly research certainly appears far greatly brighter! In Dr Kiauta's own words dated the 20th Dec., 2017, "*The recent foundation of the Indian Odonatological Society is to be understood as the crown to the more than a century-lasting systematic odonatological research in this country, for which the scientific basis was provided by the monumental works of the late Lt-Col. Dr F.C. FRASER and the organizational framework by Professor Dr B.K. TYAGI, who is acting during the past decades as an all-inspiring and tireless initiator and promulgator of odonatological research and its organization in India. Hardly any country of the World could make a boast of so many splendid results, triggered by the initiatives of so few persons, which was made possible solely due to the multitude of highly motivated and well-qualified workers, covering a large diversity of research fields. The IDS Symposia (Conferences), scheduled at regular intervals, the newsletter* Bradinopyga *and the peer-reviewed* Journal of Indian Odonatology *will facilitate the communication within the Indian odonatological community and offer the possibility of rapid publication of high quality results of the research. The importance of this cannot be exaggerated. The best research work is bound to be lost, if the results are not properly and in time published, whereas the publication inevitably opens up new horizons, most particularly so also to the coming-up young workers. It is among the objectives of the Journal of Indian Odonatology to also provide "training grounds" to the young, preparing their first manuscripts for publication under the mentorship of the experienced workers. This is a very important educational mission, which is often lacking with scientific periodicals of the (expected) high standard of our Journal. The IDS and the present Conference are opening a new page in the long and fruitful history of Indian odonatology.....*"

The SIO National Office produced its newsletter, *Fraseria*, the name given in the honour of Col. Dr F.C. Fraser for his monumental work on Indian dragonflies (*cf.* TYAGI,1988b), the inaugural issue of which appeared in 1981. *Fraseria*, edited between 1981 and 1992 by Dr B.K. Tyagi, was for the few initial issues printed in The Netherlands under the care of Dr Kiauta, in order to smoothly establish SIO Indian National Office, from a place called Ukai in a tribal belt in south Gujarat State where Dr Tyagi had recently moved in to join his first post as the Senior Research Officer in-Charge on a Malaria Research Centre (ICMR)/National Malaria Eradication Programme (Govt. of India)/WHO-TDR research project titled, "A laboratory and field study of malathion resistance in *Anopheles culicifacies*". Since 1994, *Fraseria* had been appearing as a scientific bulletin, edited by Dr R.J. Andrew.

Dr Kiauta's search for a suitable odonatologist who could effectively run the SIO National Office in India had actually begun much earlier in late 1970s. Before putting his trust in me he had invited quite a few highly competent and active odonatologists of those days to take up the responsibility, but nothing materialized, largely because the task of running a national office posed many challenges as well as responsibilities, over and above the full-time consuming professional duties one would always have! Dr Kiauta, on the other hand, knew everything about me from our extensive correspondence since 1975, when I was pursuing my PhD research under often difficult financial circumstances. Therefore, as soon as I got employment, he asked me if I would be able to head the SIO National Office in India and he explained all its objectives. I was, of course, only too delighted with the offer. As a consequence, soon after attending the 6th International Symposium of Odonatology, Prof. Kiauta informed me on 8th September, 1981 about the SIO permission to establish a National Office in India, with me as its first National Representative-in-Charge (Fig. 13.7).

Becoming National Representative-in-Charge of the SIO National Office was a turning point in my life, too. That this was a huge responsibility I knew very well, but could it be such a decorative responsibility apt to be compared with the real Head of the Republic of India was unimaginable and never ever realized! The legend goes as I once appeared for an interview being conducted at the ICMR Hqrs, New Delhi for the post of Deputy Director in Desert Medicine Research Centre, Jodhpur, in 1995. Dr S.S. Sriramachari, Former Additional Director General and a world renowned pathologist, was the Chairman of the Selection Committee. The post was open to both the medicos and the biologists, with an obvious inclination towards the former. After my interview was over, he, looking at by *curriculum vitae*, for one last time, asked me what was that 'National Representative-in-Charge' meaning for my biodata? I benignly explained that I was representing the International Society of Odonatology and, therefore, had been appointed with the position of the National Representative in-Charge of its National Office in India. He smiled at me and coolly inquired, almost in a cajoling manner, "Do you know the other Indian national in the history of the country who was also titled as the National Representative in-Charge?" And, before I would open my mouth to utter an answer, he hurriedly chipped in quipping in the same breath, "Dr C. Rajagopalachari, the first Governor-General of free India who was also the only Indian national ever to hold the office." This comparison carried a great meaning, and I thanked him humbly for the information. After leaving the interview board room I was only thinking of this new definition and was filled with the joy the SIO National Office has rewarded me in silence. The result: I was selected in this interview against all odds, even when the post was traditionally inclined heavily toward a medical incumbent!!

University of Utrecht

Department of Animal Cytogenetics
& Cytotaxonomy

Padualaan 8, 3584 CH Utrecht The Netherlands Phone: 030-533438/533841

Dr B.K.TYAGI
Malaria Research Centre (I.C.M.R.)
UGH 61-70 / Sector VI
UKAI-394 680
Surat, G u j a r a t
I n d i a

"Appointment Letter"

Utrecht, September 8, 1981

Dear Dr Tyagi,

Returning from the Symposium in Chur, I hasten to let you know that in the SIO Plenary Business Meeting, on August 18,1981, under the (modified) Agenda item No. 10, the opening of an SIO National Office in India has been formally confirmed, and you have been appointed the first SIO National Representative in charge of it.

The document on the "General Features, Structure and Operation of the SIO National and Regional Offices" (1980-1981) has been posted and passed in the same meeting, and its provisions are fully applicable to the Office in India as from August 18,1981.

In accordance with the passed constitutional amendments, the National Representatives are automatically members of the Council of the Society.

As from the date of the receipt of this communication, you are free to use the official notepaper, sent to you earlier.

The 1981-1983 composition of the SIO Committees, as established at the meeting of the Executive Council, Chur, August 19,1981, is shown in the appended list.
The formal reorganisation of the Society's Boards was considered necessary and opportune in view of the extraordinarely rapid growth of SIO, nevertheless, it should be strongly emphasised that, as heretofore, the Society remains a "family of friends" rather than a hierarchical organisation. The "success" of the Society in the first decade of its life is mainly due to this principle, therefore the same "philosophy" should be observed also by all National Offices.

I shall be glad to help you in all organisational and other matters of the Office, whenever you would need any suggestions and/or assistance in this field. Please do not hesitate to let me know about any problems.

It is the desire of SIO to get the National Office in India fully operating by December 1, 1981, enabling the Indian colleagues to fulfill the 1982 membership fees in the Indian rather than in foreign currency. It seems opportune, therefore, that the following matters should be taken care of as soon as possible:

(1) Opening of a Bank Account under the name of the National Office;
(2) The 1982 membership fee (for individual bona fide members) is US $ 30.- (postage excl.). The Treasurer is willing to offer a reasonable reduction, hence you are requested to suggest an appropriate figure in Rp. The invoices will be then prepared in Utrecht and (for 1982) mailed from Holland.
(3) The text should be prepared for the first issue of the Newsletter, which should contain the information about the establishment of the Office in India, a note on its operation, and the known addresses of all Indian odonatologists. The Newsletter will be produced in Utrecht and copies will be sent to you for local distribution.
(4) The strategy should be outlined for the Mysore 1982 odonatol. meeting

and the registration forms should be included in the Newsletter.

(5) The first issue of the Newsletter will be sent to all Indian odonatologists, therefore a Membership Registration Form should be also included.

While preparing the text for the Nesletter, please do not forget to include some general information on SIO, the exact data concerning the Mysore meeting, and any other material of interest to the Indian workers. The newsletter is not intended as a medium for publication of scientific papers. This should be emphasised in the "Editorial".

In order to facilitate your work, during the initial period the journals will be mailed from Utrecht.

A set of all SIO publications will be supplied for the library of the National Office.

This just in haste. Please let me know anything I could do for your Office, and I do hope to hearing from you before long again.

With all good wishes and kind regards,

Yours sincerely,

B. Kiauta

Fig. 13.7. Facsimile of the Appointment Letter of Dr B.K. Tyagi as the first National Representative in-Charge, SIO National Office in India, issued by Dr B. Kiauta on September 8, 1981 soon after the organization of the sixth International Symposium of Odonatology in Chur (Switzerland) in 1981.

Since its inception, Dr Kiauta constantly guided and monitored all activities of the SIO National Office with great care for consistency and commitment in all the assignments undertaken. Under his direct guidance the SIO Indian National Office paced up its activities in an unprecedented manner. With the help of SIO Central Office it was already producing *Fraseria*, always in 4 pages, and carried brief scientific notes, dragonfly news, new memberships, advertisements etc., but often also appended by either the membership form or a symposium announcement (Fig. 13.8).

FRASERIA

Newsletter of the Indian National Office of the International Odonatological Society,

SOCIETAS INTERNATIONALIS ODONATOLOGICA (S.I.O.)

No. 1	pp. 1-4	December 1, 1981	Ukai

Editorial Address: Dr B.K. TYAGI, Malaria Research Centre, Indian Council of Medical Research, UGH 61-70 / Sector VI, UKAI-394680, District Surat, Gujarat

S.I.O. NATIONAL OFFICE OPENED IN INDIA

On May 13, 1981, the Secretary-General of S.I.O., Dr K.J.Tennessen, gave the 'green light' for the establishment of a National Office of the International Odonatological Society in India. The opening of the Office was sanctioned by the S.I.O. Plenary Business Meeting in Chur, Switzerland, under the (modified) Agenda item No. 10, on August 18, 1981. The National Office will become operative as from December 1, 1981. It is headed by the S.I.O. National Representative in India, Dr B.K.TYAGI (Malaria Research Centre, Indian Council of Medical Research, UGH 61-70 Sector VI, Ukai-394680, District Surat, Gujarat).

An S.I.O. National Office is the official representative of the Society in the country of its residence, and handles all the Society's business related to the geographic area of its jurisdiction. The head of a National Office is called 'National Representative'. He/she is acting as a liaison between the Society and the local odonatologists and organizations, and is rendering assistance in all matters related to the odonatology of the territory of the Office. The National Representative is automatically a member of the S.I.O. Council. A National Office Library is keeping a stock of all S.I.O. publications, as well as a collection of local and general odonatological literature. Through the main S.I.O. Library it renders xerox- and other services to the local membership.

The need for an S.I.O. National Office in India had long been felt both by the workers in this country and by numerous odonatologists throughout the world, interested in the Indian fauna. The opening of the Office, therefore, is a really great encouragement to the Indian odonatologists, and it is the hope of the entire S.I.O. membership that the Indian National Office will provide a firm and substantial basis for advancing odonatology in India.

The following are some of the immediate practical advantages to the local workers:

(1) The opportunity to pay the membership fees and other dues in the local currency will greatly facilitate membership registration;

(2) The Library- and other S.I.O. services will facilitate research;

(3) The publication of the newsletter and the regular organization of local meetings will facilitate the communication among the workers;

(4) A part of the National Office funds will be used to furnish small grants for sponsoring the work of young odonatologists, particularly work related to the preparation of M.Sc. and Ph.D. dissertations, and to the inventarisation of the local fauna.

MAIN TASKS OF THE S.I.O. INDIAN OFFICE

The following are the major tasks of the S.I.O. National Office in India:

(1) Administration of the local membership;

(2) Collecting of the local membership fees and other dues to the Society;

(3) Mailing of S.I.O. publications to the local membership;

(4) Sale of S.I.O. publications to the local membership and those interested;

(5) Rendering financial support to young local odonatologists;

(6) Liaison with local odonatologists and assistance in all matters concerning odonatological research in India;

(7) Assistance to non-resident odonatologists visiting India;

Fig. 13.8. Facsimile of front page of Fraseria, No. 1, published December 1, 1981. - (Source: B.K. Tyagi)

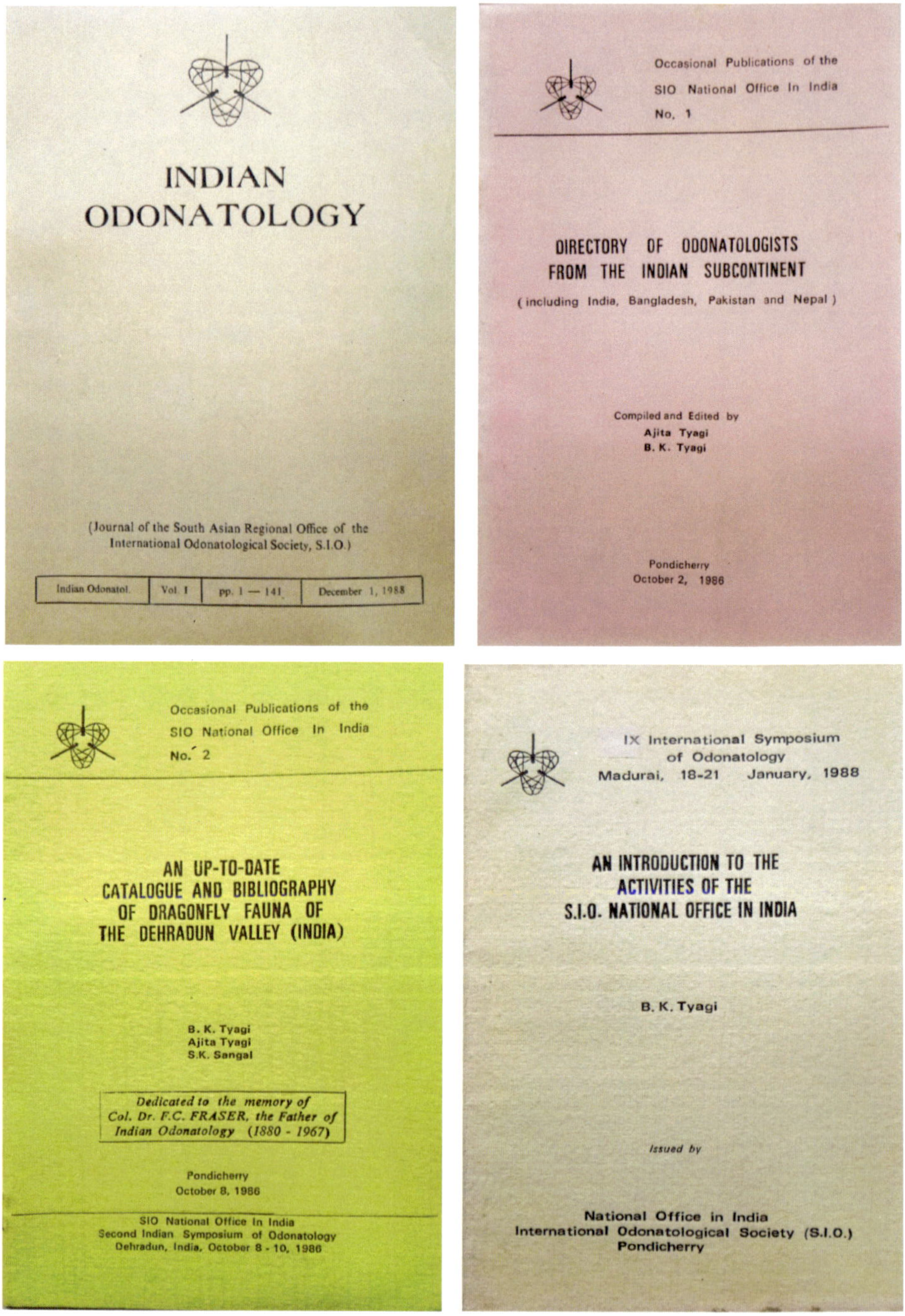

Fig. 13.9. Various publications of SIO National Office in India. – (Source: B.K. Tyagi)

In view of its phenomenal growth in a short span of time the SIO National Office was soon upgraded to Regional Office in South Asia (ROSA) in 1988, during the 9th International Symposium of Odonatology held in Madurai in 1988, with Dr B.K. Tyagi continuing as its Head. The ROSA published several periodicals and occasional publications, as mentioned below:

(1) *Indian Odonatology* (Edited by: Dr B.K. Tyagi): Commenced 1988, this was the first odonatological journal ever brought out in India (Fig. 13.9). Its Editorial Board comprised distinguished odonatologists from all over the world. About 8-10 original research articles were published per issue produced annually on December 1. Through assistance of the SIO Central Office the journal was sent to as many as 10-15 countries. The journal attracted manuscripts from many non-Indian workers.

After Dr Tyagi had relinquished from the duties of the SIOROSA in 1994, the office was renamed as SIO South Asia Council of Odonatology (SACO) during the 6th Indian Symposium of Odonatology, held in Bhagalpur, Bihar in 2003. The SACO has since been organizing symposia now and then and the latest in the series, the 9th was held in Goa, India on 15-19 September, 2017.

(2) *Abstracts of Papers read at the Indian/South Asian Symposia of Odonatology*: The SIO Regional Office in South Asia (ROSA)/South Asia Council of Odonatology (SACO) invariably published Abstracts of papers read in its biennial symposia since 1984.

(3) *Fraseria:* This is the main spokesperson of The SIO Regional Office in South Asia (ROSA)/ South Asia Council of Odonatology (SACO) activities since its inception. It was published as a newsletter between 1981 and 1992 (Editor: Dr B.K. Tyagi), and since 1994 as a scientific bulletin, in a new series, edited by Dr R.J. Andrew.

(4) *Occasional Publications of the SIO Regional Office in South Asia*: The SIO-ROSA published three issues on interesting subjects (Fig. 13.9), such as (i) Directory of odonatologists from the Indian subcontinent.Yr. 1986, No.1, pp.20 (by B.K. Tyagi, and Ajita Tyagi), (ii) An up-to-date catalogue and bibliography of dragonfly fauna of the Dehradun valley (India). Yr. 1986, No. 2, pp. 14 (by B.K.Tyagi, A. Tyagi & S.K. Sangal), and (iii) Odonatological bibliography of Frederick Charles Fraser [originally manuscripted by Basil Elwood Montgomery]. Yr. 1988,No.3, pp. 24 (by B.K. Tyagi, Editor-Author).

Additionally, the office produced an independent publication on the activities of the S.I.O. National Office in India (TYAGI, 1988).

One of the most useful activities the SIO National Office was its planning to organize a symposium on alternate years. Looking at the much unconnected and scattered odonata workers in the country/region it did not seem to be a easy venture.

However, I took the opportunity of the 69th Indian Science Congress that was to be held in Mysore University, Mysore on January 3-7, 1982. As I was a member of the Indian Science Congress Association, I took permission from the Secretary General, ISCA to organize a satellite meeting of odonatologists under the aegis of SIO National Office in India. With the permission received I circulated the agenda of an Informal Meeting to all the dragonfly enthusiasts and invited to attend the same. Only three odonatologists registered: Dr. Mahabir Prasad, Dr B. Suri Babu and Dr B.K. Tyagi. The Congress attracted over 20,000 participants, and all rooms in every department were occupied. Since no room was available I discussed with the authorities concerned and took permission to call our meeting in the open lawns of the Zoology Department. Thus was carried out the first odonatological meeting which, however small and ephemeral, certainly proved crucial to have cut the ice for future organization of our biennial symposia. The SIO ROSA/SACO has so far organized nine Indian Symposia of Odonatology;

- 1st Indian Symposium of Odonatology: Madurai Kamaraj University, Madurai, TN, India (21-23 January, 1984); Organizing Secretary, Dr. S. Mathavan.
- 2nd Indian Symposium of Odonatology: D.A.V. (PG) College, Dehradun, Uttarakhand, India (8-10 October, 1986); Organizing Secretary, Dr. S.K. Sangal.
- 3rd Indian Symposium of Odonatology: Chikkaiah Naicker College, Erode, TN, India (20-21 January, 1990); Organizing Secretary, Dr. S. Basalingappa.
- 4th Indian Symposium of Odonatology: C.M.P. (PG) College, Allahabad, Uttar Pradesh, India (10-12, 1992); Organizing Secretary, Dr. V.K. Srivastava.
- 5th Indian Symposium of Odonatology: Shri Shivaji Science College, Nagpur, Maharashtra, India (20-21 Dec., 1998), Organizing Secretary, Dr. R.J. Andrew.
- 6th Indian Symposium of Odonatology: T.M. Bhagalpur University, Bhagalpur, Bihar, India (21-22 April, 2003), Organizing Secretary, Dr S.P. Roy.
- 7th Indian Symposium of Odonatology: Hislop College, Nagpur, Maharashtra, India (2006), Organizing Secretary, Dr. R.J. Andrew.
- 8th Indian Symposium of Odonatology and Tropical Biodiversity:Hislop College, Nagpur, Maharashtra, India (17-18 August, 2014), Organizing Secretary, Dr. R.J. Andrew.
- 9th Indian Symposium of Odonatology; Nature's Nest, Goa (15-16, 2017), Organizing Secretary, Dr R.J. Andrew.

These are the maximum odonatological symposia ever in the Indian history for any single insect Order. Bringing together all odonatologists from all over the Indian Subcontinent was always a great challenge. Dr Kiauta helped in the organization of these symposia right from the stage of planning, and his communication dated 14th December, 1984 in respect with the 1st Indian Symposium of Odonatology mirrors his feelings for the success of the inaugural symposium (Fig. 13.10).

FOREWORD

The FIRST INDIAN SYMPOSIUM OF ODONATOLOGY marks an important step in the evolution of Odonatology in India, in the Indian Sub-continent, and, indeed, in the life of SOCIETAS INTERNATIONALIS ODONATOLOGICA in general.

It is extremely promising to see what this active group has achieved. A bare two years after the inauguration of the Indian National Office of the S.I.O. our Indian colleagues have been able to bring together more than 60 workers to discuss the whole spectrum of biological disciplines for which the Odonata have proved to be eminently suitable objects: taxonomy, embryology, life history, population dynamics, ecology, behaviour, physiology, cytotaxonomy, histology, insecticide effects, pest control, etc.

The ecological significance of dragonflies and their importance in environmental monitoring have often been emphasized. It is, therefore, most reassuring to see the rapidly increasing interest and the volume of primary research on this subject throughout India and southeastern Asia.

I am honoured to write these few lines at the launching of this first larger publication sponsored by the Indian National Office. I am sure many more are to follow. This will inevitably greatly facilitate a fruitful co-operation between our Indian colleagues and the odonatologists throughout the world, to the benefit of all.

Professor Dr.B. Kiauta, Ph.D., D.Sc.,
Executive Editor of Odonatologica
Chairman of the Standing Committee for the Organization of International Symposia of Odonatology

Head of the S.I.O. Commission for Coordination of National Offices and Affiliated Associations

Utrecht, 14 December, 1984

Fig. 13.10. Facsimile of the 'Foreword' by Dr B.Kiauta dated 14th Dec., 1984.
(Source: B.K. Tyagi)

In a vast country like India organizing odonatological symposia without any confirmed financial resources was never an easy task. Dr Kiauta understood the dilemma I was going through and discussed the prospects of helping with some seed money by the SIO National Office in India which I immediately accepted. The 1st Indian Symposium of Odonatology was thus convened by Dr S. Mathavan in the

School of Biological Sciences, Madurai Kamaraj University, Madurai, during January 23-25, 1984, with a major financial help from the University Grants Commission, New Delhi. About 30 papers were presented and some 45 scientists actively participated in the symposium (Fig. 13.11). On January 23, Professor J. Ramachandran, Vice-Chancellor of the Madurai Kamaraj University, inaugurated the meeting. Papers presented pertained to taxonomy, cytotaxonomy, population dynamics, endocrinology, physiology, reproductive biology and toxicology. Plenary talks on selected topics were presented by Prof. T.N. Ananthakrishnan, Dr B.K. Tyagi, Dr D.B. Tembhare and Prof. T.J. Pandian. The participants were taken to a field trip in the Alagar Hills, some 30 km from the University Campus.

(a)

(b)

Figs. 13.11. The First Indian Symposium of Odonatology, Madurai: (from left to right) (a) The Chair, Prof. S. Krishnaswamy (delivering the Opening Address), Prof. N.B. Nair. Prof. J. Ramachandran, Dr B.K. Tyagi (Head of the SIO National Office in India), Prof. TJ. Pandian, and Dr S. Mathavan (Convener); — (b): The participants, during the Opening Address. (Source: S. Mathavan)

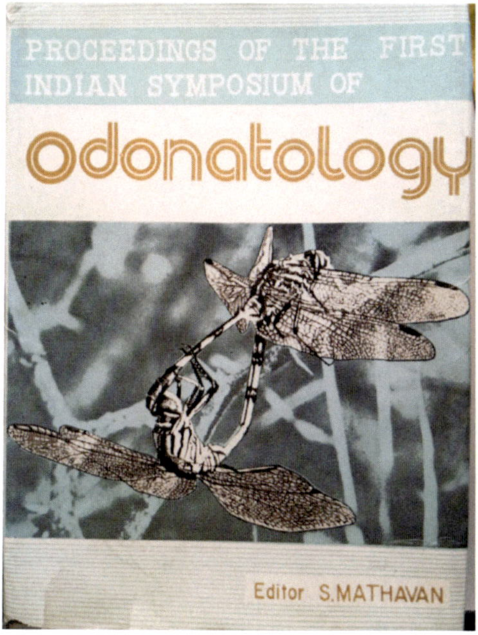

(a)

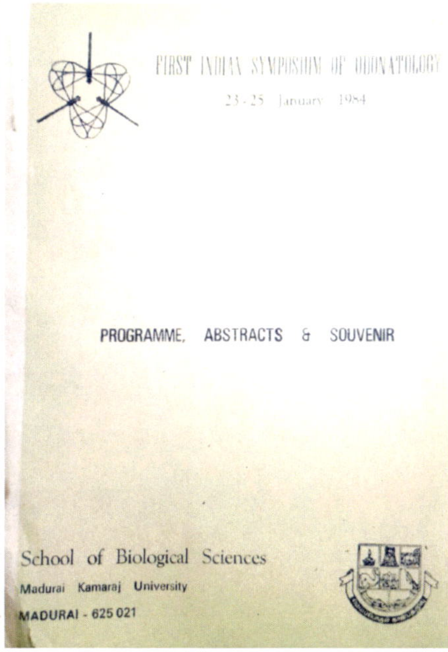

(b)

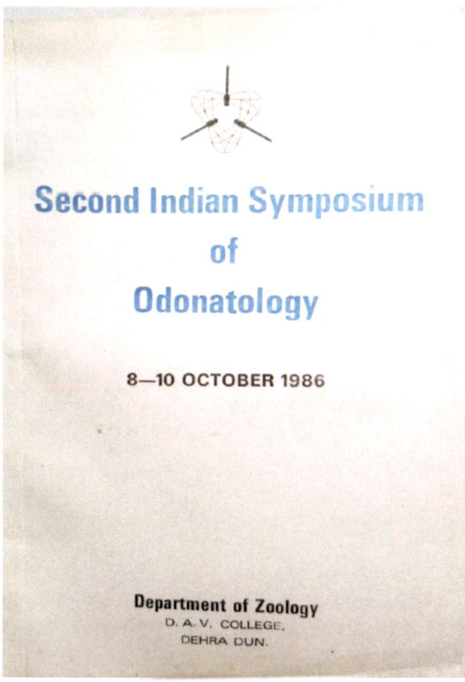

(c)

Fig. 13.12 *(a-c):* Facsimiles of the first page of the Abstracts booklets of the initial three Indian Symposia of Odonatology. – (Source: B.K. Tyagi)

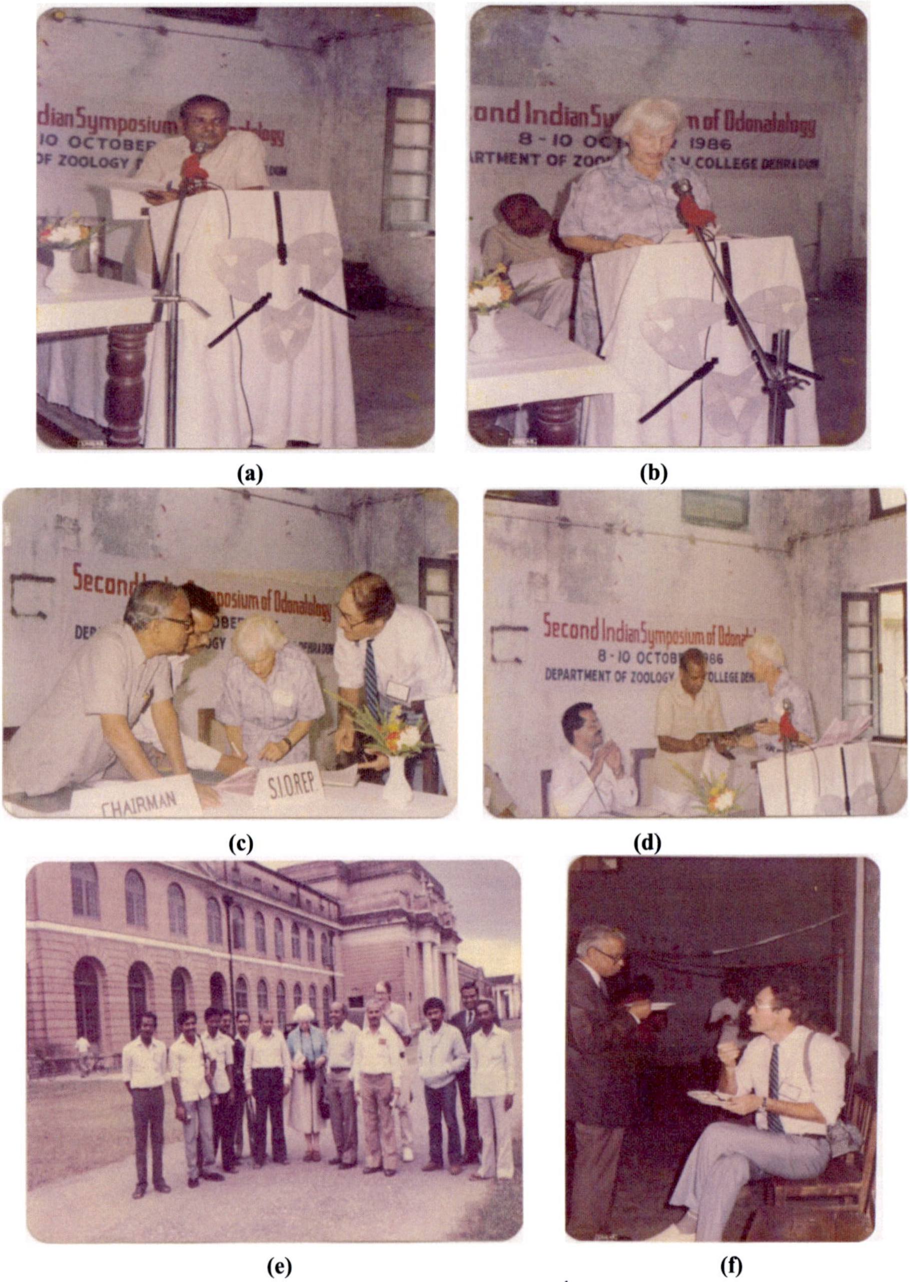

Fig. 13.13. *(a-f):* Dr S.K. Sangal, Convener, opens up the 2nd Indian Symposium of Odonatology, (b) Prof. van Brink, Chief Guest, addresses the participants, (c) Prof. Van Brink, Dr Tyagi and Dr Hämäläinen signing a Certificate of Appreciation for Dr S.K. Sangal, as Dr. S.K. Kulshrestha looks on, (d) Prof. van Brink presenting Dr S.K. Sangal a dragonfly book at the inaugural session, and (e) A dragonfly field tour including a visit to the world famous entomology museum in the Forest Research Institute & Colleges, Dehra Dun, (f) Dr Matti Hämäläinen confers with Dr. P.K. Sen-Sarma, during a lunch session. (Source: Dr B.K. Tyagi)

On top of all, participants enjoyed the religious environment of the temple city, Madurai, known for its world famous Meenakshi Temple. A very nice booklet of the abstracts, of which contents were enlisted in the OA No. 4466, in Odonatologica 13(2): 333-334 (1984), was produced by the organizers (Fig. 13.12). A part of the papers presented were subsequently processed for publication in a special Proceedings volume with the financial assistance of International Society of Odonatology, and by the Department of Science and Technology, New Delhi (Fig. 13.12).

The second Indian Symposium of Odonatology (1986) was organized at the Department of Zoology, Dayanand Anglo Vedic (Post Graduate) College, Dehra Dun in Uttar Pradesh, northern India (Organizing Secretary: Dr S.K. Sangal). The beautiful and world famous Doon Valley known for its rich entomological fauna, including dragonflies, as the site for the symposium became an instant hit and the symposium attracted participants, besides India, from Nepal (Dr Mahender Mahato attended with his family), Dr Matti Hämäläinen (Finland) and Dr Janny M. van Brink (The Netherlands), too. From India alone, a total of 40 odonatologists participated from as many as 15 States and UTs in the country. Professor van Brink and Dr Hämäläinen were the Chief Guests and Dr P.K. Sen-Sarma, Director of Biological Research at the FRI & Colleges delivered the presidential lecture. On this occasion Prof. van Brink gifted a personally signed beautiful dragonfly book to Dr Sangal. A half-day dragonfly field tour was organized to the famous Sahastradhara (= a million streams) or Sulphur Springs (the site of discovery of *Calicnemia doonensis* Sangal & Tyagi, 1984) and to the monumental entomological museum of the Forest Research Institute & Colleges (Fig. 13.13). The SIO National Office distributed among participants copies of its various *Occasional Publications.*

Highlights of the symposium were (i) the chief-guest lecture by Prof. J.M. van Brink who spoke broadly on the science of odonatology, the International Odonatological Society (SIO) and the functions of its various National Offices across the world, the future International Symposium of Odonatology in Madurai, and establishment of the unique SIO International Odonata Research Institute in Florida, USA, (ii) a special guest lecture by Dr Matti Hämäläinen who enlightened on the interesting taxonomy and distribution of dragonflies in South Asia, (iii) a lecture by Dr P.K. Sen-Sarma who recalled how, in the late 1950s, Dr Sangal, influenced by Col. Dr F.C. Fraser's *magnum opuses* on Odonata, pondered on writing his doctoral dissertation on dragonflies and the varied constraints he encountered in carrying out the work. He highlighted the significance of a mentor in guiding research to budding odonatologists, which reverberated the immortal chantings of saint Kabir in my ears:

गुरु बिन ज्ञान न ऊपजे, गुरु बिन मिले न मोक्ष

गुरु बिन लिखे न सत्य को, गुरु बिन मिटे न दोष

Transcription:

Guru bin gyan na upje, guru bin mile na moksha

Guru bin likhe na satya ko, guru bin mite na dosh

Translation

"Without the Mentor no one obtains knowledge or achieves salvation,
Without the Mentor no one can see truth or have his doubts removed."

The symposium was a great success in all aspects and concluded with the announcement of meeting again in the picturesque south Indian city of Erode in 1990! Dr Kiauta immediately shot back a highly appreciative communication full of encouragement! His encouragement only substantiated my resolve to take the SIO National Office in India to far greater heights.

The Third Indian Symposium of Odonatology was held in the Department of Zoology, Chikkaiah Naicker College, Erode, Tamil Nadu during, 1990 (Organizing Secretary: Dr S. Basalingappa). The symposium was attended by 35 participants from 9 States and UTs in the country (Fig. 13.14). A dragonfly tour in the paddy fields of rural environs was organized. The symposium concluded with the announcement of the fourth Indian Symposium of Odonatology (Organizing Secretary: Dr V.K. Srivastava) at the Zoology Department, C.M.P. College, Allahabad (Uttar Pradesh, India) during 1992.

(a)

(b)

(c)

Fig. 13.14 (a-c): (a) Dr B.K. Tyagi delivering the Presidential Speech during the 3rd Indian Symposium of Odonatology, Erode, TN (1990), (b) Dr. Varadaraj honouring Dr B.K. Tyagi, and (c) The participants. - (Source: B.K. Tyagi).

The future SIO-ROSA symposia were as spectacular, whereas the national office also geared up to organize two SIO International Symposia of Odonatology; the 9thin Madurai (1988; Organizing Secretary: Dr S. Mathavan) and the 18thin Nagpur (2013; Organizing Secretary: Dr R.J. Andrew).

From 5th Indian Symposium of Odonatology onward all ensuing symposia were organized under the aegis of South Asia Council of Odonatology (SACO) with no fixed frequency. Concomitantly, the SACO also organized its First National Workshop on Odonatology in association with *dragonflyiIndia* and *Indiabiodiversity* Portals. A photo exhibition-cum-competition was also held on the central theme of the symposium. On this occasion, Prof. D.B. Tembhare, a seasoned odonatologist with expertise on neurobiology and a noted author on insect sciences, was felicitated for his four decades long research on dragonflies. During the symposium and workshop a series of field trips including about 55 participants was arranged at the Linga Lake, the Futala Lake and the Ambazari Lake in and around Nagpur city. A total of 16 species were sampled at the Linga Lake, notable of which included *Ictinogomphus rapax, Tholymis tillarga, Paracercion calamorum, Pseudagrion decorum, P. hypermelas, Rhidischnura nursei* etc.

The SACO-*Dragonflyindia* has recently organized a highly memorable 9th Indian Symposium of Odonatology in Goa, India on September 15-16, 2017 (Fig. 13.15). About 40 participants from over a dozen States and Union Territories from India, besides those from Sri Lanka and Bhutan, participated in the diverse proceedings of the symposium.

(a) **(b)**

Fig. 13.15 a-b: (a) Inauguration celebration of the 9th Indian Symposium of Odonatology (Goa; Sep. 15-16, 2017), and (b) the participants. – (Source: B.K. Tyagi)

Significantly enough, Indian Dragonfly Society too has conducted two major events recently, viz.:

(1) *The Odonata Survey in the Periyar Tiger Reserve*. It was carried out on 27-29 October, 2017. The highlights of the survey included recording of a total of 80 species (46 Anisoptera and 34 Zygoptera; app. 50% of the

reported 161 species from Kerala) which is twice the quantum of species reported earlier from the PTR (Fig. 13.16).

(a)

(b)

Fig. 13.16 (a,b): PTR Odonata Survey being (a) inaugurated by Ms Silpa V. Kumar, IFSDD, PTR, along with Prof. Dr B.K. Tyagi, President of Indian Dragonfly Society and Mr C.G. Kiran, Secretary General, IDS, and others, (b) Odonata Survey participants. – (Source: B.K. Tyagi)

(2) *First Indian Conference of Odonatology.* IDS's 1st Indian Conference of Odonatology (Central theme: Dragonfly (Insecta: Odonata) bioecology and distribution dynamics v/s environment amelioration, with special

reference to global warming and climate change) was held in the Dolphin (PG) Institute of

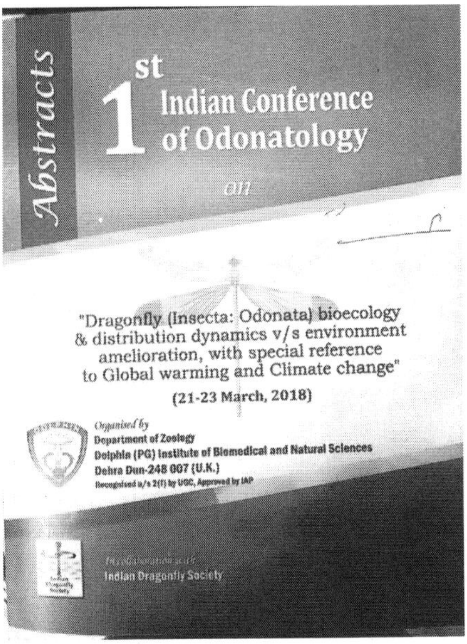

(a)

(b)

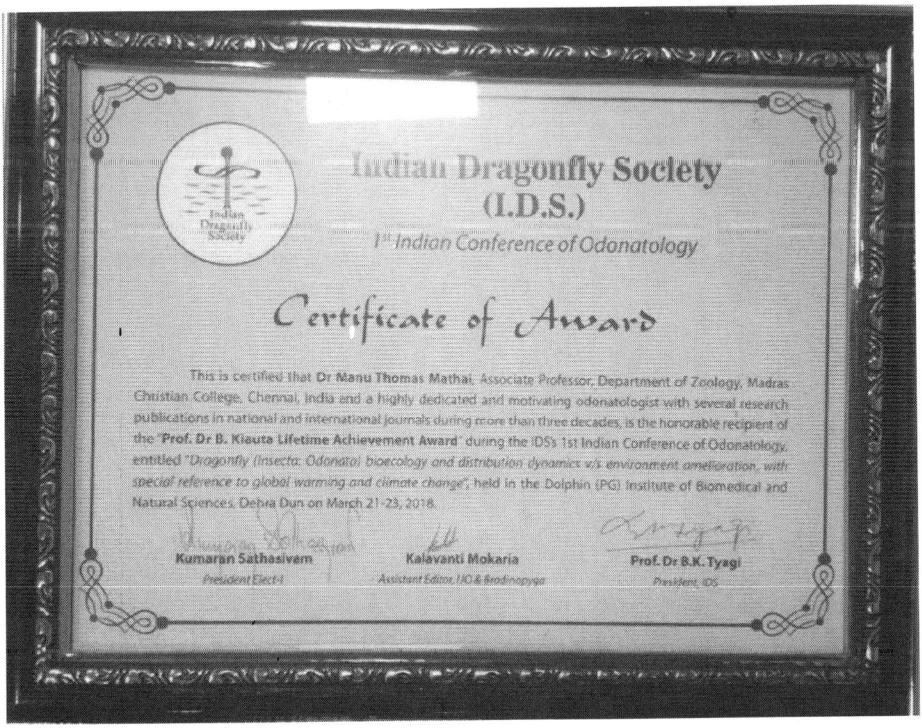

(c)

(d)

Fig. 13.17: (a) The 1st IDS Conference Abstracts booklet, (b,c)Prof. Dr B. Kiauta Lifetime Achievement Award Medal & Certificate, and (d) Inauguration of the 1st IDS Symposium and release of the Abstracts book. – (Source: B.K. Tyagi)

Biomedical and Natural Sciences, Dehra Dun (Uttara Khand, India) on 21-23 March, 2018 (Fig. 13.17). The conference was outstanding in many ways: (i) this was the first IDS Conference, (ii) A total of 32 Abstracts were received and printed in an exquisite Abstracts Booklet (46 pp.), (iii) Quite many dragonfly researchers representing nine States/UTs participated (e.g., Delhi, Gujarat, Himachal Pradesh, Kerala, Punjab, Rajasthan, Tamil Nadu, Uttara Khand and Uttar Pradesh), besides of course one odonatologist from Germany, (iv) First time in the history of science of Odonatology three prominent Awards were given away with a Citation and a Medal, namely, (a) Prof. Dr B. Kiauta Lifetime Achievement Award (to Prof. Manu Thomas Mathai, MCC, Chennai), (b) Col. Dr F.C. Fraser Oration Award (to Dr Arun Kumar, Director, DIBMS, Dehra Dun), and (iii) Dr S.K. Sangal Young Scientist Award (to Ms Kalavanti Mokaria, GEER Foundation, Gandhinagar, Gujarat), (v) the first General Body Meeting of the IDS Council was conducted and the its Constitution & By-Laws were approved, (vi) A total of 18 lectures, attracting huge interest from audience, were delivered (vii) Dr Arun Kumar's Bio-bibliography (authored by Dr B.K. Tyagi) was released, (viii) The elegant symposium Abstracts booklet with watermark of *Neurobasis chinensis* on title page was released, and (ix) the most heartening, educative and fulfilling field tour was conducted to the famous Asan Bird Sanctuary, app. 42 km west of Dehra Dun.

●●●

14

International Symposia of Odonatology

"Educationists should build the capacities of the spirit of inquiry, creativity, entrepreneurial and moral leadership among students and become their role model."

— A. P. J. Abdul Kalam

Organization of SIO biennial symposia execute the same function as the lungs do in the human body. They provided the needed freshness of thoughts and ideas both to the science of odonatology and to the Societas Internationalis Odonatologica, the SIO. Began initially as the European symposium of odonatology (Ghent, Belgium, 1971), however, the overwhelming support of odonatologists from all over the world forced to transform its very basic organizational structure and needs, with the European symposium officially counted as the First International Symposium of Odonatology. Dr Kiauta served throughout the period of all the eighteen SIO symposia as its 'Chairman of the Standing Committee for the Organization of the SIO Biennial Symposia'. Barring the Madurai (9th, 1988), Hong Kong (17th, 2006), and Nagpur (18th, 2008), all the symposia were held in odd years, commencing in 1971. Traditionally, the organizers of all SIO symposiums produced each time 3 booklets: General Information (always blue cover), Abstracts (cover green) and Field Trips (always pink or yellow), all booklets of A5 size. They were commercially available, Standing Orders accepted and served. Later, the Symposia organization was combined with the recently 'speciated' Worldwide Dragonfly Association (WDA), founded in 1997, and their frequency was changed to 3-year intervals but the subsequent numbering was omitted. The Organizing Secretaries of all International Symposia of Odonatology are publishing a Book of Abstracts (Fig. 14.1) and, in most cases, also a Field Trip Guide was organized in the framework of the respective Symposium.

The significance of the SIO biennial symposia can hardly be ever measured in words, albeit some outstanding expression from one of the greatest dragonfly conservation scientists, Professor Norman Moore tells it all, "*Yet, perhaps the most important service that Bastiaan and his colleagues provided was the initiation of biennial International Symposia all over the world with visits to wonderful dragonfly habitats where all of us lucky enough to attend could in addition enjoy dragonflies together in the field, and form friendships which stood the test of time.*"

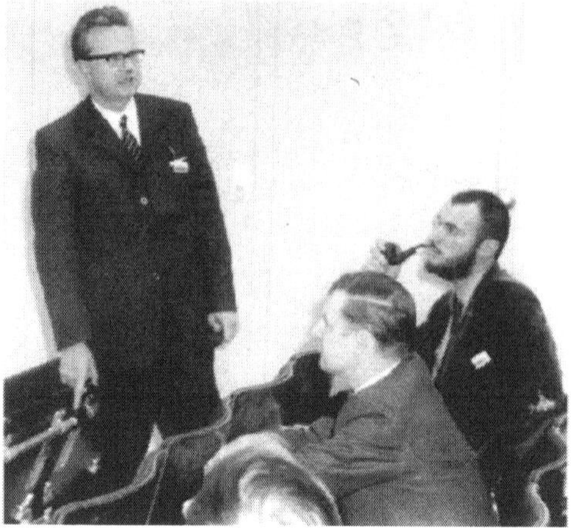

(a)

Abstracts
of Papers read at

THE NINTH
INTERNATIONAL
SYMPOSIUM
OF ODONATOLOGY

Madurai, India
January 18-23, 1988

(b)

Fig. 14.1. (a) Dr B. Kiauta with Dr G. Jurzitza and Dr E. Schmidt as one of the Charter Meeting sessions at the First European Symposium of Odonatology, Ghent, Belgium, October 1971 (Photo: Archives of Dr B.Kiauta), and (b) Facsimile of the title page of the Abstracts booklet, 9th International Symposium of Odonatology held in Madurai on January 18-23, 1988.

A chronological list of the SIO biennial symposia is bulleted below, with the Organizing Secretary's name shown in parentheses):

- (1st, 1971) Ghent, Belgium (Dr H.J. Dumont),
- (2nd, 1973) Karlsruhe, Germany (Dr G. Jurzitza),
- (3rd , 1975) Lancaster, UK (Dr P. Mill),
- (4th , 1977) Gainesville/FL, USA (Dr M.J. Westfall),
- (5th, 1979) Montreal, Canada (Dr J.G. Pilon),
- (6th, 1981) Chur, Switzerland (H. Schiess),
- (7th, 1983) Calgary, Canada (Dr G. Pritchard),
- (8th, 1985) Paris, France (Dr J. Legrand),
- (9th, 1988) Madurai, India (Dr S. Mathavan),
- (10th, 1989) Johnson City/TN, USA (Dr D.M. Johnson),
- (11th, 1991) Trevi, Italy (Dr C. Utzeri),
- (12th, 1993) Osaka, Japan (K. Inoue),
- (13th, 1995) Essen, Germany (Dr E. Schmidt),
- (14th, 1997) Maribor, Slovenia (M. Kotarac),
- (15th, 2001) Novosibirsk, Russia (Dr A.Yu. Haritonov),
- (16th, 2003) Schwerin, Germany (Dr W. Zessin),
- (17th, 2006) Hong Kong, China (Dr P.M. So),
- (18th, 2008) Nagpur, India (Dr R.J. Andrew).

Minutes of all the symposia are very meticulously archived by Dr Kiauta in his personal library. Since 1975 (3rd SIO Symposium; Lancaster, UK) he had been regularly sending me a beautifully drafted brief on each of these. He was, as usual, very opulent in his praise to the Organizers, the participants, the SIO executives in various committees for their open, frank and intelligent discourses on dragonflies on and off the stage. In one such case, after attending the 15th International Symposium of Odonatology at Novosibirsk, Russia, he wrote dated 18th August, 2001, "*The Symposium itself was splendidly organized in the "Academic Settlement", which is luxurious place close to the city of Novosibirsk. Many of the papers were highly interesting, particularly so those on taxonomy and ecology of central- and east Asian fauna, to which subject the traditional satellite Conference of the IUCN Specialist Group Odonata was also devoted. I was much fascinated by the manifold and very serious research work of the young Russian generation. Two cytogeneticists, A.E. Burgov & E. Perepelov, have also presented an interesting paper (and have another paper, on* Aeshna, *in the press in Odonatologica). Also, the Post-Symposium Tour to the Altai (close to the Mongolian border) was a much appreciated experience. Throughout the Symposium, a very strong SIO family spirit was very much felt. This makes me particularly happy.*" He went on saying, "*Of course, we have missed you there very much. Your experience and organizational work relative to the National Office in India have been repeatedly mentioned and referred to during the discussions in Novosibirsk...*"

Organizing SIO biennial symposia in different places and countries was not always easy, although everybody wanted to hold one because the great academic and research benefits were naturally harvested through them. Most of the participants possibly knew about the pleasant and hassle-free proceedings of the symposium but hardly envisioned the constraints and stumbling blocks of varied nature the organizers entered in the process, and no one else felt concerned more than Dr Kiauta. Two examples are sufficient enough to highlight his feelings: (i) Sometime in the late 1980s, possibly during the 10th SIO Symposium (1989, Tennessee, USA), on the invitation of Dr M.J. Samways (Department of Zoology and Entomology, University of Natal, South Africa), a proposal was made to hold the next symposium in Bophuthatswana. As customary opinion was commissioned from the SIO membership and discussed in the SIO Council. Because of the political instability and social turbulence some members raised security issues. After lengthy and painful discussions mainly through correspondence the SIO however considered, to somewhat disappointment of a few local odonatologists, participants' concerns of paramount importance and eventually concluded better not to hold the symposium in Bophuthatswana. Thanks to meticulous planning of Dr Kiauta, the venue of 11th International Symposium of Odonatology was decided in favour of Trevi, Italy, without any shift in the biennial nature of the SIO symposia, and (ii) Following the successful organization of the 9th International

Symposium of Odonatology when all the attendees were still enjoying knowledge-packed moments of the symposium, Dr Kiauta, though also extremely happy with the proceedings of symposium, for one exceptional count felt much pained to hear about one European participant having been denied entry in the coutry and deported by the return flight. It is his emotional character "to keep the family together" that moved me. He was always there to extend a helping hand to his colleagues anywhere in the world!

From behind the screen, and very quietly, Dr Kiauta made sure that all went fine with the symposia. To accomplish this he made extensive correspondence, costly overseas telephone calls and sometimes personal meetings with the organizers. Dwindling health (three cardiac arrests, failing eye-sight, debilitating paralytic attack etc., least to speak of similarly compelling reasons of his consort's health at the same time) and multiple shifts in residence, nothing could ever hold him back to perform his manifold duties – on dot of time! He just kept doing his duty with absolute sincerity and integrity as a truly dedicated odonatologist which reminds me of the most rewarding lines by William Wordsworth:

> "*The primal duties shine aloft like stars,*
> *And charities that soothe, and heal, and bless,*
> *Are scattered at the feet of men flowers.*"

Dr Kiauta also metamorphosed the forum of SIO international symposia for helping a large number of budding odonatologists from all over the world, particularly those hailing from the developing and the underdeveloped countries, to expose them to quality research being done elsewhere in the developed laboratories and institutions. A large number of odonatologists are on record having dedicated their success in professional career to these symposia where they freely exchanged scientific views, developed friends and colleagues with mutual odonatological interests, formulated new scientific manuscripts and research projects, and above all helped each other in any odonatologic matter. I myself had the pleasant opportunity to attend three of these international symposia which bore a huge impact on me towards understanding science, in general, and odonatology, in particular, in the best manner ever. A hard taskmaster, however, Dr Kiauta emphasized "*Do science the hardest way by your own hands (self-help), with focused attention. Repeat your observations time and again in thoroughness and perfection. Also, do not ever tend to implore about dearth of facilities or curse your destiny, and use your failures to reinvent and reinvigorate yourself to do better*". To be a scientist is like being a Goethe: "it is born in you, if you do, work twice as hard as you can. A scientist, while pursuing his goal, trusts nothing but his own quest for the Ariadne's thread to unravel the enigmatic labyrinth of his research." I tied these sermons to my vest and followed religiously all along my professional career. The essence of this teaching is so beautifully composed by Sinclair Lewis who prayed to keep *honest* in his investigations:

"God give me unclouded eyes and freedom from haste.
God give me a quiet and relentless anger against all pretence and all pretentious work and all work left slack and unfinished.
God give me a restlessness whereby I may neither sleep nor accept praise till my observed results equal my calculated results or in pious glee I discover and assault my error.
God give me strength not to trust God!"

The SIO Symposia were characterized by the phenomenal growth in both the participants and the representativeness of various different countries across the world in one after another symposium. Odonatologists were having boundless opportunities to travel all over the globe, making new friends, scientific groups of their choice or specialization, crafting future collaborations, facilitating exchange of material, reprints and knowledge etc. etc. In every manner, therefore, the Seventh International Symposium of odonatology held in Calgary, Alberta, Canada on 14-21 August, 1983 (Organizing Secretary: Dr G. Pritchard), was a great success, with one of the largest number of participants on record. During the specially organized field trips and post-symposium tours considerable collecting was undertaken at various different localities around Calgary, in the Rocky Mountains, in Banff National Park and in the Columbia River Valley of southeastern British Columbia. In addition, several participants collected in adjacent Canadian territory while enroute to and from the symposium (Fig. 14.2). A few of the more notable captures included: *Lestes unguiculatus* Hagen (Lestidae), *Coenagrion interrogatum* (Hagen), *Enallagma anna* Williamson (Coenagrionidae), *Aeshna constricta* Say, *A. palmate* Hagen, *A. subarcticta* Walker, *A. tuberculifera* Walker, *A. junius* (Aeshnidae), *Cordulegaster dorsalis* Hagen (Cordulegastridae), *Somatochlora cingulata* (Selys), *S. minor* Calvert, *S. walshii* (Scudder), *S. whitehousei* Walker (Corduliidae), and *Sympetrum madidum* Hagen (Libellulidae).

The success of the Calgary Symposium was to a great extent a result of outstanding organizational skill of Dr Gordon Pritchard, ably assisted by two of his colleagues, Drs Rob Cannings and Sid Cannings, brothers and established entomologists. Dr Pritchard was a premier Canadian entomologist and educator and an expert on many subjects from the biology of crane flies to evolutionary questions associated with the development of insects, especially aquatic ones. But his overwhelming entomological love was the study of the Odonata and he was a significant force in the international research organizations that focused on this insect order.

Gordon was born on 9th February, 1939 in Burton on Trent, Staffordshire, a city of breweries in the English Midlands. When he was nine or ten, Gordon, without a net, captured his first dragonfly, a Brown Hawker (*Aeshna grandis*) – the trick, he said, was to throw your jacket over whatever interested you. Gordon had been inspired by his biology teacher at grammar school and wanted to be a biology teacher himself. But, instead, in 1957 he enrolled in Imperial College, University of London, what he later would consider the best entomology department in the world. He received his BSc (Honours) in zoology in 1960 and, after winning a Commonwealth Scholarship that year, he came to the University of Alberta in Edmonton, Alberta, Canada. In those days the entomology department was a small place, with three faculty members – Brian Hocking (the chairman), George Evans (ecologist and Gordon's supervisor) and George Ball (systematist) – and only eight graduate students. But the place hummed with excitement and new ideas. Gordon's doctoral research focused on the life of larval dragonflies in the boreal forests of Alberta, and he produced his first major publications on the Odonata stemmed from this work. Gordon finished the field work in only one season, a sign of his efficiency and concentration. He was awarded his PhD in entomology in 1963. On 2nd February that year Gordon married June Dalby in Edmonton.

Fig. 14.2. Dr Kiauta and Marianne Kiauta at Eagle River, BC after 1983 Symposium, using the traditional Japanese collapsible and foldable nets (Photo Source: Rob Cannings).

After graduating, Gordon's first job was as a research scientist with CSIRO (Commonwealth Scientific and Industrial Research Organization) in Sydney, Australia, where he studied fruit flies. He returned to the University of Alberta, in 1966, and filled in for George Evans while George was on sabbatical. Next year he took up a one-year teaching appointment in insect physiology at the Calgary branch of the University of Alberta. The job was offered to him by Jim Cragg, Head of the Department of Biology, who had been Gordon's external examiner at Imperial College. By the late 1970s, he returned to studying Odonata, being intrigued by *Argia vivida* living in the warm pools at Banff.

Gordon was an energetic and involved teacher, researcher and administrator at the University of Calgary. By 1976 he was made Full Professor and in 1976-78 he served as an Assistant Dean of Science. In 1997 he retired, only to be later engaged by the university to continue to teach the senior courses as Professor Emeritus. He was curious about absolutely everything. This made him an insatiable reader and collector of books. Dragonflies took him to many wonderful places, from the hot springs of Canada and the US, to tropical streams in Costa Rica, to conferences above the Arctic Circle in Sweden and to the deserts of Namibia. For most of his career, he worked tirelessly in the several societies that promoted and coordinated international dragonfly research. Among much other work he organized the Seventh International Symposium of Odonatology in Calgary in 1983 and served as the President of the Worldwide Dragonfly Association from 2007 to 2009. The successful organization of the Calgary Symposium was a reflection of his inexhaustible energy, calm wisdom and negotiating skill – attributions that made him such a wonderful friend about whom Dr Kiauta always felt highly proud to associate with.

When the 8th SIO International Symposium of Odonatology was to be held in Paris in 1985, I was posted as Senior Research Officer at Vector Control Research Centre in Indian Council of Medical Research, Puduchery (then Pondicherry) – a south Indian Union Territory with French background. Within the SIO, already I was serving as the National Representative in-Charge of its Indian National Office and made very good progress there; with one informal meeting of odonatologists in Mysore (1982) and the 1st Indian Symposium of Odonatology (Madurai; 1984) as well as announced the 2nd Indian Symposium of Odonatology at Dehra Dun in northern India (1986). I attended the Paris symposium, being my life's first conference overseas, at the invitation of Dr Kiauta. In Paris I went straight to the office of Dr J. Legrand, the Organizing Secretary, who was extremely courteous. He lodged me in a cosy hotel near the famous Natural History Museum and on walking distance from the historical river Seine. There I had the privilege and honour to meet with three highly energetic and seasoned odonatologists, Prof. A.B.M. Machado, Dr Eberhard Schmidt and Dr D.A.L. Davies, also staying in the same hotel. Next day morning, during a brief introductory discussion at breakfast, I learnt with amazement that Prof. Machado, a highly respected odonatologist the world over, was actually a medical scientist, and he had no parallel in odonate taxonomy in whole of Brazil. He also had an immense interest in children's fables from rural environs, with a tinge of dragonflies in the background (Fig. 14.3).

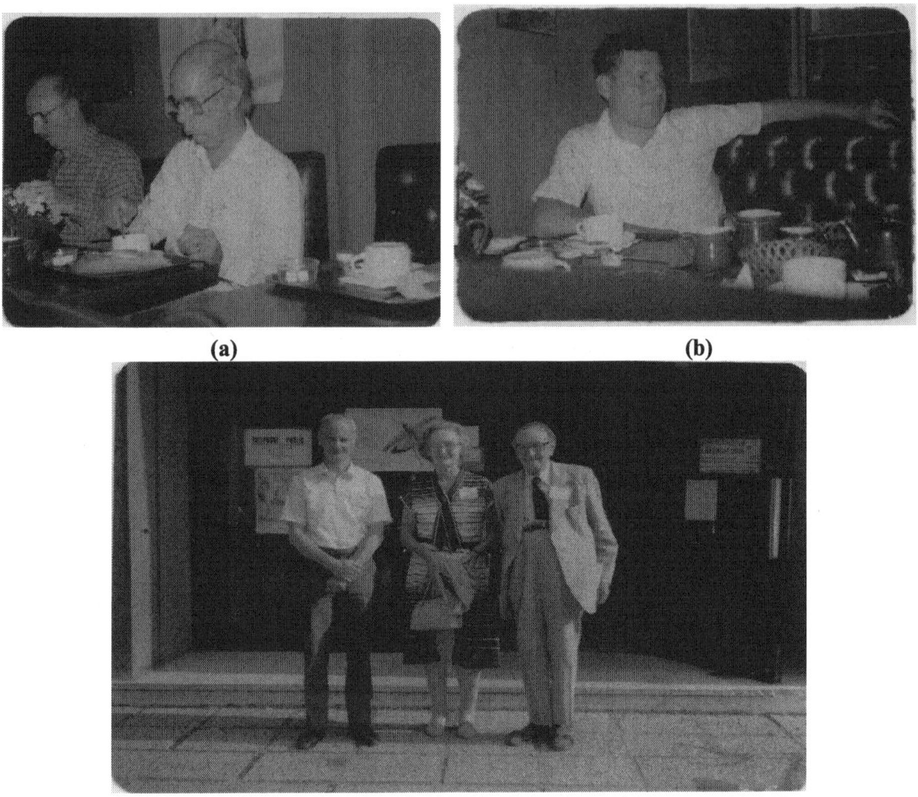

(a) (b)

(c)

Fig. 14.3. (a) Prof. D.A.L. Davies and Prof. A.B.M. Machado, (b) Prof. E. Schmidt, and (c) from left to right: Dr P.J. Mill, Mrs Gambles, Dr R. Gambles, at the Paris meeting in 1985. – (Source: B.K. Tyagi)

During the symposium I had the opportunity and privlilege of meeting not only my teacher and mentor, Prof. Dr B. Kiauta and his wife Mrs Marianne Kiauta, but also a large number of great odonatologists such as Drs J.M. van Brink, Rob Cannings, Sid Cannings, P.S. Corbet, D.A.L. Davies, D.C. Geijskes, M. Hämäläinen, G. Jurzitza, J. Legrand, A.B.M. Machado, P. Mill, N.W. Moore, M.J. Parr, J.-G. Pilon, E.C.G. Pinhey, G. Pritchard, R. Rudolph, F. Schaller, M. Schorr, C. Utzeri, J. van Tol, J.A.L. Watson, M.J. Westfall, to name a few only. All the lectures were highly inspiring, especially those dealing with taxonomy, ecology, geographical distribution etc. Dr. Rüppell showed my life's first dragonfly video film which everybody enjoyed with often lofty laughters when in the film the dragonfly, a prey, dodged repeatedly a merciless frog's capturing manoeuvres, leaving the pitiful anuran sadly retreating! This symposium exhibited uninhibited, free and frank scientific discussions and left an indelible imprint on my mind for its highest quality of scientific discourses (Fig. 14.4).

Fig. 14.4 (a-f): (a) Prof. Van Brink, Mrs Marianne Kiauta, Prof. Kiauta and Mrs. Westfall seem to be planning for the next scientific session, (b)Dr Philip S. Corbet offering a keynote lecture during the 8th International Symposium of Odonatology, held in Paris, France (1985), with Dr Kiauta is all ears to hm. (c) During one of the session breaks, free and frank discussions among participants – the hallmark of all SIO Symposia, in its true spirit of philosophy, "family of friends", (d) Prof. Minter J. Westfall in a serious discussion with Prof. Corbet, (e) Dr Rüppell with Prof. F. Schaller, possibly recalling how dragonflies escaped ambush of their predators, and (f) Dr Legrand with Dr Michael Parr.– (Source: B.K. Tyagi)

Even though Dr Legrand was the Organizing Secretary of the Paris Symposium, I could always feel the minute-to-minute review and follow up by Dr Kiauta, particularly regarding the conduction of scientific sessions. The Organizing

Committee had also generously assigned me a responsibility with the Chairmanship of one of the scientific sessions which was to start just after the lunch, at 1.30 PM sharp. The entire gathering of participants was invited to have their lunch at the restaurant of a nearby railway station called Gare d'Austerlitz, one of the six large terminus railway stations in Paris, and situated on the left bank of the Seine in the southeastern part of the City, in the 13th arrondissement. Dr Kiauta knowing that I was a staunch vegetarian got me some dishes of boiled brinjal and potatoes with green salad and some curd. As I was enjoying my sumptuous lunch, Dr Kiauta asked if the food was fine and before I could respond he was quick in pointing out that I was the Chairman in the next session. I saw my watch which clearly told that I had but 5 minutes left to reach the lecture hall. I left my food then and there in the middle, excused myself and dashed toward the destination, all the way running berserk. When I reached the lecture hall still in time with a few seconds left, I found to my pleasure with a great sense of learning that all the participants had already returned to the conference hall and were ready to commence the session (Fig. 14.5). It was entirely Dr Kiauta's mentoring and motivational generosity that I had, for the first time in my life, an opportunity to chair a scientific session and delivered an invited lecture as well during an international conference abroad.

(a)

(b)

Fig. 14.5. Eighth international Symposium of Odonatology, Paris (France), 1985: (a) Dr. B.K. Tyagi chairing a scientific session, as Dr Michael Parr offered his speech, and (b) He delivered also an invited lecture, with Drs Peter Miller and J. Legrand acting as Chairman and Co-Chairman.– (Source: B.K. Tyagi)

Back home, during my entire service spanning over there-and-a-half decades in Indian Council of Medical Research, I must have organized several dozens of symposia, conferences, workshops and training programmes, but I invariably made it a point to commence an event always on the dot of scheduled time, with all dignitaries in seat already! My symposia/conferences have been considered a benchmark in disciplinedly organizing them in their content and time management, thanks to Dr Kiauta's early teachings!

One of the spectacular highlights of the Paris Symposium was a visit to the French National Museum of Natural History, known in French as the 'Muséum national d'Histoire naturelle' (MNHN), established in 1635 on the left bank of the River Seine, which is a grand établissement of higher education part of Sorbonne Universities, with a Collection size of over 68 million specimens (Fig. 14.6). I saw there my life's first fossil Odonata, the famous fossil of *Meganeura monii*, which with a wingspan of over 50 cm is the largest known species of insects that ever existed on earth. The imprint was embossed on a rock, in three pieces, which were assembled together to complete the fossil. The grand species still lives in my eyes as if it was only yesterday I had seen it!

Fig. 14.6. Participants of the 8th International Symposium of Odonatology, held in Paris, France, 1985, photographed in front of the Paris Museum (Source: J. Legrand).

A dragonfly field tour was organized and a very cosy bus transported participants to the destination where participants enjoyed 'dragonflying'. Post-conference several participants joined the Kiautas on field tours at their Swiss and Dutch homes, respectively (Fig. 14.7).

(a)

(b)

(c)

Fig. 14.7. (a) In Liechtenstein B. Kiauta and R. Cannings after the 1985 Symposium. (Photo: Syd Cannings); (b) Dr Kiauta at a pond in Bilthoven, 24th May 1985. Kiauta's first field trip that summer (Source: Dr M. Hämäläinen); (c) Angelo, Philip, Bastiaan, Marianne) during a collecting trip in the mountains near his Casa d'Uors in Switzerland.—(Source: Dr A.B.M. Machado).

On the concluding day, during the valedictory session Dr Kiauta announced that the 9th International Symposium of Odonatology would be held in Madurai, India, and passed on to me the 'flag of SIO insignia', which I received on behalf of my friend, Dr S. Mathavan, the Organizing Secretary. I remember how much ecstatic all the participants at the Paris symposium had felt with the prospects of their next gathering in India – a country rich in art and culture, and the dragonflies!

The Madurai Symposium under the SIO Presidentship of Dr Kiauta was unique in more than one way, namely, (a) It was the first hosting site in entire Asia or even the Oriental Region for an SIO International Symposium of Odonatology, (b) For any insect order the 8th International Symposium of Odonatology was the first such a conference that was devoted entirely to a single order, Odonata, and (c) The SIO and IUCN had a meeting on Odonata for the first time on Indian soil (Fig. 14.8). Besides, more than 150 odonatologists from as many as twelve countries have participated in the symposium (Bangladesh, Canada, France, Germany, India, Italy, Nepal, South Africa, Sri Lanka, The Netherlands, UK, and USA). Two dragonfly field tours were arranged; one during the course of symposium to the Kodaikanal road and another post-symposium tour to the Thekkady (Periyar) sanctuary. An account of 36 species of Odonata collected in the Periyar sanctuary was later published (Mathavan, S. & P.L. Miller, 1989).

Fig. 14.8. Dr B. Kiauta delivering the Presidential Address at the 9th International Symposium of Odonatology (Madurai; 19-21 Jan., 1988).– (Source: B.K. Tyagi)

During the symposium Prof. S. Krishnaswamy, Vice Chancellor, MK University, Madurai and a world renowned zoologist himself, invited the foreign delegation led by Dr Kiauta for an informal discussion (Fig. 14.9). Prof. Krishnaswamy showed keen interest in the proceedings of the symposium and wished the delegation the very best. He particularly took time to discuss with Dr Kiauta on the international scenario of dragonfly research and conservation, and wished the international community of dragonfly experts to collaborate with Indian scholars for the overall healthy evolution of the science of odonatology. He extolled Dr Kiauta and the SIO's endeavours in popularizing dragonfly research worldwide.

(a)

(b)

Fig. 14.9. (a) Prof. B. Kiauta, Prof. S. Krishnaswamy, Prof. Minter J. Westfall and Dr S. Mathavan during the inauguration of the 9th International Symposium of Odonatology, (b) Prof. S. Krishnaswamy, Vice Chancellor, Madurai Kamaraj Univrersity, having a discussion with Prof. Kiauta, Prof. Pritchard, Prof. Corbet as Prof. T.J. Pandian looks on (Source: – (Source: B.K. Tyagi)

On 20th January, Dr Kiauta's birthday was celebrated. He was taken completely by surprise and had no idea who informed the participants on the date of his birth. Unaware of any kind of celebration, all the participants were invited to the main conference hall. I had the privilege to compere the programme that took everybody by surprise, pleasure and wishes galore! There was a very memorable singing of birthday songs: every singer sang his/her own birthday song, in their own local language. Dr Kiauta was much impressed, for example, by Dr Mitra (in Bengali), by Dr Tyagi (in Hindi), by Professor Georg Rüppell (in German), by Professor Carlo Utzeri (in Italian) etc. Dr Mathavan arranged a cake for the ceremony, and for all to sweetly enjoy the moment for ever in future!

The 9th International Symposium of Odonatology (Madurai, 1988) had proved a dream coming true for a large number of odonatologists, particularly young and budding ones. The indian odonatologists were particularly delighted having met in person with such a large galaxy of world renowned dragonfly researchers from so many different countries (Fig. 14.10). While Dr. Chaudhary (Bangladesh) was seen being busy getting his manuscript reviewed with Prof. Philip S. Corbet, a PhD student of Dr Mathavan was ecstatic on being invited by Prof. Schaller to join his laboratory in France. Professor Kiauta arranged a place for Mr Mahendra Mahato (Nepal) in the laboratory of Dr. Dan Johnson in Tennessee University, USA where he moved soon after the Symposium. Dr. S. Mathavan himself got a long time fellowship programme both to USA and France.

For the participants, in general, the symposium proved a great 'family get together' and a platform for informal discussion. Wives of Prof. Westfall, Prof. Kiauta, Prof. Schaller, Dr Tembhare, Dr Tyagi, Dr Mitra, Dr Lahiri and many

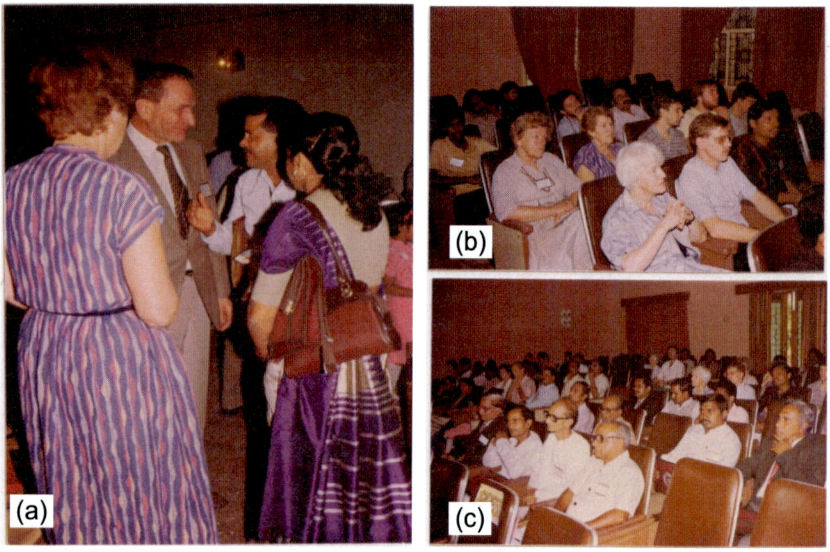

Fig. 14.10. (a) Professor Schaller, with wife, full ears to Dr B.K. Tyagi, joined by his wife, Ajita Tyagi, and (b,c) Participants during the 9th SIO Symposium (Madurai, 1988). –(Source: B.K. Tyagi)

others were often seen exchanging pleasantries. Dr Schaller happily gifted my wife a 'calculator', and Dr Kiauta and Mrs Marianne Kiauta lovingly presented to her a 'dragonfly towel' and a pair of beautiful 'dragonfly coffee cups', which we still have in our esteemed collections as wonderful momentos (Fig. 14.11).

Fig. 14.11. (a-d): Participants involved in rich odonatological discussions: (a) Dr Kiauta and Mrs Marianne in discussion with participants, (b) Dr Peter Miller seen discussing with Prof. M. Lakshmanan, (c) Prof. Van Brink, Mrs Westfall and other participants, and (d) Dr Rüppell with participants.– (Source: B.K. Tyagi)

Following the SIO Symposium Dr Tyagi invited Dr Kiauta, Mrs Marianne Kiauta and Prof. van Brink to their home in Pondicherry (now Puduchery), about 300 km west of Madurai. On the given date and early morning time they all arrived in Pondicherry after travelling overnight by bus and from the bus stop near JIPMER medical college they all walked down for about half a kilometre to reach my house within the Residential Colony of the institute, Vector Control Research Centre. We welcomed them to our home where Dr Kiauta saw my study and was pleased to see a portable typewriter there, remarking that he has a similar 'blue' typewriter. I also showed him my small library where two very old books, owned by Dr S.K. Sangal, were also present. Dr Sangal had passed on these books to me for sale. I told this story to Dr Kiauta, who without making any delay immediately purchased the books, telling that even he did not have them in his collection, hence highly precious to him. The money was soon transferred to Dr Sangal.

My wife, Ajita, prepared the lunch which they all enjoyed very much especially the Indian *chapaati* and *tadka-daal.* In the afternoon we all went for a dragonfly hunting in Pondicherry suburb. Dr Kiauta had brought his own dragonfly net and enjoyed very much hunting the dragonflies. In the evening I escorted them to the Pondicherry Bus Stand for boarding a bus to Madras (now Chennai). While I engaged an auto-riskshaw for Prof. van Brink and Ms Marianne Kiauta, Dr Kiauta delightfully accepted my request to ride pillion with me on my newly purchased '*Lambretta'* scooter. They all pleasantly departed for Madras, to pay a visit to Professor T.N. Ananthakrishnan's Entomological Research Institute, located in Loyola College, Nungambakkam, Madras. Dr Kiauta and all were always so kind and generous throughout that even if they had encountered some cumbersome moments during these journeys they certainly resolved at their level and never let anybody know about their feelings, whatsoever! I imbibed within myself all his attributions.

After returning home to Bilthoven Dr Kiauta quickly sent a note of thanks and appreciation for the successful organization of the SIO Symposium, the dragonfly field tour, and, above all, for the birthday celebration as well as memorable welcome at my home in Pondicherry. Such a kind and humble character is always going to have a deep impact on anyone, and I have learnt a lot from all these courteous and warm greetings being Dr Kiauta's inherent traits!

The Tenth International Symposium of Odonatology (Johnson City, Tennessee, USA; 1989; Organizing Secretary: Dr Dan Johnson) began with great enthusiasm and expectation of advance odonatolological learning. About 150 participants from several countries in the Americas, Europe, Asia, Australia and Africa attended. All the lectures and poster presentations were of very high standard. A dragonfly field tour was conducted in the Tennessee Valley and I saw how nicely the sylvatic and aquatic environments were maintained without any litter anywhere! During this field trip, for a while during the lunch time, I got an opportunity to walk along with Prof. Philip S. Corbet, who, knowing my professional attention toward medical entomology and especially mosquito control also by using biological agents, was narrating to me his own experiments in collaboration with Dr Sebastian in Myanmar where he used *Crocothemis servilia*, in repetitive inundative manner, to control *Aedes aegypti*, the vector for deadly dengue, chikungunya, yellow fever and the Zika virus. At the end of this talk, we sat on a bough of a felled tree partly projecting over a pond nearby. We quietly opened our packed tiffins provided by the generous organizers and settled to rejoice it. Suddenly I noticed that Prof. Corbet was all ears to the ripples created by the continuously falling tree-leaves in the otherwise placid lake. I rather hesitantly asked him if his attention to the rippling water meant something serious? He saw toward me smilingly and spoke almost murmurously,"These ripples give energy to me." I understood the essence of his philosophical words: *Movement is life* (Fig. 14.12).

Fig. 14.12. Prof. Philip S. Corbet, equipped with his dragonfly net, diary and other paraphernalia, alongside a lake in Tennessee Valley during the 10th SIO Symposium.— (Source: B.K. Tyagi)

This field tour proved a good training ground for many of us. Looking at senior odonatologists who toiled hard to catch dragon- and damselflies in the heavily vegetated lake, the message dawned upon me was clear: '*No sweat, no sweet*'. Dr Kiauta, completely oblivious of the condition of his shirt which got unbuttoned from front, was spotted vigorously chasing the dragonflies in the bushes and/or thoughtfully discussing with other keen participants about either identification of the samples collected or other associated subjects (Fig. 14.13). From India we were five scientists participating in the symposium, and Dr Kiauta's message was that we mix with others from far off countries for benefit of advancing diversity in knowledge. In that sylvatic ecosystem around the lake I witnessed that all participants were very clear about conserving the nature– neat and clean! Though food was served packed in cardboard boxes during the lunch hours, one could not see anything littered around and all the used boxes with or without leftover food were eventually disposed off in the trash collecting containers systematically kept all along the passeges in the forest. If for any reason some litter arose somewhere, the same was immediately collected and disposed off by the volunteers. I learnt that sanctity of our environment is a necessity for which all have to contribute.

(a)

(b)

Fig. 14.13 (a-b): (a) Prof. Kiauta, along with Prof. P.S. Corbet, on a dragonfly hunt in Tennessee, USA, 1989 Symposium. (Photo: Rob Cannings), (b) The towering personality of Dr. Kiauta and two other participants during a field tour organized in course of the 10th SIO International Symposium of Odonatology (Johnson City, Tennessee, USA, 1989). –(Source: B.K. Tyagi).

In my own life, dragonflies have been the creaters of ripples and, in spite of the inevitable round-the-clock bite of mosquitoes, professionally speaking, my first love was always for the dragonflies. I even used this positive energy in controlling mosquitoes by using dragonflies (*cf.* VENKATESH & TYAGI, 2013). For this I most gratefully bring it on record that it was entirely Dr Kiauta's continuous motivation and encouragement during nearly more than three decades that kept alive in me the "dragonfly fire".

The Tennessee symposium was a great success, and Dr. Dan Johnson a great and very generous host. His whole team included only four persons; himself, his wife and daughter and his research scholar Mr. Mahendra Mahato. From transporting participants from the airport, nearly 40 km away in Johnson City, to accommodating them comfortably in the university dormitories, as well as their breakfast, lunch and dinner, everything was so effectively performed that it remained an example of good teamwork all along my own life when I had occasions to organize such international meets in India (Fig. 14.14).

Fig. 14.14. The Indian and Nepalese participants on the dining table; Dr S. Mathavan, Dr D.B. Tembhare, Mr M. Mahato, V.K. Srivastava, and Dr B.K. Tyagi. – (Source: B.K. Tyagi)

The Tennessee Symposium was also a bit marred due to some unexpected scenarios which culminated in a contest by vote between two groups, for the first time. The outcome of this voting exercise was also very disappointing, with Dr Kiauta group (with myself included) had lost by a single vote! In that pindrop silence of the outcome of the voting exercise, too, I heard Dr Kiauta quietly expressing to everyone present his thanks!! What a great soul that is Dr. Kiauta – a true testimony to the democratic spirit of the SIO, "a family of friends". For me the night that day appeared longer and darker, but I knew for sure that there would be morning soon with bright light again, reminding me what Charles Haddon Spurgeon has said:

Losses and crosses are heavy to bear;
but when our hearts are right with God,
it is wonderful how easy the yoke becomes.

Once again India had the unique privilege to host an SIO Symposium, the 18th International Symposium of Odonatology, at Hislop College, Nagpur, India from November 5-9, 2008 (Organizing Secreatry: Dr R.J. Andrew). Dr. Kiyoshi Inoue, President of the S.I.O., and a noted nature photographer, led the odonatological fraternity at the Symposium (Fig. 14.14). Besides a regular Abstracts booklet, the organizers also released a dragonfly book, "*A handbook on common odonates of Central India*", by R.J. Andrew, K.A. Subramaniam and Ashish D. Tiple.

Fig. 14.14. Inauguration of the 18th International Symposium of Odonatology:(From left to right) Prof. D. B. Tembhare, President, South Asian Council of Odonatology, Mr. B. Majumdar, Principal Chief Conservator of Forest (Wildlife), Prof. S. N. Pathan, Vice Chancellor, RTM Nagpur University, Nagpur, Dr K. Inoue, President, International Odonatological Foundation (SIO), Dr (Ms.) Dipti Christian, Principal, Hislop College, Nagpur, and Dr. R. J. Andrew, Organizing Secretary, the 18th International Symposium of Odonatology. - (Adopted from R.J. Andrew)

Later, Societas Internationalis Odonatologica (SIO) and Worldwide Dragonfly Association (WDA) agreed to convene the future international symposia under the aegis of the WDA, and now referred to as International Congress of Odonatology.

●●●

15

Odonatologica—The Backbone of the S.I.O. and the Dawn of a New Era

"When you are inspired by some great purpose, some extraordinary project, all your thoughts break their bounds. Your mind transcends limitations, your consciousness expands in every direction, and you find yourself in a new, great, and wonderful world. Dormant forces, faculties, and talents come alive, and you discover yourself to be a greater person by far than you ever dreamt yourself to be."

—**Maharishi Patanchali in** *Yogasuthra*

No blood vessel system can function without a powerful heart. In the same manner, the relations between six hundred people interested in dragonflies and scattered over more than fifty countries of the Globe, like a wide-mazed network of thin blood vessels, was unthinkable till mid 20th century (*cf.* KIAUTA & VAN BRINK, 1972).

It is now more than 75 years ago that, during the Seventh International Congress of Entomology (Berlin, 1938), Drs. Maurits Anne Lieftinck, Cesare Nielsen, Erich Schmidt and Douglas St. Quentin made the first attempt to bring some coordination among several hundreds of people interested in dragonflies and scattered over more than fifty countries of the globe. Their first, though informal "odonatological colloquium" had a considerable stimulating impact on their subsequent work, therefore similar gatherings were organized by Lieftinck and St. Quentin also in the framework of the Amsterdam (1951) and Vienna (1960) International Congresses of Entomology. In Japan Dr. Syoziro Asahina went his own way and succeeded in bringing together the numerous workers of that country in a formal society centered around his quarterly *Tombo–Acta Odonatologica,* the first odonatological periodical, that started publication in 1958. It was, however, the late Professor Basil Elwood Montgomery who fully understood the paramount importance of direct personal communication and of a certain amount of coordination among the workers. His *Selysia* (since 1963) and his Purdue Odonatological Colloquia (1963, 1966) triggered the idea of setting up a worldwide odonatological society. This materialized on October 22-23, 1971, during the First European Symposium of Odonatology in Ghent, Belgium, where the SIO and *Odonatologica* were born. The Charter Meeting was chaired by Professor François Schaller. This historic decision was most commensurately reinforced by Dr Belyshev, who wrote to the participants of the Ghent Symposium, "*Every learned society is bound to die if it is not able to keep close contacts with its members by means of an own journal.*"

The journey of *Odonatologica* through past 43 years under Dr. Kiauta's editorship and almost 20 years of his odonatological life that preceded this time period would possibly make a highly motivational epic. He embarked on the study of dragonflies in 1952 and his class-mate, colleague and very close friend during the past 70 years, Professor Dr Matija Gogala (Ljubljana, Slovenia), was involved in this decision. His first dragonfly teachers were Ž.R. Adamović (Serbia) and D.E. Kimmins (UK), and he had the privilege of having been in the position to conduct some correspondence with our classic odonatologists of the past, such as P.P. Calvert (USA), F.C. Fraser (UK), F.F. Laidlaw (UK), Erich Schmidt (Germany), D.St. Quentin (Austria), K.J. Valle (Finland), E.M. Walker (Canada) and F.E. Zeuner (UK). Laidlaw's brief correspondence exercised much influence and gave him much encouragement, while the frequent and long letters from Erich Schmidt were to sow in him the first seed that later gave birth to *Odonatologica*, though in their correspondence of the 1960s they were referring to the journal as *Notulae odonatologicae*. The name, *Odonatologica*, was proposed at the First European Symposium of Odonatology (Ghent, Belgium) by Professor Dr François Schaller of the University of Strasbourg, France.

It was at the SIO's First European Symposium of Odonatology (Ghent, Belgium), held in October 1971, that the participants had entrusted Dr. Kiauta with the two major responsibilities of general overseering of the organization and editorship of *Odonatologica*. This was to appear quarterly and include, in addition to research papers, the abstracts of contemporary odonatological literature published elsewhere. Right from the day of origin of SIO, the journal *Odonatologica* was also considered its mainstay being reflective of its philosophy of functioning – of evolving the science of odonatology and odonatologists worldwide. It focused on high quality peer-reviewed publication of research manuscripts dealing with diverse subjects, both fundamental and applied, in any of the four main European Congress languages, determined the true strength of the Society and its membership. Dr Kiauta has devoted much of his active life time during past six decades in the service of the world odonatological community, by bringing together hundreds of dragonfly workers into the shape of a formal organization – the International Society of Odonatology (which Dr Kiauta fondly called as the "family of friends"), by facilitating their mutual contacts through personal discussions during the biennially organized symposia in different countries each time – extending from East to West of the world, and by helping publication of the results of odonatologists in the world's first truly international journal, *Odonatologica* – the mainstream periodical of the SIO, and, last but not least, apprising the odonatologists about research publications published anywhere in the world, irrespective of their language, through *Odonatological Abstracts* (ca.

20,000 abstracts) – an inbuilt and integral feature of *Odonatologica*, since 1971. For all the great service to *Odonatologica* all these long years the journal has become synonymous to the name of Dr Kiauta.

Because the quantum of manuscripts was continuously growing, to adjust brief notes, another regular periodical, *Notulae odonatologicae*, a semiannual bulletin was initiated in 1978, published June 1 and December 1 every year.

"What lesson have I learnt from the *Odonatologica* section of *Odonatological Abstracts*?" I once asked myself and there came the answer almost instantaneously, "Every piece of research publication, in whatever format, is invaluable for a serious researcher no matter how small or trifle-looking that matter is, even if published in a little known periodical or even when presented in absentia in a scientific colloquium etc". What is considered unimportant or less important in one's viewpoint, could be the jewel of a document in another person's consideration! The famous story behind the rejection of Albert Einstein for a position of Assiciate Professor in the University of Bern, Germany on the basis of his publications is an eye opener in this regard. The Dean of Science, vide his communication dated 6th June, 1907, did not apparently agree with Einstein's "conclusion about the nature of light and the fundamental connection between space and time", published in "*Annalen der Physik*", and considered them "somewhat radical" and "more artistic than actual Physics". The same paper, when revisited later with an open mind and focused attention, had subsequently formed the basis of fetching Einstein not only the professorship in a university but also a streak of the global laurels including the Nobel Prize in Physics, 1921 "for his services to Theoretical Physics, and especially for his discovery of the law of the photoelectric effect".

I myself have an interesting fable to report alluding to *Calicnemia doonensis* Sangal & Tyagi, 1984, which was first brought to the light only by publishing as an 'abstract' (the full version published later in 1985) in the Souvenier of the First Indian Symposium of Odonatology, held in Madurai on January 19-21, 1984 (SANGAL & TYAGI, 1984, 1985). Later on, with a difference of just a couple of months, following a tour to India Dr M.A. Lieftinck described a new species *C. carminea pyrrhosoma* from the same site in Dehra Dun from where I had described my species, i.e., Sulphur Springs, referring to my species as *nomen nudum* (LIEFTINCK, 1984). Sangal & Tyagi's taxon was the same as *Calicnemia carminea pyrrhosoma* Lieftinck, 1984, the holotype of which also comes from Sulphur Springs. Thus it became necessary to decide which name has priority. Dr Lieftinck was the greatest known taxonomist in odonatology, besides being most respectful to me personally, therefore I did not even think it an affront, albeit dismay due to disappearance of my maiden discovery of a 'dragonfly taxon'. Then, all of a sudden as if divinely decided, Dr Kiauta, Editor of *Odonatologica*, considered it appropriate to decipher the mystery surrounding my taxon. Therefore, under his advice, Dr Matti Hämäläinen set out to visit India in 1986, in connection with his participation in the Second Indian Symposium of Odonatology held in Dehradun. There he had sampled the damselfly from Sulphur Springs and returned to his department in Helsinki University, Finland. He later visited British Museum (Natural History) where type specimens of all the *Calicnemia* species were housed, and tallied the Sulphur Springs specimens with those. He reached a conclusion that my original description as *Calicnemia miniata doonensis* was actually a good species, and, therefore, upgraded the same as *C. doonensis* Sangal & Tyagi, 1984 over and obove that of Lieftinck's *C. eximia* which was made a junior synonym to my taxon (HÄMÄLÄINEN, 1989). Hämäläinen summarized his analysis as follows:

Calicnemia doonensis doonensis Sangal & Tyagi, 1984, stat. nov.

Calicnemia carminea pyrrhosoma Lieftinck, 1984, syn. Nov.

Known distribution: NW India: Dehra Dun valley.

Calicnemia doonensis carminea Lieftinck, 1984, comb. Nov.

Known distribution: Nepal.

With much indebtedness to Dr Kiauta's unique sense of judgement, my joy saw no bounds on this outcome and learnt a lesson to do science as you think it right.

The history of science is filled with myriads of such examples where not only a classic research paper in a peer-reviewed high impact factor journal has grabbed the attraction of the world, but even research publications in almost nondescript journals have fetched to the authors great laurels and appreciation, though belatedly, and one glaring example is the publication of discovery of the human malaria parasite, *Plasmodium falciparum*, by Dr Alphonse Laveran, a medical doctor in French Army, in 1880. In 1878, he was posted in Algeria (then a French territory), first at the military hospital in Bône, then in the military hospital in Constantine. Malaria was then a serious problem in the army, and this untiring doctor meticulously studied the disease's clinical aspects and its anatomic pathology – providing, for example, an excellent histologic description of cerebral malaria. His ultimate objective was to identify the causal agent of the disease and his tenacity and patience knew no bounds in his sustained endeavours. The opportunity knocked at his door in the early morning hours of November 6, 1880, when on examining the blood of a patient who had been febrile for 15 days, he saw spectacular view of "exflagellation" on the edges of a pigmented spherical body, a male gametocyte, usually to be found in the stomach of the *Anopheles* mosquito. Convinced that these motile structures belonged to a protozoan that caused malaria, Laveran sent two notes to the Academy of Medicine, in November and December 1880 on this epochal discovery of the first malaria parasite, *P. falciparum*, and later published in a rather insular periodical, *Bull. Mém. Soc. Méd. Hôpitaux, Paris*. Recognition did come to Laveran but rather belatedly, in the form of Nobel Prize in Physiology or Medicine, 1907. In essence, no scientific research publication is insignificant and all are important in the balance of time.

From the inception of the first issue of *Odonatologica*, on March 1, 1972 (Fig. 15.1) till the Volume 42, No. 4, December 1, 2013, Dr Kiauta as its Executive Editor has made several milestones in the realms of editing and publishing. He raised the standard of manuscripts accepted for publication through a rigorous process of peer-review and ensured to bring out each issue precisely on the scheduled dates of March 1, June 1, September 1 and December 1.

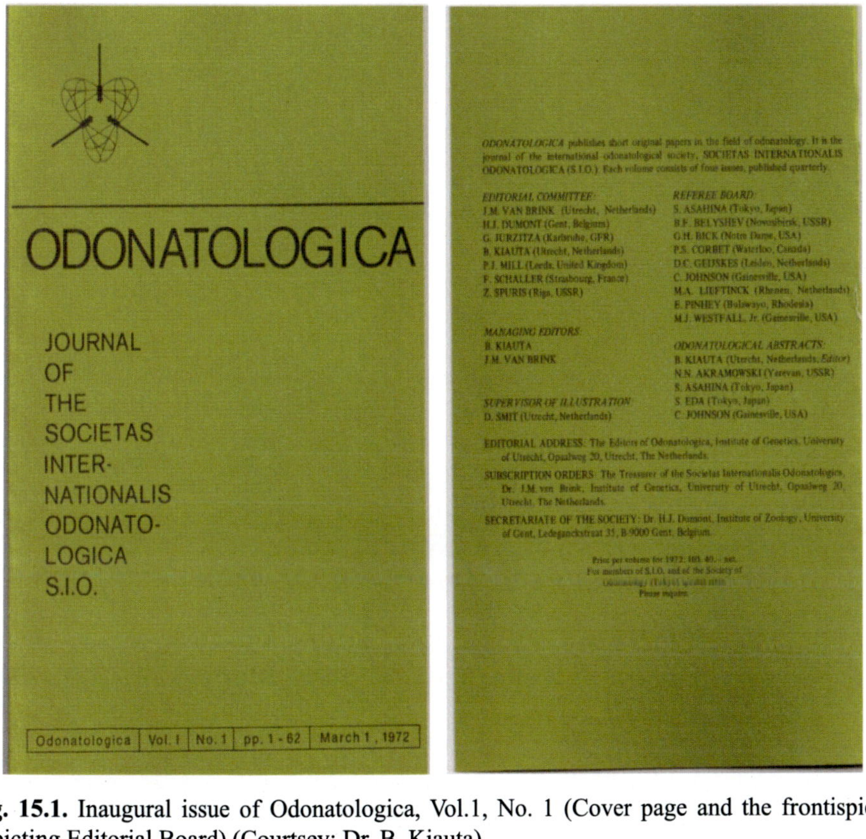

Fig. 15.1. Inaugural issue of Odonatologica, Vol.1, No. 1 (Cover page and the frontispiece depicting Editorial Board) (Courtsey: Dr. B. Kiauta)

For those fourty years of *Odonatologica*, Dr Kiauta gave everything in sacrifice; time, attention, focus, family and social life. If I were to take his words from a letter dated 15th February, 1997, Dr Kiauta wrote to Dr Rob Cannings, I could easily understand his sentiments: "(in doing Odonatologica) *It is a great sacrifice of personal freedom and time to have the journal produced punctually for 25 years now* ⌊*in Odonatologica alone over 10,000 pages, over 11,000 abstracts (all written by myself), over 820 papers, all edited by myself*⌋... ."

(a)

(b)

Fig. 15.2. (a) Dr Kiauta and Dr van Brink on a discussion table; it was a regular feature they would discuss on various SIO issues, including *Odonatologica*, and (b) Dr. Kiauta shifting the just produced *Odonatologica* from Prof. J.M. van Brink's residence to be ready for mail to the odonatologists across the world. (Source: Archive of Dr B. Kiauta).

Since the birth of *Odonatologica*, two people have been crucial for its development: without the support and much more than the manifold help from the late Professor Dr Janny M. van Brink (Utrecht/Bilthoven, The Netherlands) and his wife, Marianne, it would have been unthinkable to continue in the office for so long. For some decades, Marianne has been the backbone of the SIO and *Odonatologica*, taking care of all the technical matters and much more (Figs. 15.2 and 15.3).

Fig. 15.3. Dr Kiauta with his wife, Marianne, and Prof. van Brink, soon after the 8th International Symposium of Odonatology (Paris, 1985).– (Source: B.K. Tyagi)

Likewise, since the very first issue until today, the editorial help rendered by Dr Peter J. Mill (Leeds, UK) has been indispensable. There is hardly an issue in the production of which he has not been involved, either as a peer reviewer, style modeller or a language corrector. Kiyoshi Inoue's (Osaka, Japan) and Professor Dr Minter J. Westfall's (Gainesville/FL, USA) manifold help and continuous encouragement during the four decades were among the cornerstones of much of Dr. Kiauta's work. In recent years, first Dr Günter Bechly (Böblingen, Germany), and later his son, Adjunct Professor Dr George Kiauta-Thucydides (Vancouver, Canada) opened a new horizon by managing the website and the electronic edition of the journal. At its very onset, *Odonatologica* enjoyed the active support and assistance of well over a hundred odonatologists from all parts of the world. From among those deceased, at least a few peer reviewers, collaborators of the Odonatological Abstracts section and/or helpers in other editorial matters include; [academic titles omitted]: N.N. Akramowski (Armenia), S. Asahina (Japan), B.I. Balinsky (South Africa), J. Belle (Netherlands), B.F. Belyshev (Russia), G.H. Bick (USA), F. Capra (Italy), H.-f. Chao (China), P.S. Corbet (UK), D.A.L. Davies (UK), R.M. Gambles (UK), D.C. Geijskes (Netherlands), L.K. Gloyd (USA), A.Yu Haritonov (Russia), H. Kaiser (Germany), M.A. Lieftinck (Netherlands), C. Longfield (Eire), T.T. Macan (UK), S. Mielewczyk (Poland), P.L. Miller (UK), T.R. Mitra (India), E.B. Montgomery (USA), P. Münchberg (Germany), C. Nielsen (Italy), E.C.G. Pinhey (Rhodesia/UK), W. Piper (Germany), A. Portmann (Switzerland), G. Pritchard (Canada), M. Schorr

(Germany), Z. Spuris (Latvia), V. Teyrovský (Czech Republic), J.A.L. Watson (Australia) and M.J. Westfall (USA). *Odonatologica* is a team work. Dozens of its external peer reviewers and Editorial Board members, some of these are currently the sole (or almost so) living experts in certain narrow fields of odonatology, have been his greatest indispensable accomplices in keeping the highest standard of the journal during the past four decades. It is also due to them that the journal has been in a position to adhere strictly to the publication dates, as set-up in 1971. His perseverance and sense of punctuality soon began to pay dividends as excellent original manuscripts poured in for publication in the journal from all over the world. Dr. Kiauta enjoyed editing *Odonatologica*, and why not; he was absorbed in its editing and publishing 24×7×12×43!!

One of the strongest attributes Dr Kiauta exhibited as the Executive Editor of *Odonatologica* was his unstinted and constant encouragement to prospective authors of manuscripts and the 'never say never' attitude even to an apparently 'poorly written' manuscript! Such a magnanimous outlook bore motivational vibes powerful enough to usher in more authors and manuscripts, on one hand, and worked miracles in encouraging especially the young and brooding odonatologists, on the other. How much impact an instance of encouragement makes on a man's life can be best learnt from the life of Thomas Edison (1847-1931), one of the greatest inventors the world ever had! One day he went home from school and handed over his mother some paper which his teacher had given him and had instructed to only give it to his mother. As she read the note, tears started flowing down her cheeks. She read the note loudly to Thomas. The note said, "*Your son is a genius. This school is too small for him and doesn't have enough good teachers for training him. Please teach him yourself.*" After many years, when Edison's mother had already passed away, and he had become one of the greatest inventors in history, Edison was one day going through some old family stuff. In one of the drawers, he found this old note. Edison opened it and read it. It was the same note that his teacher had given him to give to his mother many years ago. It said: '*Your son is addled or mentally ill. We won't let him come to school anymore.*' After reading the note, Edison cried for hours and then wrote in his diary: "*Thomas Alva Edison was an addled child that, by a hero mother, became the genius of the century.*" This story tells us that if the right encouragement is given early in life this can indeed infuse great self-confidence in a person and that he or she can achieve wonders through dint of hard work and focused application. Dr Kiauta, justly remembering with deep and sincere gratitude the value of help and guidance he received from his teachers, mentors and seniors during his own evolution, always helped and encouraged others who sought it or he knew of someone in need of.

Dr Milen Marinov is one of his 'disciples', like hundreds of those who have largely been 'taught' by way of correspondence. He has recently recounted an episode of many years in past and shared his experience of publishing a research paper in *Odonatologica*, as: "*Dear Prof. Tyagi, Below is a message from March, 2011 where Prof. Kiauta encouraged me to publish one very important for me paper in Odonatologica. It appeared later the same year. It shows how effective he was in his editorial work during all those 40+ years! Kind regards, Milen*".

Dr Marinov, currently one of the leading odonatologists in the world, proudly remembers how Dr Kiauta has helped him with publication and literature in his formative days of odonatological research! He gives a passionate account of how Dr Kiauta encouraged him to publish his research findings.

On Sat, Mar 19, 2011 at 8:22 AM,

Milen Marinov <mg_marinov@yahoo.com> wrote:

Dear Prof. KIAUTA,

Here is the final draft of a paper that we were working on for a very long time. It is just the text of the draft without the figures. I could attach them later if you consider it worth publishing. As you'll see I'm contemplating of going further with this stuff and am almost ready with the next in line publication – about the spatial modeling of Odonata habitats in New Zealand. The other file I attach here is my dissertation from Lincoln University which was exactly on this subject. I can easily work on it in order to publish the results from this investigation.

With the present letter I just want to touch base with you and see if you basically agree with this publication and would like us to send you the final figures. With my best wishes and thanks in advance:

Milen"

Sooner Dr Milen received the reply:

"**From:** *Kiauta <mbkiauta@gmail.com>*
To: *Milen Marinov <mg_marinov@yahoo.com>*
Sent: *Tue, March 22, 2011 12:07:44 AM*
Subject: *Re: From_Milen*

Dear Dr Marinov,

Thank you very much for the pdf of your dissertation, which I am very glad to have.

The draft on Spatial modelling of odonate habitats in the Pacific is highly interesting indeed. When ready, I would be glad to have the manuscript for publication in Odonatologica.

With thanks again and kind regards,
Yours sincerely,
B. Kiauta"

From: Milen Marinov <mg_marinov@yahoo.com>
To: Kiauta <mbkiauta@gmail.com>
Sent: Wednesday, 23 March 2011 8:16 AM
Subject: Re: From_Milen

Dear Prof. KIAUTA,

Thank you very much for the encouraging words. I forwarded your letter to my co-authors and shall wait to hear anything back from them. When ready I'll let you know.
Best wishes,

Milen

My own story is worth mentioning here since not only Dr Kiauta's encouragement proved crucial for my 'first ever' scientific research manuscript intended for publication in *Odonatologica*, but we also evolved a teacher-taught relationship for ever! For past more than four decades his finest and critically assessed teachings in science have been my guiding force, reminding me of the famous maxim: "*A word of encouragement during a failure is worth more than an hour of praise after success.*"

I came in contact with Dr Kiauta during 1975 through my PhD guide, Dr S.K. Sangal, who had sought his expert advice on my first research manuscript before finally submitting it to *Odonatologica.* Within a fortnight Dr Kiauta's reply to Dr Sangal arrived, with a copy marked to myself, thanking him profusely for 'developing the Doon Valley into a stronghold for dragonfly research largely due to his personal influence'. Most strikingly in his response to my manuscript "*Studies on the karyotypes of three Indian species of Odonata*..." he instructed that '*for any discussion he would take up contact directly with the author* (Tyagi)'. This last sentence alone surged in me a beam of confidence, unfelt until then, and I began corresponding with him since then. He later offered his comments on the manuscript – systematically explaining the plus and minus points. Upon receiving his suggestions I appropriately revised the manuscript, restricting to the hitherto cytologically undescribed two species, viz., *Burmagomphus pyramidalis* Laidlaw (Fig. 15.4) and *Onychogomphus saundersi duaricus* Fraser. In my manuscript I had emphatically referred to, among other details, the centromere as 'localized' in both the species and made no assumptions as to the decrease of the chromosome numbers (2n, n) in *O. saundersi duaricus.* Very coolly and motivatingly Dr Kiauta apprised me about the understanding then in vogue about the centromere type in the order Odonata (*cf.* KIAUTA, 1970b) and encouraged me, if I thought otherwise, to come up with a demonstrable interpretation of the existence of localized kinetochore! He also informed how important, in his opinion, were the reduced chromosome complements, both 2n and n, in *O. s. duaricus*, and supplied a large number of photocopies of literature dealing with the subject, for my perusal. I was convinced, based on the proof of my microphotographic slides, that it was a diffuse centromere in all my dragonflies, and that the reduced chronomosome number was a result of the fusion of sex chromosome with the autosomes (Fig. 15.5 and 15.6). The manuscript was finally published in *Odonatologica* (TYAGI, 1977) to become one of the finest research publications in my formative days and laid foundation of my future thinking about quality research!

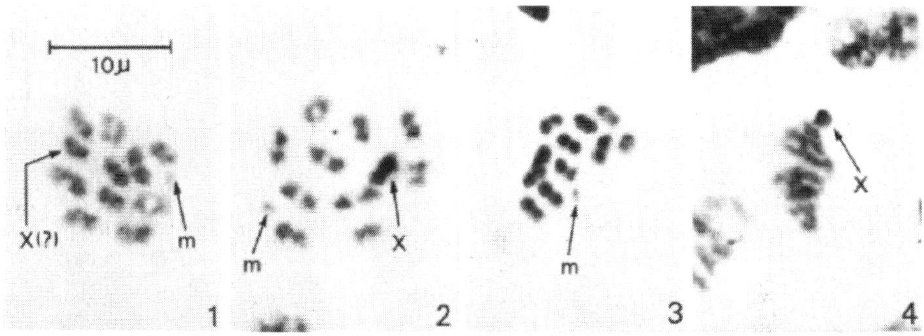

Figs. 15.4 (1-4). Spermatocyte stages of *Burmagomphus pyramidalis* Laidlaw (Dehra Dun Valley, India; carbol fuchsin squash; 1500 X): (1-2) Primary spermatocyte metaphase, polar view; —(3) Secondary spermatocyte metaphase, polar view; — (4) Secondary spermatocyte metaphase/anaphase, lateral view. (From Tyagi, 1977)

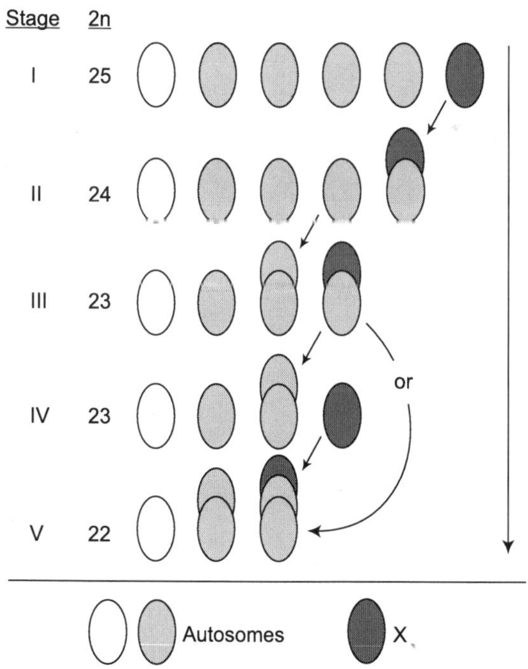

Fig. 15.5. Schematic interpretation of the evolution of sex determination and reduction of chromosome number in the Gomphidae, based on Kiauta (1969a). In the case of *Onychogomphus saundersi duaricus* Fraser the stages (I) through (III) and (V) are applicable and cytologically demonstrable.(From Tyagi, 1977)

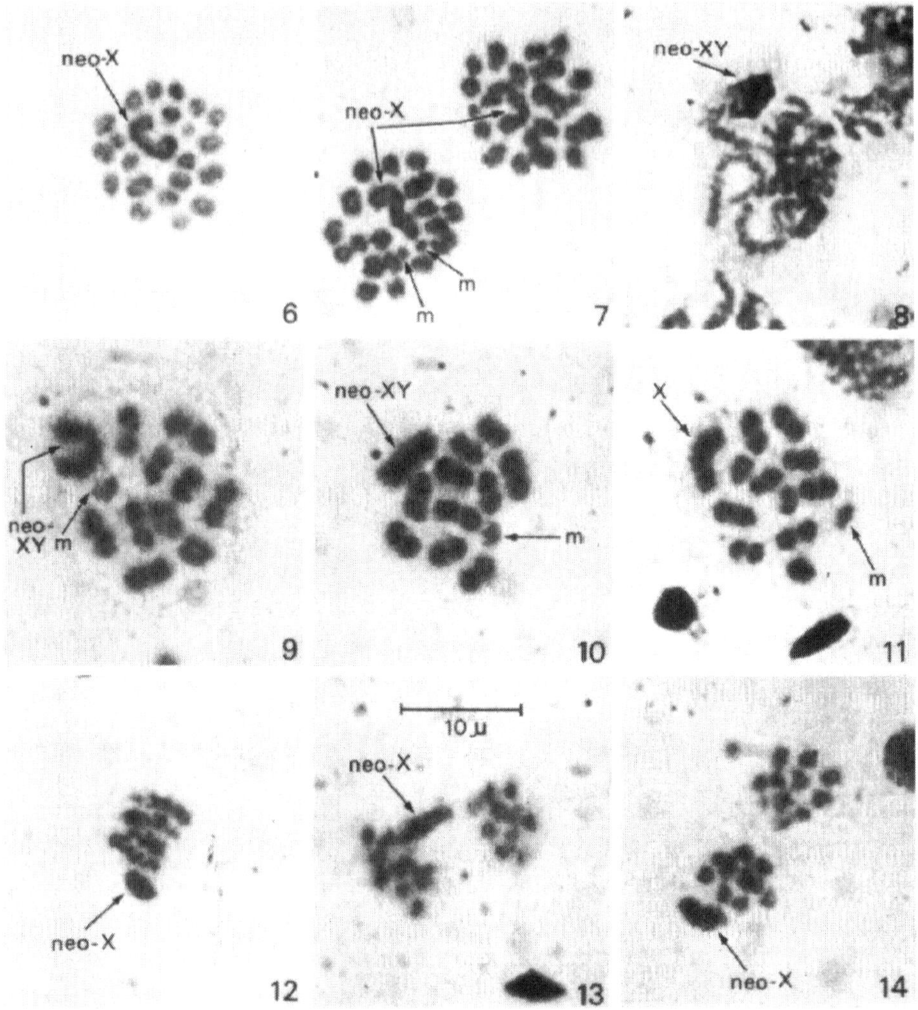

Fig. 15.6. Male germ cell chromosomes of *Onychogomphus saundersi duaricus* Fraser (Dehra Dun Valley, India; carbol fuchsin squash; 1500 X): (6-7) Spermatogonial metaphase, polar view; — (8) Pachytene,; — (9-10) Polar view of metaphase I, n=12 complement; — (11) Ditto, n=13 complement; — (12) Metaphase/anaphase II, lateral view (note the large neo-sex element); - (13) Late anaphase II (note the delayed kinesis of the neo-sex element); — (14) Telophase II, one daughter cell with, and the other without the neo-X.(Source: Tyagi, 1977)

Dr Liliana Maria Mola (Genetica, Universidad de Buenos Aires, Argentina) is, like several of us, one of the noteworthy disciples of Dr Kiauta. They met once during the symposium in Trevi. Dr Kiauta assummably has all of Dr Mola's publications, since the inception of her work in the eighties. She was apparently greatly influenced by the Odonata cytotaxonomy research of Dr Kiauta, and in all her research publications she liberally referred to his works. She seemed to be also impressed by the work of GAMA et al. (1981), where they claimed the occurrence of 4 chromatids in the meiotic bivalents of *Enallagma cheliferum*, documented by electron-microscopic photographs. MOLA (2007) knew this paper, published in Portuguese in a Brazilian journal, and compared the matter with the situation in *Luzula* and concluded with the assumption of "holokinetic nature of the (odonate) chromosomes … in mitotic chromosomes". She finally came up with a new sex-determining mechanism, X_1X_2Y, based on cytotaxonomic studies on *Micrathyria ungulata*. In karyotypic (and species) evolution the polycentric or diffuse nature of the kinetochore makes no substantial difference, and Dr Kiauta, according to whom monocentric chromosomes, with a single localised centromere, are from the point of view of evolution of course quite a different matter, nevertheless he welcomed her innovative ideas, however confounding to the established fact owing much to his own lifelong research, and encouraged her to convincingly show this "complicated situation". Dr Mola is presently one of the leading authorities on odonate cytogenetics. This being one of the finest examples of encouragement to a novial odonatologist in the realms of scientific pursuit, it is of paramount importance to quote *verbatim* inspirational words of none other than Dr Kiauta himself: "*Anyone working much, inevitably makes many mistakes. If these are discovered, corrected and/or the ideas improved during the original author's lifetime, this is to be considered as a help and gratefully acknowledged.*"

"*A journal is made by the contributing authors, rather than by the editor*", wrote Dr Kiauta once to me in context with my own journal, *Indian Odonatology*, published under the aegis of SIO National Office in India. From this he meant that authors are the all-important persons in a journal publication, that is, submission of only good and original research manuscripts largely decide the standard and fate of the journal, while the editor works as a catalyst to ensure a flawless publication. There were some hard times, too, when desired quantum of manuscripts have lacked threatening procrastination of the publication date of *Odonatologica* against his principle. On such occasions his uncanny style of incessant persuasion and encouragement helped a lot in attracting certain very good manuscripts almost at the nick of time! For example, the 3rd issue of Volume 2 appeared in time solely due to M.J. Parr's (UK) submission of his two excellent *Ischnura elegans* manuscripts. Generally, otherwise, due to a constant flow of manuscripts, occasionally overwhelming, he never had any problem meeting publication dates. Throughout *Odonatologica*'s life Dr Kiauta enjoyed the loyalty and cooperativeness of the odonatological community which has been invariably splendid, sincere, trustful and dependable.

The members of the Editorial Board were always eager to encourage and help young dragonfly students in the presentation of their work. Some of the senior members, such as, e.g., the late George Bick (USA), Philip Corbet (UK), Dirk Geijskes (Netherlands), Maus Lieftinck (Netherlands), Peter Miller (UK), Gordon

Pritchard (Canada) and Minter Westfall (USA) even rendered pre-submission guidance to the authors by reviewing the manuscript so as to help the author improve its quality. My own first research publication (TYAGI,1977) is a vivid example as to how superbly it was metamorphosed by Dr Kiauta to become a great publication to be proud of. What is more because of my financial situation, Dr Kiauta and Dr van Brink waived the page charges for my publication and even courteously provided 20 copies *gratis*! These kinds of help, however trifle they may look, bear an everlasting impact on the mind of a developing young scientist, and in my own case whenever a similar situation occurs I do invariably emulate this example to help young and needy researchers! Whenever there was any uncertainty as to the identity of the species to any odonatologist in any part of the world, Dr Kiauta would have the great Dr Lieftinck generously check the specimens systematically and identify the vouchered specimens, if provided by the authors upon request.

The Odonatological Abstracts section has been a standard feature of the journal since its inception. During the earlier years the personal computer was unknown, the facilities for copying scarce, not very good, elaborate and expensive. Dr Kiauta had to depend almost entirely on the exceptionally well-stocked library of the Netherlands Entomological Society (Amsterdam) where, during 40 years, he checked monthly the newly arrived issues of hundreds of entomological periodicals. The reprints sent more or less regularly by numerous workers were, of course, very useful. Some of his colleagues volunteered at various times and over various lengths of time, to collect for him systematically the odonatological works published in their country or a larger region, without an interruption, such as the contributions covering Australia and New Zealand (by J.A.L. Watson), China (H.-f. Chao), Germany (R. Rudolph), India (T.R. Mitra), Italy (C. Utzeri), Japan (in succession: S. Asahina, S. Eda, K. Inoue), The Netherlands (M.Th. Wasscher), New Zealand (W.J. Winstanley), Poland and eastern Europe (P. Buczyński), Russia (N.N. Akramowski, S.N. Gorb, A.Yu. Haritonov), Slovenia (M. Bedjanič), Switzerland (H. Wildermuth) and Thailand (A. Pinratana). During recent years, until mid 2012, the overwhelming help by W. Piper (Germany), covering the entire world and including considerably more than three thousand publications, made the preparation of the abstracts easier than ever before. Since that date M. Schorr (Germany) has been helping similarly both to Dr Kiauta and to *Odonatologica*.

Printing a journal of the world-class quality with an infallible frequency that no other single insect order anywhere in the world could match was a herculean task from day one!! This was apparently a great challenge throughout since Dr Kiauta had several other editorial responsibilities almost lifelong. In 1962 he was already the Editor-in-Chief of the inaugural issue of the Slovenian periodical "*Nature Conservation*", and the journal continues appearing until this very day, although he is lately not involved any more in its editorial matters. His other editorial works include:

- *CONTACTBRIEF NEDERLANDSE LIBELLENONDERZOEKERS*, Utrecht. 1970-1974, pts 1-13 + Supplements, Editor
- *ABSTRACTS OF PAPERS READ AT THE FIRST EUROPEAN SYMPOSIUM OF ODONATOLOGY*, Ghent, Belgium, 1971. Editor
- *GENETICA*, The Hague, 1971-1990, Vols 42-83. Joint Editor
- *GENEN EN PHAENEN*, Journal of the Netherlands Genetic Society, Utrecht, vols 18-19, 1972-1977. Joint Editor
- *ODONATOLOGICA*, Utrecht, 1972-2013, vols 1-42. Editor
- *NOTULAE ODONATOLOGICAE*, Utrecht, 1978-2013, vols 1-8. Editor
- *ABSTRACTS OF PAPERS READ AT THE SECOND INTERNATIONAL SYMPOSIUM OF ODONATOLOGY*, Karlsruhe, 1973. Joint Editor
- *FIELD REPORTS OF THE NETHERLANDS CENTRE FOR ALPINE BIOLOGICAL RESEARCH (N.C.A.B.R.)*, Utrecht, vol. 1, 1973. Joint Editor
- *SOCIETAS INTERNATIONALIS ODONATOLOGICA RAPID COMMUNICATIONS,* with *Supplements*, Utrecht, vols 1-4, Suppl. 1-14, 1980-1992. Editor
- *OPUSCULA ZOOLOGICA FLUMINENSIA*, Flums/Switzerland, pts 1-224, 1984-present, Editor

Yet, Dr. Kiauta's heart and soul was fully occupied with *Odonatologica.* He made it world's one of the most respected periodicals for any of the insect orders through his complete devotion, hard work and application. The bond of trust he developed with a large number of key players was crucial in *Odonatologica*'s life! As said earlier, *Odonatologica* was a collective effort of the 'family of friends' and one can not afford to remain unmoved and overwhelmed by the strikingly superb quality of its line-drawings, and labeling of figures and photographs but also general administration and logistic management. Such people, often remaining behind the scene but are inevitably indispensable in their contribution, need a special mention. Mr. Dick Smit, the artist at the former Institute of Genetics of the University of Utrecht, The Netherlands, who had designed the SIO and *Odonatologica* logo, also handled all the illustrations published in the first 12 volumes. For her reliability and her invariably high-quality work; also for her very dependable cooperation of every kind and for her strict adherence to publication dates, even when these were occasionally not easy to meet, Dr Kiauta has always thankfully banked on his page-maker and go-between with the printers, Mrs Joke M. Schippers-Kok of "De Zetspiegel" (Hardinxveld-Giessendam, The Netherlands). Finally, but not the least, in the field of administration and local distribution of *Odonatologica* much help was continuously rendered by some of the SIO National Offices: in Germany (in succession R. Rudolph, R. Jödicke), in Italy (C. Utzeri), Japan (K. Inoue), Slovenia (A. Pirnat), Thailand (A. Pinratana),

United Kingdom (first P.J. Mill, then J. Silsby, currently G. Vick), India (first B.K. Tyagi, then D.B. Tembhare/R.J. Andrew) and in the United States (M.J. Westfall, later B. Mauffray).

With the December 2013 issue edited by Dr Kiauta, the 42nd volume – the last under his Executive Editorship – was completed: the 168 issues contained about 1400 research papers and over 19,800 abstracts of works published throughout the world since 1971. The editing and the intensive daily contacts with the odonatological community became his life style, bringing much pleasure and many personal friendships. It was not easy. However, owing to advanced age and some serious eye-sight related health issues, with much regret he decided to bid adieu to the office upon the completion of the 42ndvolume. As from the 43rd volume onwards Dr. Kiauta passed on the batton of the responsibilities for the journal to Dr Florian Weihrauch (odonatologica@osmylus.com), along with its administration, production, editorship and circulation. With his profound experience in worldwide scientific communication and odonatological editorship, Dr Florian Weihrauch is perfectly qualified to navigate the journal on the waves of contemporary scientific requirements and standards and to guide it to an increasingly high scientific level. Dr Kiauta is nevertheless continuing on the Editorial Board of *Odonatologica*.

●●●

16

Worldwide Odonatological Periodicals of the S.I.O. and Dr Kiauta's Network

Societas Internationalis Odonatologica (= S.I.O., International Society of Odonatology) has grown famous phenomenally not only for its mainstay international journal, *Odonatologica*, and regularly organized biennial symposia all over the world, as well as almost a dozen national offices across the globe, but also for an arcane reason of publishing unbelievably about five dozen periodicals in countries where either of its national section or affiliation functioned, besides the SIO Central Office in the Netherlands. Nowhere in the world today or in yore would there be an instance when a scientific society, leave aside entomological, ever published or coordinated publication of so many scientific periodicals, clearly surpassing any imaginable world record of sort! "Tombo Dokokai" (= Dragonfly Lovers' Association) was the first formal odonatological organization on record (Nagoya, Japan, 1956). The Society of Odonatology, Tokyo and its periodical, *Tombo – Acta odonalologica* (1958), are the oldest organization/joumal in existence/in-print. All these periodicals that were produced during various lengths of time, between 1971 and 2013 when Dr Kiauta was at the helm of the SIO affairs, are listed here under.

A total of 60 odonatological periodicals and serials in-print, published by the International Odonatological Society (SIO) and by 29 regional, national and local odonatological associations in 20 countries are briefly annotated. After the respective title, the year of publication of the first issue is stated in brackets. Although some titles have not been published since several years, they are included if no formal notification on their suspension has been received. Some additional information such as, e.g., editor of periodical, is also appended.

Periodicals published by S.I.O. and its local sections:

I. CENTRAL OFFICE (Bilthoven, Siebengewald, The Netherlands)

1. *Advances in Odonatology* (commenced in 1982)
   1982-1994, Editors: R.M. Gambles (1982), G. Pritchard (1984) and J. Legrand (1991-1994)
2. *Noltulae odonatologicae* (1978).
   Editor: 1978→, B. Kiauta (1978-2013), A. Schröter (2014→)

3. *Odonalologica* (1972) (incl. Odonatological Abstracts).
   Editor: 1972→, B. Kiauta (1972-2013), F. Weihrauch (2014→)
4. *Reports of the Odonata Specialist Group, Species Survival Commission, International Union for Conservation of Nature and Natural Resources* (1982)
   Editor: 1982-1996, N.W. Moore
5. *Sales Catalogue of the SIO Antiquarian Department* (1985)
   Editor: 1985-1997, B. Kiauta
6. *Societas internationalis odonatological rapid Communications* (incl. *Supplements*) (1980)
   Editor: 1980-1992, B. Kiauta

   The numerous periodicals where Dr Kiauta is/was serving either as a member of the Editorial Board and/or as a regular or incidental peer reviewer, are not mentioned here. In Switzerland he is publishing since 1984 an international serial, *Opuscula zoological fluminensia*. So far 224 parts have appeared at irregular intervals, over 100 of these dealing with the Odonata.

II. SIO INTERNATIONAL ODONATA RESEARCH INSTITUTE (Gainesville, Florida, United States)
   – *Annual Report of the International Odonata Research Institute* (1988)

III. NATIONAL AND REGIONAL OFFICES, SECTIONS AND CHAPTERS
1. Canada –*Walkeria* (1985)
2. Germany –*Hagenia* (1991)
3. Iberian Peninsula –*Navasia* (1992)
4. India (Office Dr B.K. Tyagi):
   - *Indian Odonatology* (1988) (SIO Regional Office in South Asia):
   - *Abstracts of Papers read at the South Asian/Indian Symposia of Odonatology of Odonatology* (1984)
   - *Fraseria* (1981-1992 as a newsletter, since 1994 as a scientific bulletin)
   - *Occasional Publications of the SIO Regional Office in South Asia* (1986)
5. Italy –*Lindenia* (1984)
6. Japan –*Sioja* (1981)
7. Slovenia –*Erjavecia* (1995) and *Exuviae* (1994)
8. Thailand –*Malangpo* (1985)
9. United Kingdom – *Kimminsia* (1990)
10. United States -*Selysia* (1963)

IV. S.I.O. SPECIALIST GROUP FOR SYSTEMATIC AND PHYLOGENETIC ODONATOLOGY

- *Petalura* (1993)

V. ORGANIZATION COMMITTEES OF THE INTERNATIONAL SYMPOSIA OF ODONATOLOGY

- *Abstracts of Papers read at the International Symposia of Odonatology* (1971)
- *Abstracts of Papers and Posters presented at Odonatological Symposia of the Alps-Adriatic regional Community* (1994)

Regional, national and local odonatological organisations and their periodicals.

The societies listed under (4), (12), and (24) represent national/regional chapters of SIO. Those listed under (1), (10), (22) and (27) are either affiliated to, or enjoy a special status in SIO.

The Americas

(1) DRAGONFLY SOCIETY OF THE AMERICAS (Dr & Mrs T.W. Donnelly)

- *Argia* (incl. Supplements 1989)
- *Bulletin of American Odonatology* (1992)

Austria

(2) OSTERRE1CHISCHE ARBE1TSGEMEINSCHAFT LIBELLEN (Dr R. Raab)

- *Anax*, Wien (1994)

Belgium

(3) BELGISCHE LIBELLENONDERZOEKERS ODONATOLOGUES BELGES (Ms A. Anselin)

- *Gomphus* (1984)

Eastern Europe

(4) In the framework of SIO, name not yet fixed (Dr S. Gorb)

- *East European Odonatological Annual* (1995)

Eire

(5) IRISH ODONATA RECORDING SCHEME (C. Ronayne)

- *Irish Odonata Recording Scheme Newsletter* (1988)

France

(6) GROUPE DE RECHERCHE ET DE PROTECTION DES LIBELLULES "SYMPETRUM" (Ms K. Funkiewiez)

- *Le Depressiusculum*– Feuille de liaison pour Tetude des odonates dans la Crau (1992)
- *Sympetrum*, Grenoble (1987)

(7) SOCIÉTÉ FRANÇAISE d'ODONATOLOGIE (J.-L, Dommanget)

- *Letlre des Socielaires* (1994)
- *Martinia* (1985)
- *Martina Hors-serie* (1990)

Germany

(8) ARBEITSGEMEINSCHAFT LIBELLEN IM LKR. SCHWÄBISCH HALL (B. Kunz)

- *Jber. Libellen Schwäbisch Hall* (1987)

(9) BREMER L1BELLENGRUPPE (Dipl.-lng. J. Ruddek)

- *Hefte der Bremer Libellengruppe* (1989)

(10) GESELLSCHAFT DEUTSCHSPRACHIGER ODONATOLOGEN (Ms U. Krüner)

- *Hagenia* (with SIO Natn. Office in Germany, 1991)
- *Libellula* (1981)

[**Note:** Abstracts of Papers of some of the GdO Annual Symposia are incidentally published by the respective Organizers, but they do not have a common title and order address, therefore they cannot be considered a serial work.]

(11) SCHUTZGEME1NSCHAFT LIBELLEN IN BADEN-WÜRTTEMBERG (Dr R. Buchwald)

- *Sammelbericht uber Libellenvorkommen in Baden-Württemberg* (1980)

Hungary

(12) FRATERNITY OF HUNGARIAN ODONATOLOGISTS (M.O.B.K.) (Dr A. Ambrus)

- *MOBK-Communications* (= informal, numbered circulars, 1988)
- *Vizipasztor* (not yet published)

(13) STUDIA ODONATOLOGICA HUNGARICA (Dr G. Dévai)

- *Studio odonatologica hungarica* (1993)

Japan

(14) AKA-TOMBO NETWORK (Dr T. Uéda)

- *Symnet* (1994)

(15) DRAGONFLY RESEARCH ASSOCIATION, TOKYO (I. Wakana)

- *Shibuchan* (1994)

(16) HOKKAIDO ODONATOLOGICAL SOCIETY

(Prof, Dr H. Ubukata, Dept Sci. Educ., Kushiro Coll., Shiroyama 1, Kushiro, 085, Japan)

- *Bulletin of the Hokkaido Odonatological Society* (1986)

(17) KANSAI RESEARCH GROUP OF ODONATOLOGY

(K. Tani, Jizo-cho 129, Nara, 630, Japan; - foreign correspondence to: K. Inoue, 5-9, Fuminosato 4-chome, Abeno-ku, Osaka, 545, Japan)

- *Gracile* (1965)
- *Odonatological Library News* (1987)

(18) SOCIETY OF DRAGONFLIES AND NATURE (Ms M. Tanimoto)

- *Tombo to Bunka* (1985)

(19) SOCIETY OF ODONATOLOGY, HYOGO (S. Nishu)

- *Sympetrum, Hyogo* (1993)

(20) SOCIETY OF ODONATOLOGY, TOKYO (Dr Eda)

- *Tombo –Acta Odonatologica* (1958)

(21) TOMBO KENKYUKAI (= Odonatological Society of Osaka) (A. Muraki)

- *Aeschna* (1974)
- *Macromia* (1975)

The Netherlands

(22) NEDERLANDSE LIBELLENONDERZOEKERS (W.J.A. Hoeffnagel)

- *Contactblad Nederlandse Libellenonderzoekers* (1981)
- *Libellennieuwsbrief* (1993)

Scandinavia

(23) Informal organisation set up at the 1994 Rakkestad (Norway) Regional Symposium (Dr H. Olsvik)

- *Nordisk odonatologisk Forum* (to commence in March 1995)

Slovenia

(24) SLOVENSKO ODONATOLOŠKO DRUŠTVO (= Slovenian Odonatological Society) (M. Bedjanič)

- *Erjavecia* (1995)
- *Exuviae* (1994)

Switzerland

(25) GROUPEMENT DES ODONATOLOGUES DE SUISSE – VEREINIGUNG DER SCHWE1ZERISCHEN L1BELLENKUNDLER - GRUPPO DEGLI ODONATOLOGI DI SVIZZER– –GRUPPAMENT OILS ODONATOLOGS SVIZZERS) (Dipl.-Zool. C. Meier)

- solely a standard section, "Anax", in *Nouvelles Cent, suisse Cartogr. Faune* (1991)

(26) ZÜRCHER LIBELLENFORUM (Dipl.-Zool. C. Meier)

- Zürcher Libellenforum (1980; status after 1988 unknown)

United Kingdom

(27) BRITISCH DRAGONFLY SOCIETY (Ms J. Silsby)

- *Journal of the British Dragonfly Society* (1983)
- *Newsletter [of the] British Dragonfly Society* (1983)

(**Note:** Occasional publications on selected topics are published incidentally, but they have neither a common serial title, nor a consecutive issue numbering.)

(28) ODONATA RECORDING SCHEME (R. Merritt)

- *Odonata Recording Scheme Newsletter* (1977)

United States

(29) OHIO DRAGONFLY SURVEY (B. Glotzhober)

- *Dragon-Flier* (1992)

●●●

17

Some non-S.I.O. Odonatological Organizations and their Periodicals Influenced by S.I.O. or Dr B. Kiauta

The Japanese journal, *Tombo - Acta odonatologica* (1958), founded and edited by the doyen of Japanese odonatology, Dr. S. Asahina, is the first truly odonatological journal in the world, while *Selysia*, initiated in 1963 by America's one of the greatest odonatologists, Prof. Basil Elwood Montgomery, was the first bulletin or the newsletter of any odonatological association. Both these periodicals existed before the Societas Internationalis Odonatologica (=International Society of Odonatology, S.I.O.) was founded in 1971, with their key journal, *Odonatologica*, also became available since March 1972. The SIO was in fact the first international odonatological society, represented by nearly 600 members in all the continents of the world, and with its national and/or regional offices in about a dozen countries as well as manifold scheduled academic and/or research activities such as the biennial international odonatological symposia and affiliation with the International Union for Conservation of Nature and Natural Resources (I.U.C.N.). The SIO cast an enormous influence on the diversification of odonatological activities in a large number of countries, any of which ventured to bring out their own journals, bulletins and/or newsletters. These periodicals though did not bear any affiliation to the SIO, nevertheless, they did aparently emerge under its motivation and encouragement largely spearheaded by its general overseer, Dr B. Kiauta. Among all the continents Europe emerged most prolific with close to a dozen journals or bulletins, some of which have been doing well for 20 or 30 years, sometimes even longer, reflecting the increased interest in dragonflies.

1. *Crenata* - Finland

The journal *Crenata* has been published in print format each year since 2008 and thereafter as electronic version, freely downloadable at www.sudenkorento.fi . The older issues are also planned to be made available for download from the same site at, http://www.sudenkorento.fi/kwiki/Crenata .

2. *Odonatrix* –Poland

Odonatrix (2005), published twice a year originally and after 2016 appearing in electronic format annually, is a journal devoted to the study of dragonflies and damselflies (Odonata) of Poland, now downloadable (both old and new issues as separate PDF) at, http://www.wazki.pl . The idea of the journal was initiated by Paweł Buczyński. It publishes faunistic data and establishes a group of amateur researchers studying odonates, now significantly increased under the banner of Odonatological Section of the Polish Entomological Society which has already organized 13 annual seminars and/or conferences in various localities in Poland. Contributions published in *Odonatrix* contributed to '*A distribution atlas of dragonflies (Odonata) in Poland*' published in 2009. *Odonatrix* is a peer-reviewed journal and included in AGRO and Index Copernicus.

3. *Martinia* – Journal of the Société Française d'Odonatologie (SfO)

The French odonatological society (SfO) is an association born on 23 April, 1991. It plays an expert role in developing odonatological studies in collaboration with the National Museum of Natural History (MNHN), the French Ministry in charge of the Environment, the Office for the insects and their environment (Opie) and other institutions among which are wildlife managers. The SfO organizes annual meetings but also international conferences, such as the 3rd European Congress on Odonatology which was held in 2014 in Montpellier (southern France). Targeting the conservation of dragonflies and their habitats, the SfO also offers education and training sessions and edits books and field guides.

The SfO publishes *Martinia*, the association's scientific journal which is dedicated to the French odonatologist René Martin (1846-1925). With two issues a year, *Martinia* aims to provide short communications and articles dealing with ecology, behavior, distribution, conservation and management of the odonates. Both French and English texts are accepted currently. *Martinia* is even older than the association itself and was created in 1985 by a small group of dragonfly enthusiasts, led by Jean-Louis Dommanget who was the first editor. *Martinia* was therefore the first journal providing information about dragonflies from France and helped substantially to organize and energize the national distribution survey of odonates which had been launched in 1982. From 1985 up to now, more than 810 papers have been published in *Martinia*, which represents 6,700 pages. A repository of these papers (1985-2011) is available on the SfO website [www. Libellules.org].

4. *Libellula* – Journal of the GdO e.V. (Gesellschaft der deutschsprachigen Odonatologen/Society of German-speaking Odonatologists)

Libellula (Editor: Dr Christoph Willigalla), published semiannually since 1981, is the journal of the GdO (Gesellschaft der deutschsprachigen Odonatologen e.V. (Society of German-speaking Odonatologists), comprising scientific papers on biology, ecology, distribution, and conservation of Odonata. For subscription, a membership with the GdO is obligatory. The annual membership fee is €35 plus additional shipping costs. The membership fee also includes subscription to the *Libellula Supplement*, published at irregular intervals, and which has contributed to the publication of such important odonatological monuments like the *Atlas of the Odonata of the Mediterranean and North Africa* and the *Atlas of the Odonata of Germany* were published within this series. Manuscripts in both German and English texts are acceptable to *Libellula*, and may be submitted to its editor, Dr Christoph Willigalla [christoph@willigalla.de].

5. *Boletín ROLA* – Journal of odonatologists from Andalusia (Spain)

Boletín ROLA is the journal of the network of dragonfly recorders and enthusiasts in Andalusia (southern Spain). It was launched in 2012 as a consequence of the activities of a freshly emerged group of local odonatologists who are great Spanish amateurs and professional scientists. Currently edited by Florent Prunier, it is freely downloadable from the site: http://aeaelbosqueanimado.blogspot.be/p/boletin-rola.html.

6. *Journal of the British Dragonfly Society*

The *Journal of the British Dragonfly Society* (1983), currently edited by Dr Peter J. Mill [p.j.mill@leeds.ac.uk], is the official journal of the Society and is published twice a year, in April and October. It includes articles on odonates that have been recorded from the UK; also articles on European species written by members of the British Dragonfly Society. It contains some of the latest research into all aspects of dragonfly biology. Species reviews are included and, to date, 10 have been published. Members to the BDS also receive, in addition to the journal, the newsletter, *Dragonfly News*, published twice a year and the annual *Darter Magazine*, which is devoted to recording odonates in the U.K. The journal is included in the BDS membership which is open to anyone. The membership application form is downloadable from the Spociety's website [www.british-dragonflies.org.uk/].

7. *Brachytron* – Journal of the Dutch Dragonfly Society

Brachytron is the journal of the Dutch Dragonfly Society (Nederlandse Vereniging voor Libellenstudie, NVL) and the Flemish Dragonfly Society (Libellenvereniging Vlaanderen, LVV). It has been published twice a year since 1997 and includes to date 188 scientific papers on the distribution, biology, ecology and conservation of Odonata, chiefly concerning the fauna of the Netherlands and Belgium, but also from other parts of Europe. Most of the papers can be downloaded from the website, http://www.brachytron.nl/Publicaties.html. Besides the regular two issues a year, supplements are irregularly published such as on the Odonata of Turkey, or field guides to the damselflies and dragonflies of New Guinea, or a special issue on *Sympecma paedisca*. The journal is included in the membership of the Dutch or Flemish Dragonfly Society. *Brachytron* accepts manuscripts in Dutch or English (Editor: Dr Geert De Knijf,<redactie@brachytron.nl>).

8. ***Bradinopyga*** **– the Semiannual Newsletter of the Indian Dragonfly Society**

Bradinopyga, commencing January 1, 2018, is the semiannual newsletter of the Indian Dragonfly Society (IDS), founded recently on 15thMay, 2017. It publishes all kinds of news, advertisements, circulars and a couple of research notes as well. English texts are to be submitted to the editor, Mr. Abraham Samuel K. [Abrahamcms@gmail.com], or, from Vol. 2 onward, the Chief Editor, Prof. B.K. Tyagi (abktyagi@gmail.com). or the Assistant Editor, Ms Kala Mokaria [kala303@gmail.com].

9. ***Indian Journal of Odonatology*** **– the annual periodical of the Indian Dragonfly Society**

Indian Journal of Odonatology, to be commencing from January 1, 2019, is an open access annual journal of the Indian Dragonfly Society (IDS). It is general policy all manuscripts are critically peer-reviewed. It will publish original research and review works as well as short communications. Manuscripts in English are to be submitted to the Editor, Kumaran Sathasivam [kumaran.sathasivam@gmail.com] or the Assistant Editor, Ms Kala Mokaria [kala303@gmail.com] or the Chief Editor, Prof. B.K. Tyagi (abktyagi@gmail.com).

10. ***Erjavecia*** **- Bulletin of Slovene Odonatological Society**

Highly rich in dragonfly fauna, distributional studies were initiated in a systematic way by Prof. Bastiaan Kiauta over six decades ago. Erjavecia (commenced 1985) is the national odonatological bulletin of Slovene Odonatological Society which has been active for more than two decades.

Erjavecia is named after Fran Erjavec (1834–1887), a Slovene writer, naturalist and teacher. In 1864, he introduced the expression 'kačji pastir' which was the name used for dragonflies in the first Slovene biological schoolbook. Since 1996 the bulletin has been edited by Matjaž Bedjanič.

For more details about *Erjavecia* please read it elsewhere in the book and/or the website.

●●●

18

International Odonata Research Institute (I.O.R.I.)

Dr. Zandis Davidovich Spuris in his Open Letter to the Ghent Symposium was the first to launch the idea of an "International Odonatological Documentation Center". His suggestion had ever since occupied the minds of many of the odonatologists in the SIO. While the SIO "family" and its organizational structures and activities have grown rapidly, the scope and tasks of such a Center were defined and redefined countless times until in August 1981 Professor Minter Westfall, Jr. presented a formal, specified outline in the Plenary Business Meeting at the Sixth International Symposium in Chur, Switzerland, during the celebration of SIO's tenth anniversary.

In this outline, approved by the SIO Executive Board in their last-minute meeting in the library of the Natural History Museum in Chur, Professor Westfall did not dwell too long on the financial aspects of the venture. At that stage the Executive Board did not consider it opportune to lay undue emphasis, in the Business Meeting, on financial "uncertainties", they were of the opinion that an extensive discussion of this subject could easily frustrate all further planning at the most crucial initial stage. The philosophy prevailing in the Executive Board was largely identical to that underlying the initiation of SIO itself, ten years earlier: "Let us start, even if it has to be without means, and we shall definitely manage somehow. A very modest start is better than none at all and, in view of our lack of experience with management and administration of a permanent institute, is probably even better than starting immediately with a fully-fledged Institute". The financial aspects were left to be tackled by the Preparatory Committee to be appointed, while the initial statement primarily highlighted the general importance of a "Center", and gave a brief outline of its organizational scheme, scope and tasks. Presented in this way, the project was immediately received with overwhelming enthusiasm by the membership, and Dr Westfall and Dr Kiauta were asked to work out the details and report to the 1983 Symposium in Calgary, Canada.

Their task was not easy, but it was most pleasant to let their imagination fly freely. They were well aware of the financial limitations and technical difficulties, but their ultimate goal was clear: SIO should develop a permanent International Odonata Research Institute, harbouring the large library, their archives, a world collection, the central editorial offices, providing research facilities to workers in any field of Odonatology and rendering (paid) services to museums, run in a similar way to institutions sponsored by such illustrious organizations as, for instance, the Smithsonian Institution in Washington, DC or the German Oriental Society. Without financial means, the way to the goal described was inevitably long, and in their "day-dreaming" they had started at the end. In Casa d'Uors – Mrs Marianne and Prof. Kiauta's home in Flumserberg, where Prof. Westfall and his wife, Margaret, stayed during the Chur Symposium, Prof. Westfall and Dr Kiauta prepared the final draft and the related documents 'in the kitchen of the Kiautas' Swiss home for the foundation of the SIO International Odonata Research Institute (IORI), to be located in Gainesville/FL, USA, as well as that of a 'Memorandum of Understanding' with the Florida State Collection of Arthropods (Fig. 18.1). They had the sketch of the Center's building ready on August 26, 1981. Professors Philip S. Corbet, Angelo B.M. Machado, Jean-Guy Pilon, Gordon Pritchard and Janny M. van Brink were also their guests during the symposium and have contributed their ideas. [The original idea was to construct a building in Gainesvile FL on land secured by the then existing officials of the SIO…and it was to be flanked by a separate dormitory building with a capacity sufficient to house up to 4 guest research workers for any length of time.]

Fig. 18.1. At Bastiaan's home, Casa D'Uors, at Flums, Switzerland, after 1985 Symposium. From left to right: Dr Bastiaan Kiauta, Mrs Marianne Kiauta, Dr Minter Westfall, Dr Robert Cannings, Dr Stefan Mielewczyk, Dr Janny M. van Brink and Mrs Margaret Westfall. – (Photo: Syd Cannings).

Dr. Westfall took the drawing to Gainesville and had the costs estimated by an expert. The price for the 3360 sq. ft. building of the Institute would amount (early September 1981) to about US $ 35-40 – per sq. ft, meaning a total of some US $ 117,000 to 135,000. This of course could only be realized with the help of some benefactors. Unfortunately, these were few and far apart, and none were known to them. This being so, it was clear that they had to find other ways. Unexpectedly soon Prof. Westfall came up with the plan to set up the Institute within the newly created (Florida) Center for Arthropod Systematics. This would allow an immediate start. In exchange for SIO services, they would be given free housing and various kinds of facilities, while retaining the freedom to move out to their own premises as soon as SIO would be financially able to do so. It was an opportunity they could not afford to miss. On August 20, 1985, in one of their meetings during the Eighth International Symposium of Odonatology in Paris, the SIO Council agreed on a 'Memorandum of Understanding" with the Florida State Collection of Arthropods, relative to the foundation of the Institute, and authorized Dr M.J. Westfall, Dr. S.W. Dunkle and Dr. G.H. Bick to sign on behalf of the Society. They did so upon Prof. Westfall's return to Gainesville in September 1985, and the Institute was formally opened on January 28, 1986, the 70th birthday of its first Director. In November 1986 (filed on December 5, 1986) the Institute was registered under the laws of the State of Florida as an "Incorporation not for Profit". Until his death on July 20,2003, at the ripe age of 87, Dr Westfall served as the Director, whereas during 1986-2003 Dr Kiauta was a member of its Policy Board.

The SIO, thus, had a home then. It is but very small and modest and it is not yet able to fulfill all the tasks of the powerful heart of a worldwide society. However, they are firmly convinced that it will grow rapidly with SIO. Its work and vitality will depend entirely on the support and enthusiasm of each and everyone of the 600 SIO members.

The IORI is now actually I.O.R.I., Inc, due to the fact that it became incorporated in the State of Florida on 5th Dec., 1986. Thus, it is now able to receive tax-deductible gifts of cash, books, specimens, and other materials – as donations!

The IORI is operating in facilities generously provided by the Florida Division of Plant Industry in a large substantial brick building. While the area IORI occupy is relatively small, they do have research desks and microscopes, a library, a collection, and archives. It also has many other benefits, at no cost to SIO, by being in the facilities of the Florida State Collection of Arthropods (FSCA), such as: the large FSCA Odonata Collection, fumigation, automatic fire protection, security protection, and water, electricity, temperature control, telephone, and restrooms. In addition, expert taxonomists on other insect and arachnid groups are available for consultation, as is the Division of Plant Industry Library, which has extensive holdings on insect taxonomy. The IORI curates the Odonata collection of the FSCA in return for the use of these facilities. Last but not least, a small cafeteria is available. Its staff includes the General Manager, a position currently held by Dr Bill Mauffray.

The IORI provides a depository for Odonata specimens, offering full curatorial services, as well as a library and archive for personal historical Odonata data and notes of deceased or retired Odonata specialist worldwide. It has archival materials/donations from a large number of deceased odonatologists, some of which are Philip Calvert, Robert Hodges, B. Elwood Montgomery, James Needham, Septima Smith etc. Its collections of varied nature continuously growing, designing of a database was also strongly felt. The database, designed and written by Bill Mauffray in 1993 originally with Microsoft Access 1.0, has since been upgraded several times and finally converted to Office 2000 access in 1999. Now, of course, it is friendly with the latest programme development. The IORI has a website, too.

●●●

19

A Teacher, Scientist and Philosopher

गुरु ब्रह्मा गुरु विष्णु गुरुर्देवो महेश्वरः।
गुरुरेव परं ब्रह्मा तस्मै श्रीगुरवे नमः।।

Transliteration–

(Gurur-Brahmaa Guru-Vissnnur-Gurur Devo Maheshvarah,
Gurure Iva Param Brahma Tasmai Shrii-Gurave Namah.)

Translation–

(The Guru (or the teacher) is Brahma, the Guru is Vishnu,
the Guru Deva is Maheswara (Shiva),
The Guru is verily the Para-Brahman (Supreme wisdom); Salutations to that Guru.)

Teaching is hard work, and a great responsibility in national building. Execution of great responsibilities also demands the investment of an enormous amount of energy and labour. All great teachers have some indispensable traits, though not all inclusive, in common: They work tirelessly to create a challenging, nurturing environment for their students, have deep knowledge of their subject and are good communicators, and above all they are warm, accessible, enthusiastic and caring. A great teacher respects students, hears them attentively and find solutions to their queries – *hook, line and sinker*! The best part of every man's education is that which he gives to himself. The education received at school or college is but a beginning, and is valuable mainly inasmuch as it trains the mind and habituates it to continuous application and study. Dr Kiauta's family was blessed with a legacy of love for higher education and, therefore, the environment in his home was always charged with discussions on great patriots, academicians, scientists, musicians, nature, animals and plants, and so on. Aside of his parents and the family, quite a few persons had a profound influence on his choice for the study of nature. In High School, it was Pavel Kunaver, the teacher of geography, who opened to him the mountain and the cave world. At University, Professor Dr Jovan Hadži triggered and much enhanced his interest in invertebrate systematics and particularly in invertebrate phylogeny. At his first professional post, as a bio-speleologist at the Institute of Karst Research in Postojna, the incidental working visit of Professor James Kezar of the University of Oregon (USA) turned his attention to cytogenetics and cytotaxonomy, to which disciplines he remained true throughout his subsequent life. The odonatology he has learnt while going, but his principal teacher and mentor in this field was Dr M.A. Lieftinck. The knowledge he acquired from long discourses with his parents and grandparents had sown the seeds for an ever inquisitive mind that would not cease to think unless a proper solution to a given problem is found (Fig. 19.1). He inherently became very sharp in picking up many a foreign language – trait that came to be very handy when he started editing *Odonatologica.*

The habit of rising early, completing his daily homework before retiring to sleep, and attending to all his needs related to school, including polishing his shoes and sharpening his pencil, was ingrained into him by the parents at a very early age. He hardly looked upon anybody for any help in his daily chores pertaining to preparation for the school. Dr Kiauta thus cultivated the most important of all human characters, the 'self-help', at a novial age of five, the time when he first caught his dragonfly. In the process he also developed into a voracious reader.

(a)

(b)

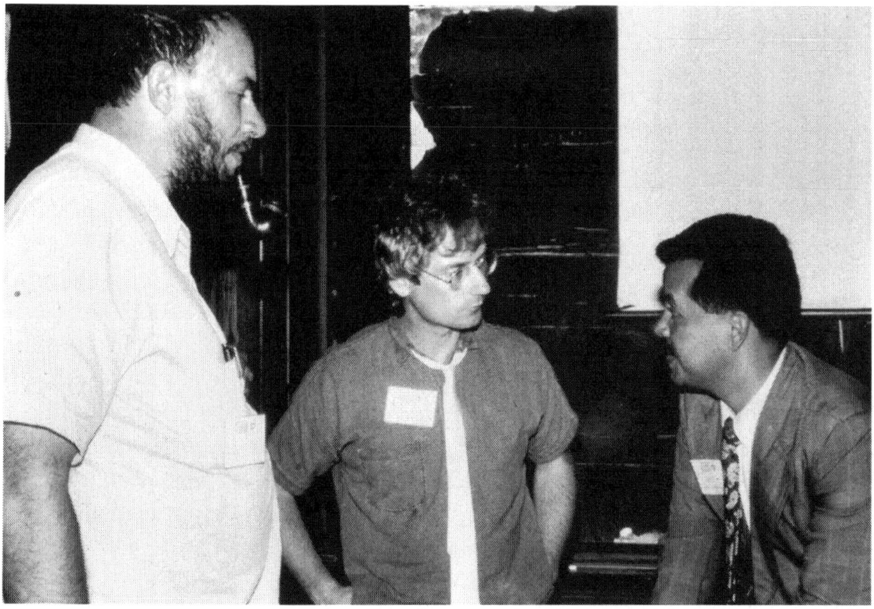

(c)

Fig. 19.1. Prof. Dr B. Kiauta in three moods; as (a) a teacher, (b) a scientist, and (c) a philosopher. (Sources: Archive B. Kiauta and B.K. Tyagi)

Like all great men of science, literature, art and music – apostles of great thoughts and masters of the great heart, he worked out his way in the hardest and sincerest possible manner maintaining throughout 'simple living with high thinking'. The road to success was never easy, and was comparable to like treading a double-edged sword. No amount of difficulties apparently the most insuperable proved obstacles in his way. These very difficulties, as would turn out later in his life, helped him surmount many challenges in the long life of both the SIO and *Odonatologica*, by evoking their powers of industry and endurance. In the life of Dr Kiauta the instances of obstacles thus surmounted, and of triumphs thus achieved, are indeed so numerous, as almost to justify the proverb that "with will one can do anything." While maintaining highest standards of teaching in the University, he still devoted his leisure to the study of natural and life sciences in all their branches and his researches in connection with the odonate cytotaxonomy were soon rewarded by several novelties including discoveries, hypotheses and theories which have all stood the test of time!

A towering personality, how then could he obtain such a high level of energy to support him work consistently for about 16 hours a day? Even though he was never an addict to athletics and team sports, yet he was lifelong a passionate alpinist and mountain wall climber and a (tour) skier. Even as an octogenarian, he remains a passionate long-distance swimmer. In his younger years he participated sporadically in regional ski matches, but otherwise he was never the type of a competitive sportsman. Cycling and (horseback) riding were also much exercised, though rather for the sake of work requirements than as a passion for its own sake. Horses remain his passion to this very day and his wife and himself are running an asylum for old, disabled or discarded horses and other animals (SITAR, 2012b).

Like all great teachers, Dr Kiauta relied more upon *training* than upon *telling*, and sought to make his pupils themselves active parties to the work in which they were engaged, thus making teaching something far higher than the mere passive reception of the scraps and details of knowledge from the classroom lessons learnt by rote. He always wants to see his students surpassing his own level as he once wrote to me, "*A good teacher has to produce some students and/ or young foreign associates, whose work and excellency will surpass his own work. This is the only way to guarantee the development and progress of our science.*" He has a phenomenal memory, too. He remembers almost everything he has ever seen, heard, or read, and never forgets a colleague's or student's name. He is truly a teacher par excellence. Today, his students, research scholars and younger protégé of odonatology are spread all over the world. He is held in high veneration by all his students. For example, Dr Milen Marinov, in an email to me, paid tribute to Dr. Kiauta on the occasion of his 70th birthday, "*In fact he (Dr Bastiaan Kiauta) is the one who supported me in the beginning of my as researches with dragonflies and gave me every important advices and magazines as well.*"

Sherely by dint of his vast odonatological knowledge, not to say in the least about other orders of Insects, and also many other invertebrate groups, topped invariably by his indomitable spirit of extreme humility, great camaraderie and infinite benevolence, Dr Kiauta has influenced and mentored a very large number of contemporary odonatologists as well as budding scientists in other disciplines from all over the world. The world odonatological community have dedicated several species' names (#14) in his honour, namely, *Drepanosticta kiautai* Zhou, 1986; *Elasmothemis kiautai* (De Marmels, 1989) [orig. *Dythemis kiautai*]; *Risiocnemis kiautai* Hämäläinen, 1991; *Paragomphus kiautai* Legrand, 1992; *Macromia kiautai* Zhou, Wang, Shuai & Liu, 1994; *Navicordulia kiautai* Machado & Costa, 1995; *Telebasis bastiaani* Bick & Bick, 1996; *Argia kiautai* Steinmann, 1997 [syn. of *Rhipidolestes okinawanus* Asahina, 1951]; *Rhipidolestes bastiaani* Zhu & Yang, 1998; *Priscagrion kiautai* Zhou & Wilson, 2001; *Austroagrion kiautai* Theischinger & Richards, 2007; *Neoneura kiautai* Machado, 2007; *Teinobasis kiautai* Theischinger & Richards, 2007; *Pericnemis kiautarum* Orr & Hämäläinen, 2013. His community- and dragonfly research work were acknowledged by rendering him various awards and honorary positions, such as, e.g. the Membership of Honour of the Societas Internationalis Odonatologica (Chur, 1981), the Golden (dragonfly) Medal of the Netherlands Dragonfly Society (Zwolle, 2004) and the Membership of Honour of the Slovenian Dragonfly Society (Ljubljana, 2015) (Fig. 19.2). Beatrix, the Queen of the Netherlands, appointed him, with a particular reference to his odonate work, as a Knight in the Order Oranje Nassau (2007), and the voluminous *Festschrift* was published in celebration of his 70th birthday (TYAGI, 2007a), in which 38 authors, from 17 countries collaborated.

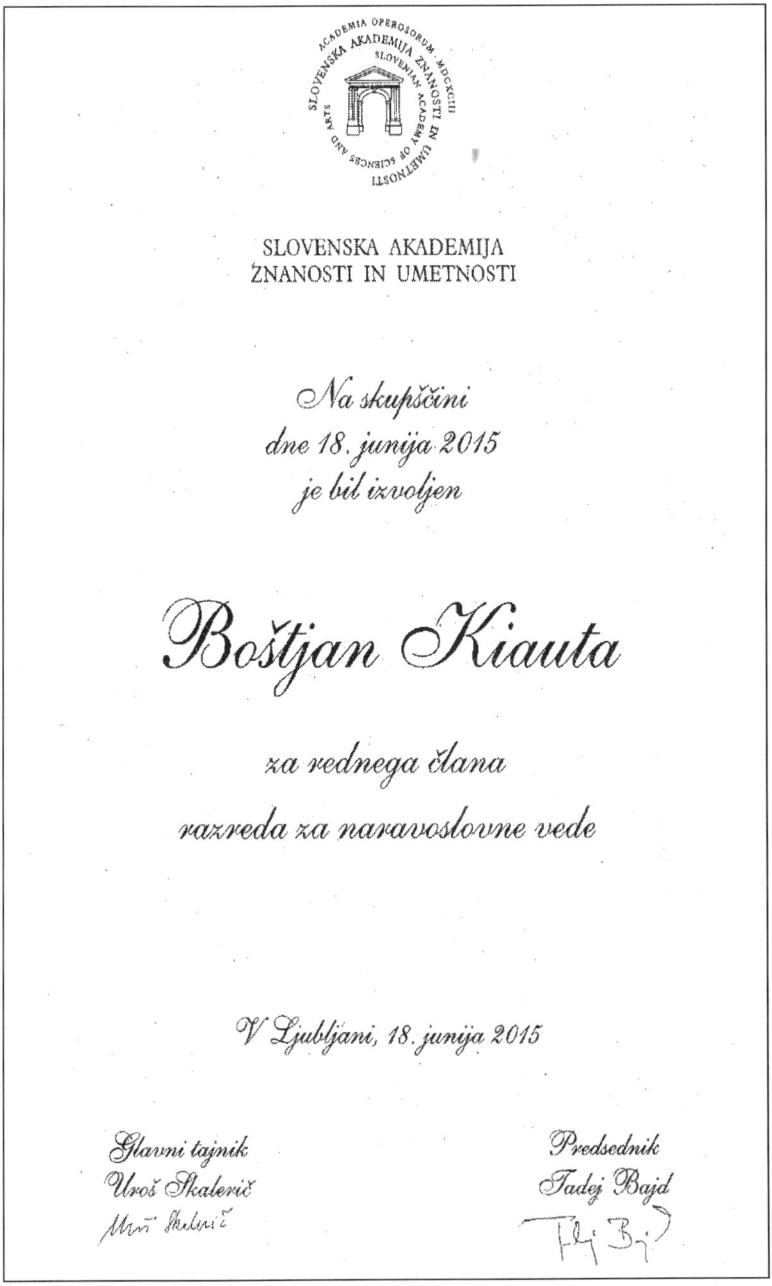

SLOVENSKA AKADEMIJA
ZNANOSTI IN UMETNOSTI

Na skupščini
dne 18. junija 2015
je bil izvoljen

Boštjan Kiauta

za rednega člana
razreda za naravoslovne vede

V Ljubljani, 18. junija 2015

Glavni tajnik
Uroš Skalerič

Predsednik
Tadej Bajd

Fig. 19.2. Document on the B. Kiauta: election as a full member of the Slovenian Academy of Sciences and Arts, Ljubljana; dated 18 June 2015. – (Source: B.Kiauta)

He is a complete teacher: with vast and deep knowledge, sharp and photographic memory, great articulation and prowess to narrate, friendly, affable and mentoring!

A teacher is considered above all and the path to enlightenment passes through him alone!

गुरु गोविन्द दोऊ खड़े, काके लागूं पांय।

बलिहारी गुरु आपने गोविन्द दिया बताय।।

(Transliteration: Guru Govind dou khade, kake lagoon paay, Balihari Guru aapno, Govind diyo batay.)

(Translation: "The Teacher and the God are both standing in front of me, and the question before me is to whom I must go first to take blessings by touching the feet. The decision is difficult but I prefer to first of all approach the teacher because he alone can show the path to enlightenment, the God.")

As thoroughness and accuracy are two principal attributes applied in teaching, Dr Kiauta – during his more than six decades of career as a teacher (professor) and scientist (odonatologist) left no stone unturned in his endeavours to study with a definite aim and object all the related books and literature in every form available (including microfilm, microfiche, xeroxed and cyclostyled documents, internet etc. etc.). During Dr Kiauta's active life, he had of course met hundreds upon hundreds of eminent scientists. At the 16th International Congress of Entomology (Kyoto, Japan) he met the famous cytologist, Professor Dr M.J.D. White and he was in regular touch with the world-wide acknowledged cytologist Professor Dr J.W. Boyes of McGill University (Montreal, Canada), with whom he also has a joint paper (BOYES et al., 1980b). As to the odonatologists, he did have the privilege of some correspondence with the great workers of the past, such as, e.g., Drs P.P. Calvert, F.C. Fraser, F.F. Laidlaw, Erich Schmidt, K.J. Valle and many others. Since the 1960s, he was in touch with almost all contemporary odonatologists, many of whom he also visited in person, met at the Symposia, or they visited with him. In the process he also undertook travels to interact with intellectuals in many countries across the world such as, e.g., Canada (Drs R. & S. Cannings, G. Pritchard), most European countries (meeting very many local workers), India (Drs B.K. Tyagi, S. Mathavan, D.B. Tembhare, T.N. Ananthakrishnan), Japan (Drs S. Asahina, K. Inoue, S. Eda and numerous others), Nepal (Drs M. Mahato, T.C. Majupuria), Russia: Siberia (Drs A.Yu. Haritonov, E.I. Malikova, O.L. Kosterin, S.N. Borisov, O.N. Popova and many others), Taiwan (Dr J.C. Lien), Thailand (Drs A. Pinratana, M. Titayavan), The Philippines (Drs E.B. Akiew, G. Braeckman, M.L. Ligeralde), United States (Drs M.J. Westfall, G. Bick, C. Williams, S. Dunkle and many others).

A huge responsibility like producing an international journal (*Odonatologica*) warrants free accessibility to the fast flowing world literature, from all countries in different languages! Dr Kiauta devised two different modes to fulfil these requirements; first, he began to update his own library, meticulously developed over many years of his enviable bibliophilic and bibliopolic habits, and, secondly, he developed an unfallible network of tens of hundreds of tributaries of communication (in far earlier days of *Odonatologica*, only the letters by surface or air mail; he invariably preferred the latter) largely to update the Odonatological Abstracts section in *Odonatologica.* Soon his library and archive swelled with the current and old and rare literature. I had a chance to look at his library in 2007 when he was still living in Bilthoven, near his *alma mater*, the Utrecht University.

Dr Kiauta's library is a very busy room, occupied by one or another work of odonatological literature, with hardly an inch of space available, either on walls, racks or even floor. In a corner I saw also that famous typewriter from which, before the advent of the computer, he used to send his scores of letters, several times running into several pages and all immaculately typed. Since his library was also his study room, I found his working table full of several books and other literature systematically stacked up near the latest issue of *Odonatologica.* At present, though the entire literature he has is organized in at least 3 places (e.g., his homes in Slovenia, Switzerland and The Netherlands), but to quench his day-to-day requirement of readily available literature particularly during his nearly four-and-a-half decades of regular editing and management of *Odonatologica* and *Notulae odonatologicae*, two of SIO mainstream journals, besides of course very many other assignments on constant basis, he does have a massive library at his home in The Netherlands which amounts to some 300 shelf metres in his archive. At worst, if he could not collect a sample copy of a given literature for any reason, he would make sure that he at least gets an opportunity to go through that, e.g. Thomas Moufet's *Insectorum sive minimorum theatrum* (1634), Johann Swammerdam's *Historia insectorum generalis* (1669), Johann Weichard von Valvasor (1685), Antoni van Leeuwenhoek's *Arcana naturae detecta* (1695), John Ray's *Historia insectorum* (1710), E.L. Geoffroy's *Histoire abrégée des insectes* (1752), and Carl von Linne's 10th edition of *Systema Naturae* (1758). He made an allround study about dragonflies, and once started reading a book he would not leave it without finishing and, in the process, expanded his knowledge in scope and richness. He always appreciates the works of his predecessors, acknowledging very well what the great physicist Issac Newton famously said, "*You can see farther because you stand on the shoulder of a giant.*"

Dr Kiauta picked up every opportunity to study from the works of all the great odonatologists of yore, such as, for example T. de Charpentier (1840), J.C. Fabricius (a disciple of Linnaeus who in 1793, gave dragonflies separate, ordinal status under the name 'Odonata', and earlier in the 1775 edition of his *Systema entomologiae* he had included 30 species of dragonflies assigned to three genera *–Aeshna, Agrion* and *Libellula*, and by 1798 his list of described species of dragonfly stood at 69), Kirby, Drury, Baron Edmond de Selys-Longchamps (1840, 1850), who discovered that wing venation provides a reliable guide to relatedness at several hierarchical levels, and he applied this knowledge with great success to the classification of most families of Odonata, P.L. Vander Linden, who soon in 1825 produced his treatise on the Dragonflies of Europe, H.A. Hagen (1840) etc. It was Selys' great contribution to provide the hierarchical classification that has served odonatologists so well ever since. Selys assembled a magnificent personal collection of dragonflies, and with great perseverance and skill produced a succession of monographs on different subfamilies (now mostly regarded as families), leaving at his death two to be completed by Rene Martin (1906, 1908-1909) and the formidable Libellulinae, completed by F. Ris (1909-1916). Selys exerted a profound influence by placing the taxonomy of the Order on a firm footing. Indeed, almost all subsequent work in odonatology owes something, directly or indirectly, to the system he devised; so it is fitting that the international newsletter of odonatology, started in 1963 by B. E. Montgomery, should be called *Selysia* (*cf.* CORBET, 1999).

The 20th century could be possibly called a golden period since it saw great odonatological activity by experts like Drs Ris, Lucas, Popova, Needham & Westfall, Walker, Santos, Lieftinck, Asahina etc. Dr Kiauta has read about and/or corresponded with them extensively. Dr Kiauta was particularly motivated by Dr M.A. Lieftinck, a Dutch odonatologist who was regarded unequivocally one of the greatest taxonomists ever. He influenced development of odonatology as a science by the inspiration provided for more than half a century and by setting high standards of taxonomic work and giving unstinting help to many odonatologists. Dr Kiauta had the privilege to associate with him for a long time, learn taxonomy the hard way Lieftinck was known for, and considered him his teacher and mentor! With taxonomic expertise on a sound footing, and with attractive regional monographs available to provide the incentive, Dr Kiauta was by then absolutely convinced about Odonata being his first choice for lifelong research.

Before deciding to undertake research on dragonflies, Dr Kiauta however reviewed all other options of his long term interests in butterflies (Lepidoptera), beetles (Coleoptera), wasps, bees and ants (Hymenoptera), bugs and cicadas (Heteroptera and Homoptera) – all insects, but also other arthropodan groups such as spiders. Though, he had fairly good understanding and interests especially in Lepidoptera and Coleoptera, the decision clearly weighed in favour of Odonata, largely owing to three reasons, viz. (i) Dragonflies, as an order have a very special place in the class Insecta, forming along with the mayflies (Ephemeroptera) the Paleoptera, thus having a significant role in phylogeny and evolution, a subject that was extremely close to Bastiaan's heart, (ii) The startling physical and behavioural uniqueness as well as the speed with almost incredible flight control manoeuvres had always attracted him most and he used to look upon them with a feeling of wonderment if these elegant creatures had in their biology the Ariadne's thread to some of the man's own quests embroiled in a labyrinth of a streak of enigmas which remained to be unravelled yet, and (iii) He had, following a rigorous exercise of worldwide reference search, found that dragonflies were by far the least studied group when compared to Lepidoptera, Coleoptera, Hymenoptera and Heteroptera, clearly sending a message in his mind that there was a huge opportunity awaiting to explore these wonderful insects. He had fallen in love with the dragon- and damselflies, and he remained focused to this group, Odonata, throughout his life! Following is what he had made a picture of dragonflies in his mind:

"Dragonflies (Odonata) comprise one of the most important groups of aquatic animals. Being voracious predators in both immature (aquatic) and adult (aerial) stages they are important elements of all, except the drier (or high alpine)

environments in temperate and tropical regions, occupying a position at the apex of the food chain of invertebrate life. In addition to their biological importance in such environments, these relationship render them the ultimate accumulators of persistent compounds in polluted waters. They are the ideal organisms to be used as indicators of water pollution and contamination. Many species serve as intermediate hosts of fluke parasites of birds, and thus are important in the transmission of parasitic diseases, especially of domestic poultry and wild ducks. Because of their unique morphology and physiology, dragonflies are used extensively in the study of many biological phenomena. In insect phylogeny dragonflies also occupy a unique position. Since the Upper Paleozoic they have not changed essentially from the structural point of view and are considered as "living fossils" par excellence. For these obvious reasons dragonflies have also been one of the extensively researched insects."

Although well versed with the taxonomy of many best-known orders of insects, he took keen interest in the taxonomic studies of dragonflies, which is a continuing process, and this branch of odonatology remains an active one, producing descriptions of new taxa, revisions, monographs and catalogues galore. He particularly reviewed certain publications, such as, for example "*A reclassification of the Order Odonata*" by Tillyard & Fraser (1938-1940) and by FRASER (1957), and, in fact, encouraged the catalogues of world genera (numbering 630) by DAVIES (1981) and of world species (numbering 4875) by DAVIES and TOBIN (1984, 1985), and TSUDA (1986). His own "*Annotated catalogue and bibliography of taxa introduced in Odonata from 1971 to 1980*" became a highly useful reference for a serious student of odonate taxonomy (KIAUTA, 1981a) (Fig. 19.3).

Dr Kiauta soon emerged as an odonatologist blessed with so many virtues casting an indelible imprint of greatness and influence on the global platform. His innovative approaches provided the contemporary odonatology with an ever enriching wisdom that became his landmark to redefine the science of odonatology at a global level. He began writing about history of odonatology, and one of the first he compiled authoritatively is that from Switzerland (KIAUTA, 1978b). Soon he came up with a unique compilation of taxa introduced in Odonata from 1971 to 1980 (KIAUTA, 1981a). Even his wife, Marianne, discovered poetry in odonatology (KIAUTA, 1985). He suggested Professor P.S. Corbet, a long-time friend, to compile the history of odonatology and provided all necessary help to complete and publish the seminal assay in *Advances in Odonatology*, (Vol. 5, pages 21-44), bringing on record a massive information on dragonfly research in the past! Dr. Corbet summed up one of the most profitable aspect of Dr Kiauta's life as: "*At his home in The Netherlands, Dr Kiauta maintains a rather small but useful dragonfly collection of about 500-600 species of all continents. All taxa are identified and a few appear new and are still undescribed. He never had much time to work on the collection and he does not have a list of species*."

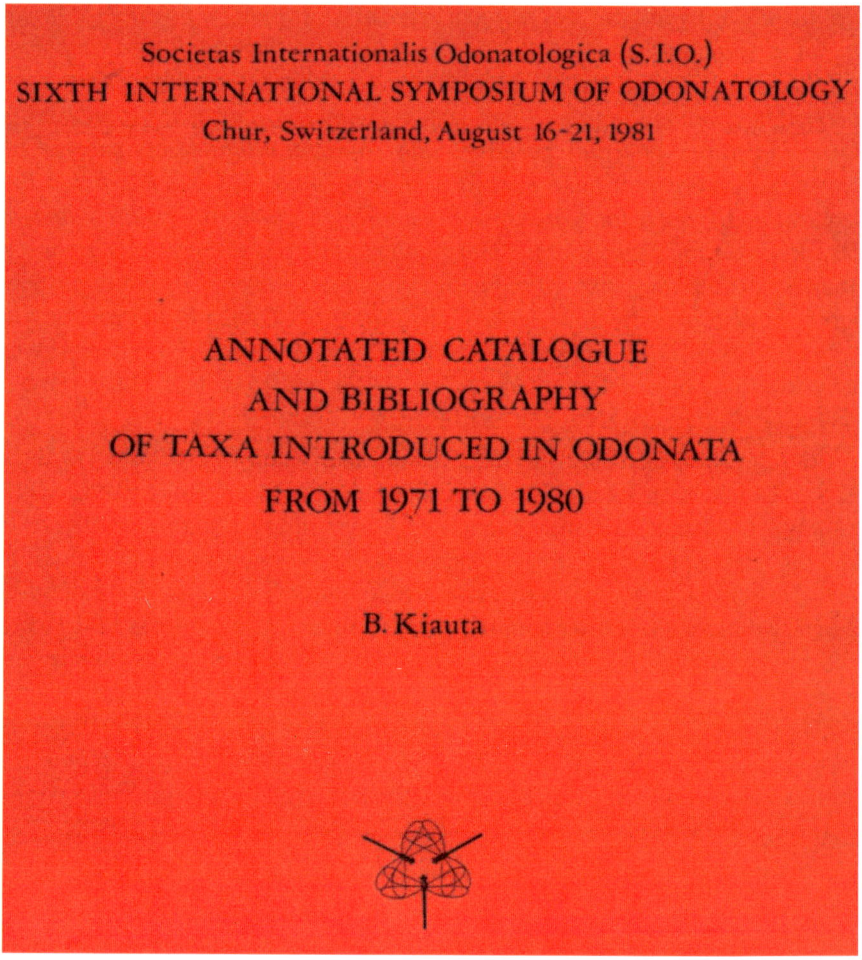

Societas Internationalis Odonatologica (S. I. O.)
SIXTH INTERNATIONAL SYMPOSIUM OF ODONATOLOGY
Chur, Switzerland, August 16-21, 1981

ANNOTATED CATALOGUE
AND BIBLIOGRAPHY
OF TAXA INTRODUCED IN ODONATA
FROM 1971 TO 1980

B. Kiauta

Fig. 19.3. B. Kiauta (1981 a): Annotated catalogue and bibliography of odonate taxa, introduced during 1971-1980. A slim volume, preceding the monumental work of C.A. Bridges (1991, enlarged editions in 1993 and 1994). - (Source: B. Kiauta)

Dr. Kiauta's wife, Marianne, actively participated in almost all his field work. Aside of her technical and other assistance, her skillful management of the logistics was indispensable particularly in the Himalayan, SE Asian and the Philippine missions, where her familiarity with some of the local languages and scripts was often a valuable asset also. Dr Kiauta's phenomenal rise to a great teacher, scientist and philosopher has in fact a clear imprint of inexorable support lent by Mrs Marianne Kiauta.

●●●

20

Societas Internationalis Odonatologica (S.I.O.) and International Union for Conservation of Nature and Natural Resources (I.U.C.N.)

"At the edge of the pond ducks flapping their wings a wave of leaves".
—Marianne Kiauta
(Dedicated to Norman W. Moore,
from the cycle "The first crocus",
Mondena, Grosuplje & Ursus, Bilthoven, 2000;
earlier published in Odonatologica, vol. 32, No. 1)

The International Society of Odonatology (Societas Internationalis Odonatologica, S.I.O.) is committed to research and educate about conservation of both the dragonfly species and their habitats – an objective prominently emphasized in her Constitution and By-Laws. Dragonflies, being voracious predators in both immature (aquatic) and adult (aerial) stages, are important elements of all, except the drier (or high alpine), environments in temperate and tropical regions, occupying a position at the apex of the food chain of invertebrate life. In addition to their biological importance in such environments, these relationships render them the ultimate accumulators of persistent compounds in polluted waters. They are the ideal organisms to be used as indicators of water pollution and contamination. One odonatologist who championed the cause of conservation of dragonflies and their habitats was Professor Dr Norman W. Moore. As the SIO Representative and Chairman of the Odonata Specialist Group of the Species Survival Commission in IUCN, he waged all along his life a fierce battle to allow a proper recognition to dragonflies despite the immense complexity of problems which confronted him in achieving practical conservation action on behalf of large, poorly known invertebrate groups in conservation world, such as Odonata, which could serve as a robust model value for monitoring chastity of the wetlands already on the cross-roads of contamination due to excessive pesticides in agriculture and health sectors (Fig. 20.1). His work led to a ban on pesticides such as DDT, and so helped populations of dragonflies, along with birds of prey, to recover from a near extinction!

Norman Winfrid Moore was born on 24th February, 1923, in London to Mary (nee Burrows) and Sir Alan Moore, a medical officer of health and a baronet. From an early age he was always happiest in the field as a practising naturalist, and the matrix of complexity, elegance and vulnerability of natural systems was the unchanging focus of his life. At prep school in Hampshire the young Moore was more often out in the fields studying insects and birds than in his proper place in class. He was educated at Eton College, and Trinity College, Cambridge. He joined the army in 1943 and trained as a mountain gunner in the Cairngorms. He graduated during World War II, and then served in the Royal Artillery in the last two years of the war, reaching the rank of Lieutenant. He saw action in the Netherlands and Germany, was wounded, and became a prisoner of war. When the Second World War ended, he became an assistant lecturer at Bristol University, where for his PhD he carried out a detailed ecological study of dragonflies, which were a lifelong focus of fascination and research.

Fig. 20.1. Professor Dr Norman Winfrid Moore (Source: Mrs Charles Moore, published in *Odonatologica* 32(1): 1-8.)

From 1960 to 1974 he was Head of the Toxic Chemicals and Wildlife Division at Monks Wood Experimental Station where he studied the effects of toxic chemicals on wildlife, in particular the adverse effect of organochlorine pesticides on raptors. This work led to him developing the "precautionary principle" exemplified by his recommendation that the use of such pesticides should be phased out even though the extent of the harm they caused was not yet fully known. His pioneering work on nature conservation and his pesticide research led to requests for advice from governmental and other scientific organisations in Europe, India, Australia and the United States. It was his work on dragonflies and conservation that led to him coining the term "the birdwatcher's insect", aiming to raise public interest in the role of insect monitoring in ecosystem conservation. Due to his background in dragonfly research and conservation, Dr Moore was invited to chair the Odonata specialist group of the IUCN Species Survival Commission. This international group first met in 1980, and produced a world plan for dragonfly conservation in 1995, which was published in 1997. Recipient of many international awards and honours, he even co-authored a book in the New Naturalist series: *Dragonflies* (1960). His book, "*Oaks, Dragonflies and People*" (2002) charted the creation of a nature reserve and dragonfly pond at his home in Cambridgeshire. The British Dragonfly Society administers an award in Moore's honour, called the 'Norman Moore Award Fund'. In addition to this, several species of dragonflies and damselflies are named after Moore.

Dr Moore was globally regarded as one of the most influential figures in nature conservation over half a century. Having been the first British biologist to spend his entire professional career in research, planning and negotiation directly related to practical nature conservation, he retired in 1983 and became chairman of the Farming and Wildlife Advisory Group, which he had helped to set up 20 years earlier to give environmental advice to farmers. He also continued to chair an IUCN specialist group dealing with the identification, ecology and survival of dragonflies.

The IUCN/Species Survival Commission is committed to communicate important species conservation information to natural resource managers, decision-makers and others whose actions affect the conservation of biodiversity. The SSC's Action Plans, Occasional Papers, news magazine (Species), Membership Directory and other publications are supported by a wide variety of generous donors. The Species Survival Commission (SSC) is one of six volunteer commissions of IUCN – The World Conservation Union, a union of sovereign states, government agencies and nongovernmental organizations. IUCN has three basic conservation objectives: to secure the conservation of nature, and especially of biological diversity, as an essential foundation for the future; to ensure that where the earth's natural resources are used this is done in a wise, equitable and sustainable way; and to guide the development of human communities towards ways of life that are both of good quality and in enduring harmony with other components of the biosphere.

The SSC's mission is to conserve biological diversity by developing and executing programs to save, restore and wisely manage species and their habitats. A volunteer network comprised of nearly 7,000 scientists, field researchers, government officials and conservation leaders from 188 countries, the SSC membership is an unmatched source of information about biological diversity and its conservation. As such, SSC members provide technical and scientific counsel for conservation projects throughout the world and serve as resources to governments, international conventions and conservation organizations. The IUCN/SSC Action Plan series assesses the conservation status of species and their habitats, and specifies conservation priorities. The series is one of the world's most authoritative sources of species conservation information available to nature resource managers, conservationists and government officials around the world.

In early 1980s the SIO had appointed Professor Moore as its permanent representative to the Odonata Specialist Group (IUCN). The Odonata Specialist Group of the IUCN's Species Survival Commission was created in 1980 through the initiative of the late Sir Peter Scott. It meets once every two years on the occasion of the International Symposia of Odonatology. In addition, the members keep in touch by correspondence and through a small journal *"Reports of the Odonata Specialist Group, Species Survival Commission, International Union for Conservation of Nature and Natural Resources (IUCN)"* – edited by the Chairman and published under the aegis of the journal *Odonatologica* (Fig. 20.2).

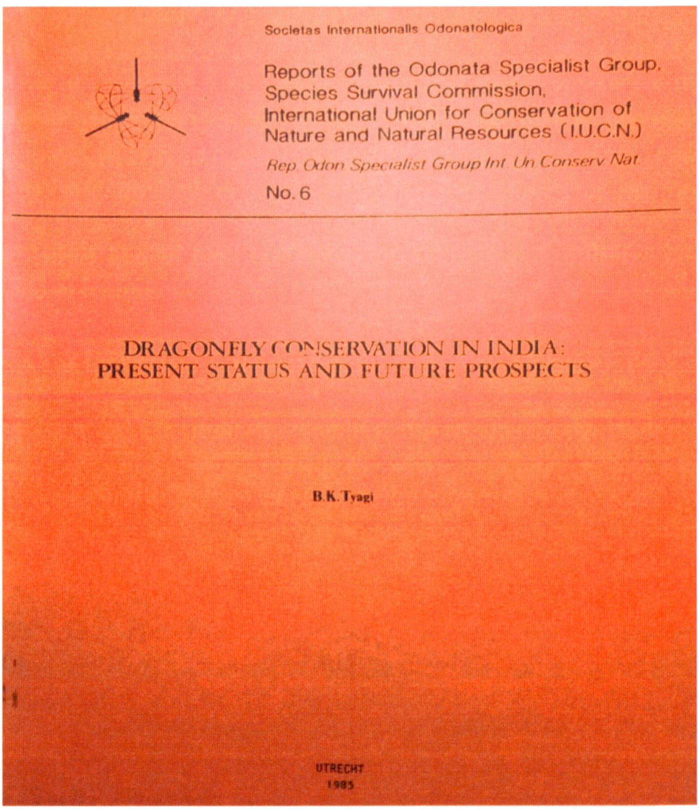

Societas Internationalis Odonatologica

Reports of the Odonata Specialist Group, Species Survival Commission, International Union for Conservation of Nature and Natural Resources (I.U.C.N.)

Rep. Odon. Specialist Group Int. Un. Conserv. Nat.

No. 6

DRAGONFLY CONSERVATION IN INDIA:
PRESENT STATUS AND FUTURE PROSPECTS

B.K. Tyagi

UTRECHT
1985

Fig. 20.2. Facsimile of title page of a issue of the SIO Odonata Specialist Group (SSC/IUCN). – (Source: B.K. Tyagi)

Although for most part of action in his life, Dr Moore – a gentle giant in the field of wildlife and conservation policies – remained a lone crusader in his fight to save dragonflies and their ecosystems, nevertheless, his humility and sense of respect for fellow colleagues was immense, in true spirit of a *Japanese haiku.*

"An ear of rice is bent because a
Dragonfly is perched upon it."

Fig. 20.3. The first report on conservation of Odonata prepared by Prof. N. Moore for the IUCN/ SSC Odonata Specialist Group.– (Source: N.W. Moore)

He was a team leader in every sense. While submitting his report to the IUCN/ SSC Odonata Specialist Group on the "Status Survey and Conservation Action Plan: Dragonflies" (MOORE, 1997), he generously commented with admiration "the huge support and help provided by a large number dragonfly experts through correspondence and at the meetings of the Group. Other odonatologists who are not members of the Group, but who regularly attend its meetings, have also been most helpful. Thus this action plan owes much to numerous people, but particularly to past and present members of the Odonata Specialist Group – S. Asahina, H. Bick, H.J. Dumont, S.W. Dunkle, J.I.R. Furtado, M. Hämäläinen, B. Kiauta, A.B.M. Machado, M.N. Parr, D.R. Paulson, R. Rowe, M.J. Samways, E.G. Schmidt, Z. Spuris, B.K. Tyagi, J. van Tol and J.A.L. Watson." (Fig. 20.3).

Propelled by the SIO Charter to work in the field of dragonfly conservation, Prof. Moore had updated the endangered species worldwide in this Report (see Table 20.1 below; *Source:* Moore, 1997):

Table 20.1. Dragonflies in the 1996 Red List of Threatened Animals listed under countries in which they occur.

Country	Species
Angola	*Aciagrion rarum* (Coen) DD
	Monardithemis flava (Lib) VU(B1, 2c)
Armenia	*Calopteryx syriaca* (Cal) EN(A2c)
	Onychogomphus assimilis (Gom) VU(B1,2c)
	Cordulegaster mzymtae (Clg) VU(B1,2c)
Australia	*Hemiphlebia mirabilis* (Hem) VU(B1,2c)*
	Acanthaeshna victoria (Ae) VU(B1,2c)*
	Petalura pulcherrima (Pet) EN(B1,2c)*
	Austrocordulia leonardi (Cord) CR(B1,2c)
Azerbaijan	*Onychogomphus assimilis* (Gom) VU(B1,2c)
	Cordulegaster mzymtae (Clg) VU(B1,2c)
Bangladesh	*Lyriothemis tricolor* (Lib) EN(B1,2c)
Belize	*Amphipteryx agrioides* (Amph) EN(A1 c)*
Brazil	*Mecistogaster asticta* (Pst) VU(A1c)
	Mecistogaster pronoti (Pst) CR(A1c)
Cameroon	*Argiocnemis umbargae* (Coen) EN(B1,2c)
	Enallagma camerunense (Coen) EN(B1,2c)
Canada	*Somatochlora brevicincta* (Cord) VU(D2)
China	*Macromia urania* (Cord) EN(B1,2c)
	Libellula angelina (Lib) CR(A1,2c)
Colombia	*Amphipteryx agrioides* (Amph) EN(A1c)*
Congo	*Aethiothemis watuliki* (Lib) DD
Costa Rica	*Thaumatoneura inopinata* (Meg) EN(A1,2c)*
	Palaemnema chiriquita (Plst) EN(A1,2c)
	Palaemnema gigantula (Plst) EN(A1,2c)
	Palaemnema melanota (Plst) EN(A1,2c)
	Palaemnema reventazoni (Plst) EN(A1,2c)
	Epigomphus camelus (Gom) EN(A1,2c)
	Epigomphus subsimilis (Gom) EN(A1,2c)
	Epigomphus verticornis (Gom) EN(A1,2c)
	Diaphlebia pallidistylus (Gom) EN(A1,2c)
Cuba	*Hypolestes trinitatis* (Meg) EN(A1,2c)*
Democratic Republic of the Congo	*Pseudagrion quadrioculatum* (Coen) DD
	Anectothemis apicalis (Lib) DD
	Congothemis longistyla (Lib) DD
Dominican Republic	*Hypolestes clara* (Meg) EN(A1c)*
Egypt	*Paragomphus sinaiticus* (Gom) VU(A2c)
	Calopteryx syriaca (Cal)
El Salvador	*Amphipteryx agrioides* (Amph) EN(A1 c)*
Equatorial Guinea	*Cornigomphus guineensis* (Gom) EN(A1 c)*
	Trithemis hartwigi (Lib) EN(A1c)
Ethiopia	*Urothemis thomasi* (Lib) DD
France	*Gomphus graslini* (Gom) VU(B1,2c)
	Macromia splendens (Cord) VU(B1,2c)
Gabon	*Palpopleura albifrons* (Lib) CR(A1c)
Georgia	*Onychogomphus assimilis* (Gom) VU(B1,2c)
	Cordulegaster mzymtae (Clg) VU(B1,2c)
Guatemala	*Amphipteryx agrioides* (Amph) EN(A1 c)*
	Hetaerina rudis (Cal) EN(B1,2c)
	Heteragrion eboratum (Meg) EN(A1c)
	Heteragrion tricellulare (Meg) EN(A1 c)
	Epiogomphus clavatus (Gom) EN(A1 c)
	Progomphus risi (Gom) EN(A1 c)
Guinea-Bissau	*Brachythemis liberiensis* (Lib) CR(A1 c)
Haiti	*Phylolestesethelae* (Syn) VU(A1c)
	Hypolestes trinitatis (Meg) EN(A1c)*
	Hypolestes clara (Meg) EN(A1 c)*
Hawaii (USA)	*Megalagrion adytum* (Coen) VU(B1,2c)
	Megalagrion leptodemas (Coen) EN(B1,2c)
	Megalagrion nigrohamatum (Coen) LR
	Megalagrion nigrolineatum (Coen) VU(B1,2c)
	Megalagrion oahuense (Coen) VU(B1,2c)
	Megalagrion oceanicum (Coen) VU(B1,2c)
	Megalagrion pacificum (Coen) VU(B1,2c)
	Megalagrion xanthomelas (Coen) VU(B1,2c)
Honduras	*Amphipteryx agriodes* (Amph) EN(A1 c)*
	Heteragrion eboratum (Meg) EN(A1c)
India	*Epiophlebia laidlawi* (Ep) VU(B1,2c)*
	Cephalaeschna acutifrons (Ae) VU(B1,2c)
	Burmagomphus sivalikensis (Gom) CR(A1c)
	Lyriothemis tricolor (Lib) EN(B1,2c)
Iran	*Brachythemis fuscopalliata* (Lib) VU(B1,2c)
	Onychogomphus macrodon (Gom) EN(B1,2c)
Iraq	*Brachythemis fuscopalliata* (Lib) VU(B1,2c)
Israel	*Onychogomphus macrodon* (Gom) EN(B1,2c)
	Brachythemis fuscopalliata (Lib) VU(B1,2c)
Jamaica	*Hypolestes clara* (Meg) EN(A1c)*
Japan	*Rhinocypha ogasawarensis* (Chl) CR(B1,2c)
	Rhinocypha uenoi (Chl) EN(B1,2c)
	Indolestes boninensis (Lest) CR(B1,2c)
	Rhipidolestes okinawanus (Meg) EN(B1,2c)*
	Boninagrion ezoin (Coen) CR(B1,2c)*
	Mortonagrion hirosei (Coen) EN(B1,2c)
	Oligoaeschna kunigamiensis (Ae) ENM(B1,2c)
	Asiagomphus yayeyamensis (Gom) EN(B1,2c)
	Leptogomphusyayeyamensis (Gom) EN(B1,2c)
	Chlorogomphus iriomotensis (Clg) EN(B1,2c)
	Hemicordulia ogasawarensis (Cord) EN(B1,2c)
	Hemicordulia okinawensis (Cord) EN(B1,2c)
	Macromidia ishidai (Cord) EN(B1,2c)
	Macromia kubokaiya (Cord) EN(B1,2c)
	Macromia urania (Cord) EN(B1,2c)
	Boninthemis insularis (Lib) CR(B1,2c)*
	Libellula angelina (Lib) CR(A1,2c)
	Lyriothemis tricolor (Lib) EN(B1,2c)
Jordan	*Calopteryx syriaca* (Cal) EN(A2c)
	Onychogomphus macrodon (Gom) EN(B1,2c)
Lebanon	*Onychogomphus macrodon* (Gom) EN(B1,2c)
Liberia	*Brachythemis liberiensis* (Lib) CR(A1c)
Korea	*Libellula angelina* (Lib) CR(A1,2c)
Madagascar	*Isomma hieroglyphicum* (Gom) VU(A1c)*
	Libellulosoma minuta (Cord) VU(B1,2c)*
Mauritius	*Platycnemis mauriciana* (Plcn) CR(B1,2c)
Mexico	*Amphiperyx agrioides* (Amph) EN(A1 c)*
	Hetaerina rudis (Cal) EN(B1,2c)
	Heteragrion tricellulare (Meg) EN(A1 c)

Country	Species	Country	Species
	Paraphlebia zoe (Meg) EN(A1c)	Tanzania	*Amanipodagrion gilliesi* (Meg) EN(B1,2c)
	Palaemnema paulicoba (Plst) EN(A1c)		*Aeshna meruensis* (Ae) EN(B1,2c)
	Aeshna williamsoniana(Ae) EN(A1c)	Taiwan	*Macromia urania* (Cord) EN(B1,2c)
	Epigomphus crepidus (Gom) EN(A1 c)		*Lyriothemis tricolor* (Lib) EN(B1,2c)
	Epigomphus paulsoni (Gom) EN(A1c)	Turkey	*Calopteryx syriaca* (Cal EN(A2c)
	Progomphus risi (Gom) EN(A1 c)		*Onychogomphus assimilis* (Gom) VU(B1,2c)
Mozambique	*Ceriagrion mourae* (Coen) DD		*Onychogomphus macrodon* (Gom) EN(B1,2c)
Myanmar	*Cephalaeschna acutifrons* (Ae) VU(B1,2c)		*Cordulegaster mzymtae* (Clg) VU(B1,2c)
	Lyriothemis tricolor (Lib) EN(B1,2c)		*Brachythemis fuscopalliata* (Lib) VU(B1,2c)
Nepal	*Epiophlebia laidlawi* (Ep) VU(B1,2c)*	USA	*Calopteryx angustipennis* (Cal) LR
	Cephalaeschna acutifrons (Ae) VU(B1,2c)		*Nehalennia pallidula* (Coen) VU(B1,2c)
New Zealand	*Antipodochlora braueri* (Cord) LR		*Enallagma recurvatum* (Coen) VU(A2c)
Niger	*Paragomphus sinaiticus* (Gom) VU(A1 c)		*Ischnura gemina* (Coen) N(A2e)
Oman	*Paragomphus sinaiticus* (Gom) VU(A2c)		*Aeshna persephone* (Ae) VU(B1,2c)
	Urothemis thomasi (Lib) DD		*Erpetogomphus lampropeltis* (Gom) EN(B1,2c)
Pakistan	*Burmagomphus sivalikensis* (Gom) CR(A1 c)		*Gomphurus consanguis* (Gom) VU(B1,2c)
Panama	*Thaumatoneura inopinata* (Meg) EN(A1,2c)*		*Gomphurus lynnae* (Gom) EN(B1,2c)
	Aeshna williamsoniana(Ae) EN(A1c)		*Gomphurus modestus* (Gom) LR
Philippines	*Rhinocypha hageni* (Chl) EN(B1,2c)		*Gomphus diminutus* (Gom) LR
	Rhinocypha latimaculata (Chl) EN(B1,2c)		*Gomphus hodgesi* (Gom) LR
Portugal	*Gomphus graslini* (Gom) VU(B1,2c)		*Gomphus sandrius* (Gom) EN(A1 c)
	Macromia splendens (Cord) VU(B1,2c)		*Gomphus septima* (Gom) (Gom) VU(B1,2c)
Principe			*Hylogomphus geminatus* (Gom) LR
(Sao Tome)	*Trithemis nigra* (Lib) CR(B1,2c)		*Hylogomphus parvidens* (Gom) LR
Rodrigues			*Ophiogomphus acuminatus* (Gom) VU(A2c)
(Mauritius)	*Argiocnemis solitaria* (Coen) CR(B1,2c)		*Ophiogomphus anomalus* (Gom) LR
Saudi Arabia	*Paragomphus sinaiticus* (Gom) VU(A2c)		*Ophiogomphus edmundo* (Gom) CR(D2)
	Urothemis thomasi (Lib) DD		*Ophiogomphus howei* (Gom) VU(A2c)
Sierra Leone	*Argiagrion leoninum* (Coen) EN(B1,2c)		*Ophiogomphus incurvatus* (Gom) LR
	Allorhizucha campioni (Lib) EN(A1c)		*Progomphus bellei* (Gom) VU(A2c)
Somalia	*Urothemis thomasi* (Lib) DD		*Stylurus potulentus* (Gom) VU(A2c)
South Africa	*Chlorolestes apricans* (Syn) CR(A1a)		*Stylurus townesi* (Gom) VU(A2c)
	Chlorolestes draconicus (Syn) LR		*Zoraena sayi* (Clg) VU(B1,2c)
	Ecchlorolestes nylephtha (Syn) LR		*Macromia margarita* (Cord) LR
	Ecchlorolestes peringueyi (Syn) VU(B1,2c)		*Somatochlora calverti* (Cord) LR
	Enallagma polychromaticum (Coen) DD		*Somatochlora hineana* (Cord) EN(B1,2c)
	Metacnemis angusta (Plcn) DD		*Somatochlora margarita* (Cord) VU(D2)
	Orthetrum rubens (Lib) DD		*Somatochlora incurvata* (Cord) LR
	Urothemis luciana (Lib) DD		*Somatochlora ozarkensis* (Cord) LR
Spain	*Gomphus graslini* (Gom) VU(B1,2c)		*Williamsonia lintneri* (Cord) VU(A2c)
	Macromia splendens (Cord) VU(B1,2c)		*Libellulajesseana* (Lib) VU(A2c)
Sudan	*Paragomphus sinaiticus* (Gom) VU(A2c)	Vietnam	*Macromia urania* (Cord) EN(B1,2c)
Syria	*Calopteryx syriaca* (Cal) EN(A2c)	Zambia	*Monardithemis flava* (Lib) VU(B1,2c)
	Brachythemis fuscopalliata (Lib) VU(B1,2c)		
	Onychogomphus macrodon (Gom) EN(B1,2c)		

Note:
Each species name is followed by the abbreviation of its family (see Table 2.1), its threat category (DD = data deficient, EN = endangered, LR = lower risk - near threatened, VU = vulnerable) and the criteria used in selecting its threat category (see IUCN 1994 and 1996).
Note that several species occur in more than one country.
* = Priority Species including those of monotypic genera confined to one country.

Some country entries have been added since the publication of the 1996 Red List. Three European species *Coenagrion mercuriale* (Coen) VU (A2c); *Oxygastra curtisii* (Cord) VU(A1 c); and *Aeshna viridis* (Ae) LR occur in many countries and have been omitted to save space. Their distribution and the threats to them are sufficiently well known for confidence to be placed in their present categorisation.

As to the conservation scenario of the Indian taxa, it was foremost at the insistence of MITRA (1982) that *Epiophlebia laidlawi* Tillyard, the first and only anisozygopterous odonate from India's Darjeeling Himalaya, was included in the revised list of the animals in Schedule One of the Wildlife (Protection) Act 1972 of the Government of India. Until then *E. laidlawi* was the only Indian dragonfly declared protected constitutionally. Next taxon to get attention for protection was the endemic *Burmagomphus siwalikensis* Laidlaw in the Siwalik hill ranges bracing the Doon Valley in the foothill Himalaya in the Uttara Khand State, India. The Doon or the Dehradun Valley in the northern Indian State, gifted divinely with thick forest cover all around, several rivers and rivulets traversing across in a network, and scores of cascades etc., is one of the famous strongholds for dragonflies, earning it commensurately the sobriquet "Dragonfly Valley" – the abode of *B. siwalikensis*, *Agriocnemis corbeti* Kumar & Prasad, 1978 and *Calicnemia doonensis* Sangal & Tyagi, 1984. However, as in case of many fast-developing towns and cities, the Doon Valley, with Dehradun subsequently becoming the capital of the newly carved out state of Uttara Khand, too lost its sheen due largely to unplanned extensive deforestation, on one hand, and establishment of a large number of quarrymines, lime kilns and mushrooming construction activities. This resulted in a great deal of devastation of the ecosystem the Doon Valley was known for, and virtually converted in the "living graveyard" for the above dragonflies. Due to an interruption of the active citizens of the Doon Valley the matter was taken to the court who gave verdict by early 1985 by banning denudation of forests, mountain quarry-mining and lime kilns with immediate effect. But the damage had occurred by that time and *B. siwalikensis,* in particular, had suffered an irreparable loss of its habitat in the Sulphur Springs region. Last time I encountered *B. siwalikensis* was in the Robber's Cave region along the Tons river. Encouraged by both Dr Kiauta and Prof. Moore I reported my observations in the SIO Reports of the Odonata Specialist Group (SSC/IUCN) (TYAGI, 1985).

Professor Moore had a very special bond with Dr Kiauta who, upon his turning 80th on 24th February, 2003, celebrated by publishing an appreciation note in his honour in *Odonatologica*, vol. 32(1): 1-8 (KIAUTA & INOUE, 2016a), and lauded Prof. Moore in the following most saluting manner:

"Dear Norman,

No doubt, if they could, dragonflies would enthusiastically celebrate today's 80th birthday of the man who spent more than six decades of his life studying their habits and, above all, promoting the protection of their habitats, in a life devoted to conservation. They are perhaps not aware of the way many of their homes would look, and of the grave destiny many of their kin would have suffered – if it were not for your conservation initiatives throughout the world. Therefore, on behalf of dragonflies and on our own behalf, the world community of dragonfly students presents you with this issue of Odonatologica, as a token of thanks for everything you have done, and are still doing, for the protection and conservation of their world.

Your attention to morphological detail, as apparent from your early publications (1939, 1942), would seem to indicate a young potential taxonomist. But your life and interests took a different turn – and your research on odonate territoriality (1952a) was one of the highlights both of your studies and of the science of odonatology of that time. This is an absolutely classical publication, and a firm basis for most work in dragonfly ethology and population biology ever since. Various aspects of population biology have had your continuous attention. For the duration and precision of systematic recording, your studies in Woodwalton Fen (1991a, 2001a) will certainly remain unmatched in odonatological literature for a long time to come; the two publications represent an important reference for studies of community succession in the Northern Hemisphere.

In 1938 you started recording British dragonflies. Your professional work with the Nature Conservancy Council took you to various parts of the country, where unprecedented changes in agricultural practices have caused the loss of aquatic habitats, while water supply schemes and increased gravel and clay extraction have produced many new water bodies. Some man-made ("dragonfly") ponds have also been constructed in some nature reserves. These developments have triggered appreciable changes in the odonate fauna and in the status of some species. In your 1976 publication you reviewed the position of British dragonfly conservation and provided suggestions for the necessary policies, management and research in this field. Your Swavesey Dragonfly Reserve served as a model for the planting programme at Ashton Water. The success of the Sanctuary, largely based on your advice, led to the foundation of the National Dragonfly Museum, as R. Mackenzie Dodds recalls elsewhere in this issue.

Your initiatives in dragonfly conservation have not been confined to the United Kingdom; they have encompassed the entire globe. In 1980, at Kyoto in Japan, you founded, within the framework of the IUCN Species Survival Commission, the Odonata Specialist Group (OSG). For two decades, under your inspirational chairmanship, the OSG performed unprecedented conservation work on a global scale. Some of its achievements and suggestions, as discussed at the biennial OSG meetings, are summarized in the eleven issues published to date of the Reports of the Odonata Specialist Group, Species Survival Commission, IUCN – a periodical you founded in 1982, and have edited ever since. The highlight of the OSG achievements is your world action plan for Odonata conservation (1993a, 1997a), an outline of a realistic strategy, with the objective of integrating global dragonfly conservation with that of other organisms and habitats. The book is an indispensable reference work for dragonfly conservation in any part of the world.

In May 1987 you have visited the Japanese 'Dragonfly Kingdom ' in Nakamura, established in 1985. The encouragement and technical suggestions you have rendered on that occasion were of decisive importance for the development of this world-famous dragonfly sanctuary.

As a result of your research, conservation initiatives, organisational achievements and publications, you have created a lasting monument. Even more precious are the living jewels that are to perpetuate your name, as from this very day: Drepanosticta moorei van Tol & R.A. Midler, *Neoneura moorei* Machado *and Notogomphus moorei* Vick, *are flying, respectively, in the Philippines, in Brazil and in Africa.*

On behalf of the world odonatological community and on behalf of the dragonflies, a big THANK YOU, and many happy returns!"

Environmental philosophers have debated at length the facets of conserving wild nature through utilitarian value and intrinsic value to the significance of literally feeling the pulse of nature. Norman Moore has straddled all these facets. It has come deep from within his being, and is a rare glimpse of humanity at its best. Professor Moore passed away 21st October, 2015 at a mature age of 92 years, whereupon KIAUTA (2016) respectfully and befittingly dedicated an obituary published in *Odonatologica*, vol. 45(1/2).

The IUCN/SSC Dragonfly Specialist Group focuses on the conservation of damselflies, dragonflies and their freshwater habitats. The DSG has a global volunteer network of over 50 expert members across the globe. Main focus of the work is the gathering and dissemination of information on distribution and threats of the nearly 6.000 damselflies and dragonflies. Currently the DSG is working on distribution databases and an assessment of the threat status of all damselflies and dragonflies. The Global Dragonfly Assessment will be an important tool in the conservation of freshwater habitats and the SIO's voice in the SSC/IUCN will be resonated time and again, just as the love for dragonflies is hummed by naturalist odonatologist, Mr Pankaj Koparde, in '*The Dragonfly*":

Today I saw a dragonfly
just an insect soaring high
and also a damselfly in grasses
hidin n' blendin in masses

I felt pity looking at them
mere insects with no names
cleaning their heads while perching
the big eyes were really searching

The nameless insect flew
caught a poor fly and chewed
swift and accurate was its fly
like the evolution's marvel fly

I can't fly but sort of try
I dream of being a dragonfly...
I see a dragonfly with my name
and I don't see its lame!

●●●

21

Global Mentorship

Sherely by dint of his vast odonatological knowledge, not to say in the least about other orders of insects, and also many other invertebrate groups, topped invariably by his indomitable spirit of extreme humility, great camaraderie and infinite benevolence, Dr Kiauta has influenced and mentored a large number of odonatologists in all continents of the world, particularly young and budding scientists in the science of odonatology. Owing to his prolific linguistic prowess, inexhaustible energy and rare insight into the science of dragonflies, Dr Kiauta emerged much earlier in his professional career as a champion of odonatological researchers across the globe. Consequently, when a group of celebrated scientists conceived the idea of founding an international society of odonatology, the Societas Internationalis Odonatologica (SIO), for the advancement of both the dragonfly knowledge and the researchers, he was naturally voted unanimously for the most challenging job of the Executive Editor of *Odonatologica*, the society's newly commenced quarterly journal, in addition to much enviable responsibilities of the Coordinator for organizing the SIO's biennial international odonatological symposia and Chief Executive of nearly half a dozen SIO National Offices in various different parts of the world – all at a young age of 32 years! Later on, he added up to these, the editorial and publication responsibilities of equally significant *Notulae odonatologicae* (semi-annual), *Advances in Odonatology* (biennial), and the time-specific *SIO Rapid Communications* and Odonata Specialist Group series of the Species Survival Commission under the International Union for Conservation of Nature & Natural Resources, IUCN, with which the SIO collaborated on conservation issues.

Adherence to deadlines of publication dates of so many international periodicals, particularly *Odonatologica* and *Notulae*, was never an easy task but Dr Kiauta, through his highly disciplined life, has made it possible largely due to his habits of honouring the inexorable principles of time and accomplishing all commitments once made. A resolute will to stick to a work timeline is of paramount importance not only for the performance of difficult duties, but in order to go promptly, energetically, and with self-esteem, through the thousand difficult things which come in almost everybody's way. His sense of valuing time is phenomenal and forms an institution by itself. He has been, in fact, never late in meeting his commitments, come what may! I quote here two examples where I remained in the scene to witness this one of the greatest virtues of a scientist.

First, it occurred during the 10th SIO Symposium held in Johnson City, Tennessee State, USA, where to resolve some serious management matters Professor Kiauta and other senior members of the SIO Council such as, e.g., Drs Westfall, Corbet, Johnson, Pritchard, van Brink, Miller, Moore, Dunkle, Tyagi etc. had continued through late night hours discussing on a new SIO Council and the Executive Committee members. After past mid-night while most of the members had retired to their rooms, he and I remained behind still picking up strands of many isolated activities of the SIO National Office in India. I remember it was well past 2 am when we had to conclude in hurry our discourse per force by remembering that he had also got to be ready for delivering in the next few hours his 'Presidential address' to the participants in the symposium hall sharp at 8 am. That morning the SIO symposium did begin at 8 am sharp, with Dr Kiauta very much there on the podium to deliver the 'all important' speech in his inimitable style, smiling as ever, and without a trace of fatigue or exhaustion! The lesson I learnt from this episode is that doing our duty to accomplish all undertakings involves an immense responsibility. When this sense of responsibility is blunted, the entire life goes to ruin. On the contrary, the sense of duty smooths our path through life. Fulfilling a commitment is among the most priceless of all traits of human life. I have followed this principle all my life – thanks to Dr Kiauta's immortal teachings.

The second instance took place during the 8th SIO Symposium, held in Paris in 1985. During a lengthy discussion between myself and Dr Kiauta, I casually revealed that, after the symposium, I would be returning to India with a brief stopover in Geneva. To this Dr Kiauta, who has a home in Flums (Switzerland), generously invited me to pay a visit there, since he had also planned to spend some days there following the Paris symposium. I was indeed very pleased with this invitation and responded in affirmative. He guided me about the date and approximate time of arrival of the train on a particular station where they (both Dr & Mrs Kiauta) would be waiting to receive me. I remember that he had cautioned me to adhere to the date and time since they would drive a long way to come to receive me on that station. I, of course, nodded in conformity. Alas! In spite of my best intention to travel down to meet with them, I couldn't make it, even though I had wandered unproductively at the Geneva Central Station, under heavy repair work at the moment, in search of ticket windows so that I could buy a ticket and proceed to board a train to Flums!! I, after spending a couple of hours at the Station, returned to my hotel in a state of absolute chagrin and disappointment, overburdened further by the thought of Dr and Mrs Kiauta having had the great trouble of a long road drive and then fruitless waiting at the railway station etc.!! I was truly shattered by these thoughts.

Fig. 21.1. Dr Kiauta and Mrs Marianne Kiauta, just after inviting Dr Tyagi to their home in Flums, Switzerland, after the Paris symposium. – (Source: B.K. Tyagi)

A few days later when I arrived at my residence in Pondicherry (now Puduchery), I saw an air-mailed letter from Dr Kiauta waiting for me. I thought it would certainly contain a message of rebuke on my indolent act and insincere behaviour. I, somehow mustering in enough courage, opened the letter to read. The letter contained the biggest and most pleasant surprise; in the letter Dr Kiauta has only greatly thanked me and praised my active participation in the Paris Symposium and various other SIO meetings including that of the IUCN/Species Survival Commission's Odonata Specialist Group of which I was the SIO Member Representative in South Asia. The letter, as characteristic of Dr Kiauta, mainly concentrated on the SIO activities and future development, but somewhere also indicated, marginally though, that they (Dr and Mrs Kiauta) had come to the Railway Station on the date and time fixed to pick me up, as arranged, and had been waiting train after train for several hours for me to arrive, but in vain!! Obviously, after reading that only sentence, I had my eyes full of tears, and even cried feeling a guilt in my heart of not standing by my promise to Dr Kiauta and that they had had to endure so much pain because of me only!

This pain and anguish kept on boiling within my heart for ever, until at least in 2007, when I had an opportunity to travel to Amsterdam for attending the 5th European Congress of Tropical Medicine & International Health and, for obvious reasons I called on Dr Kiauta to seek an appointment at his home in Bilthoven. He was extremely happy to hear my voice on the phone I made from Amsterdam, and quickly guided me as to how I should reach Bilthoven railway station. I, of course, had the extra bonus *en route* to Bilthoven, of having passed by the Utrecht railway station. I must have changed trains at least twice before reaching Bilthoven whereupon, to my greatest delight, I glanced that both Dr and Mrs Kiauta were already waiting to receive me, with a smile. I touched their feet and sought blessings. It was a good long drive to their home where my stay was extremely memorable, and would always remain imprinted on my mind for the rest of my life for their love, warmth, and hospitality, filled with a parents-like care.

In that brief, a day-long stay at their home, I had learnt many valuable things associated with the life of a scientist. Besides having had the unique pleasure of discussing with Professor Kiauta on varied odonatological subjects, I had a chance to also peep into his library. In his welcome room I saw a huge visitor's book where ultimately I had penned down the essence of my visit to the Kiautas. I found in the guest book entries by many prominent odonatologists from different, sometimes very far-off overseas, countries. I also saw a very voluminous telephone directory near the landline in the room but what perplexed me was that there was also kept a big-handled handlens to assist in viewing the numbers. Then, for the first time, I knew that Dr Kiauta was a having a problem with the sight, particularly in one of the eyes. In my mind I wondered how then the professor was still able to produce those four volumes of *Odonatologica* in a year, absolutely flawless and on time! I was learning with a sense of duty.

Soon in the later part of the morning, Mrs. Kiauta prepared breakfast for us, which among many delicious cookies contained also roasted groundnuts. She noted my inclination towards nuts and instantly replenished the bowl with double the quantum of the roasted groundnuts. I remember having acknowledged by saying, "Groundnuts are my weakness". Pat came the reply from Mrs. Kiauta, "A pleasant and healthy weakness!" Great people have always a great sense of humour.

At the stroke of 1 pm, Dr Kiauta inquired if they were going out to rejoice the lunch? And, in few minutes we all were leaving for a wonderful Chinese restaurant nearly 10 km away. On way two things happened which attracted my attention. First, Dr Kiauta informed that Mrs Kiauta would drive the car since 'he knows nothing but dragonflies.' Oh God, I thought, only this kind of focused research on Odonata could have made him one of the greatest odonatologists ever! I learnt a lesson on the importance of 'focused' research which alone can bear desired fruits of novelties, discoveries, innovations and inventions. Secondly, while driving, we once came across a big water body, which he informed is a part of the lake and that he daily comes there for swimming. Nearby the lake a few horses were also seen grazing in the grassland about which he had told that he very much likes horse-riding. They are both good exercises and energise him much to work long. I quickly understood the reason behind his inexhaustible energy enabling him daily to work for over 16 hours a day, producing in course a quarterly, 100 pages thick, issue of the journal, *Odonatologica* (with tens of odonatological abstracts), a half-yearly bulletin, *Notulae odonatologicae*, management of about a dozen of SIO National/Regional Offices across the globe, coordination for organizing a biennial SIO International Symposium of Odonatology, each time in a different country worldwide, dozens of letters a day, and many more other matters. I learnt a lesson in perseverance and application, with the power of positive thinking.

Eating at the Chinese restaurant was a fabulous experience; first we chose our items, which were then roasted or fried by the cooks, and then we served ourselves on a good big table in the cosiest corner of the restaurant. I did not stop after first round of eating and wished to go for the second which Mrs Kiauta graciously joined to help me select the specific food items, since I am a staunch vegetarian. On my completing the lunch, I thanked Dr and Mrs Kiauta for their grace to host such a highly sumptuous lunch, adding hesitatingly on being so gluttony that it was the first time in past three days to have a lunch at all. I later explained that the Organizers of the 5th European Congress of Tropical Medicine & International Health had expressed inability, albeit a registration fee charged for US$ 800.- per person, to provide any lunch or dinner, and, at the same time, to my dismay, I was unable to find a vegetarian hotel or restaurant. I was so grateful to them that day! "You are most welcome", the Kiautas have greeted, with a parental tenderness and solicitude.

It must have been around 5 pm when Dr and Mrs Kiauta drove me back to the railway station. When I boarded the train we waived at each other saying 'so long', and at that moment of departure somewhere in one corner of my mind I was for the first time feeling a sense of some kind of relief as if to have freed myself from the shackles of guilt and pain committed two decades ago!

Dr. Kiauta is the humblest human being I ever met in my life! This whole cycle of learning about the value of words changed my life, too; never ever letting anything once committed slip away without being fully accomplished, be it my profession or personal family life. I felt growing wiser in the true spirit as radiated from the saintly advice of Maharishi Patanjali in the Yogasuthra;

"When you are inspired by some great purpose, some extraordinary project, all your thoughts break their bounds.

Your mind transcends limitations, your consciousness expands in every direction, and you find yourself in a new, great, and wonderful world.

Dormant forces, faculties, and talents come alive, and you discover yourself to be a greater person by far than you ever dreamt yourself to be."

He has always gratefully remembered the help and guidance received from his mentor, Dr M.A. Lieftinck. In fact, even before he came in contact with Dr Lieftinck, he had quite some other geniuses who during his early times of profession and research had paved him the right path of action. Two examples of effective mentorship are worth mentioning here. First his decision to take up dragonflies as his life objects of research, under the friendly mentoring of his lifelong friend, Professor M. Gogala. His decision to choose odonate cytotaxonomy for his doctoral dissertation was influenced also by his mentor Professor Hadzi, who suggested him to look into holokinetic centromere and post-reduction type of meiosis. Dr Kiauta accomplished his PhD thesis with much appreciation.

It is likely that because of such mentoring Dr Kiauta developed the habit of deep delving into the then evolving new science of insect cytotaxonomy and phylogeny, with special inclination toward dragonflies for whom his love as a group was unmatched. Dragonflies became his lifelong companion – the object of his studies and the objective of his irrepressible love and attention for their worldwide protection and conservation. Soon he began to win for him exotic research projects of great challenges. When he conducted the Alpine Biological Expeditions to Nepal, during early 1970s, he wrote several significant monographs as teaching aids in insect cytotaxonomy which, along with several other publications particularly his annotated dragonfly chromosome catalogue, continue to be the reference works for a serious research student even today! His writings, like his persona, were open schools for mentoring and guiding dedicated budding odonatologists. In Nepal, during his expeditions, he helped Mr. Mahendra Mahato much to find a PhD research position in the United States.

Inspired by odonate cytotaxonomy work of Professor Bastiaan Kiauta, Dr Matija Franković (Department of Animal Physiology, University of Zagreb, Croatia) recalled in his paper titled, '*Odonatology in Croatia*', published in *Selysia* (1993), Vol. 22, No. 1, p. 6, how in 1987 he had finished his studies of Biology with a Master thesis on cytogenetics of *Libellula depressa.*

Dr Stephen Butler (United Kingdom) is yet another odonatologist who profoundly respectfully think of Dr Kiauta for his encouragement during his initial odonatological days. In his email dated the 13th October, 2016 he writes,"*Dear Dr Tyagi, I have put together a short appreciation of what Bastiaan has meant to me as a very amateur dragonfly worker*", and his brief appreciation is in fact speaking volumes of admiration for Dr Kiauta (*vide infra*):

> *"Like many dragonfly students, my first memory of Bastiaan was a request for an obscure publication of mine, which I was pleased and rather flattered to send him. From then on correspondence developed between us – mainly because I requested help with obtaining publications on larvae, or asking for contacts and general advice. He never failed to reply both promptly and courteously, to someone who was but a very minor player on the odonatological stage.*

Also like others I was privileged to stay with Bastiaan and Marianne in their beautiful house Casa d'Uors in Switzerland in April 1995 and spent several days being both entertained and stimulated (see attached photo). Of course there was an odonatological twist to the visit as Bastiaan had arranged for me to visit the known Coenagrion hylas *sites, where I spent several hours, before a snowstorm forced me to beat a retreat to the safety of their home, one of the highlights of my "career."*

I have no doubt that I could never have experienced such amazing visits to many parts of our fascinating world, nor made so many good dragonfly friends, had it not been for his kind and selfless encouragement. I salute a good friend – Bastiaan – and his lovely wife Marianne."

Similarly, Dr Gunther Theischinger, an internationally reputed dragonfly taxonomist and a senior member of editorial board of *Odonatologica*, has urged me in his mail dated the 9th March, 2017 to express his feelings about Dr Kiauta in this biography as, "*Ever since I met Bastiaan at the Symposium in Karlsruhe (1973) when I was 33, he was a great support and friend to me in matters Odonatology. I am glad to have the opportunity to officially thank him, and I hope he will be around and enjoy doing what he has been doing so well, for many years to come.*" How true. Recently, on 11th November, 2017, when Mr Kumaran Sathasivan, an entrepreuner naturalist-odonatologist inquired if photographic evidence could be used in describing a new species, Dr Kiauta, whom I had refered the matter for clarification, stated vide his email dated the 23rd November, 2017, in clear terms as reproduced below,

"Dear Professor Tyagi,

subsequent to my letter on the above subject, dated on the 13th instant, I am appending here a PDF of the KRELL & MARSHALL paper, published last month and concerning the new ICZN declaration that was also just published.

As you see, the description of a new species, without the name-bearing holotype, is permissible now, though under very strict conditions. In case of the proposed Protosticta, such circumstances are not fulfilled, therefore the holotype is needed and it is also absolutely required in order to include the new species in the existing key of the genus.

When I wrote to you on the 13th, I was not familiar with the new ICZN declaration, hence this additional information – just for good order only.

With kindest regards,

Yours ever,

B. Kiauta".

Mr Kumaran felt extremely grateful by this information and is now on his way to prepare an appropriate manuscript.

I can never forget my own early days of learning under his watchful patronage. Early in 1976 I had just stumbled upon an interesting phenomenon of sex-determining mechanism in two dragonflies, *Onychogomphus saundersi duaricus* and *O. schmidti*, but had no clue as to how the decreased number of chromosomes in both mitotic and meiotic stages of development could be explained! Dr Kiauta then came to my help; he studied my chromosome slides, photographed and developed pictures in his laboratory, critically perused my manuscript submitted to *Odonatologica*, suggested a whole lot of changes in my write-up (by virtually having pains of re-writing the whole manuscript) and, allowed publication of my first ever scientific (odonatological, of course) paper absolutely *gratis* (because I was unemployed and had no money to pay the costs of either the publication or the reprints)! This research paper, published in 1977, became a milestone in my career and continues to be the source of great inspiration of learning science in the finest tradition.

That a mentor is a father figure is best exemplified by Dr Kiauta. In my case, I owe to him all of success in my much chequered professional career in the field of vector-borne diseases and medical entomology. As if providence had decided, three events occurred simultaneously in my life soon after I ushered in the arena of doctoral dissertation research in odonate cytotaxonomy: selection of the Doon Valley – the world famous dragonfly stronghold – as site of research, dragonflies as the object of research, and, most importantly, Dr Kiauta's mentoring, during the PhD period and thereafter to date! Inspired by Dr. Sangal's work on dragonflies, particularly, *Crocothemis servilia* Drury, I came in contact with some of the greatest entomologists and/or cytogeneticists, such as Prof. P.K. Sen-Sarma, an internationally acclaimed termite expert and the Director, Biology Research, Forest Research Institute & Colleges, Dehradun, Mr. Pratap Singh, Forest Entomologist, *ibid.*, Prof. T.N. Ananthakrishnan, an unparalleled thysanopterist and the Director, Zoological Survey of India, Prof. G.P. Sharma, an internationally reputed cytogeneticist and Head, Zoology Department, Punjab University, Chandigarh, Prof. J.S. Yadav, a celebrated geneticist on Coleoptera, in particular, and Principal, Choturam College, Jind in Haryana, Dr. S.K. Kulshrestha, a colleague of Dr. Sangal in the same department, Dr. Santokh Singh, noted high altitude entomology expert and Head, School of Entomology, St. John's College, Agra, Dr. H.N. Baijal, a famous collembologist and Head of Zoology Department, Agra College, Agra, Prof. J.M. van Brink, Editor of *Genetica* and *Genen en Phaenen*, and a senior professor at the Utrecht University's Institute of Genetics, and many more...! But, one person who made the most significant impression on my persona is Prof. Bastiaan Kiauta, also the Executive Editor, *Odonatologica* as well as the Head, Societas Internationalis Odonatologica (SIO) and the Director, Khumbu Himal Research Expeditions of the Netherlands Centre for Alpine Biological Research (N.C.A.B.R.), Utrecht in collaboration with Nepal Research Centre, Kathmandu. These were the early teachings of Prof. Kiauta which had not only guided me towards accomplishing successfully my doctoral dissertation titled "*Studies on the chromosomes of Odonata of Dun Valley (Dehradun, India)*", but also taught me the lessons in performing science in the best tradition! After submitting my thesis to the university I presented a copy for his perusal. He not only praised my research output but also published an abstract of the thesis in the *Odonatological Abstracts* (*cf.* OA No. 2683), in the following manner:

> *"The comprehensive volume is based on original observations on 45 spp., the main cytotaxonomic data on 13 of which were subsequently published elsewhere (cf. OA No. 2665), leaving as new* Anisopleura lestoides *[n=13, m] only on a thorough literature study. It is organized into thirteen chapters, incl. a comprehensive bibliography and a synopsis of chromosome numbers recorded in the order up to 1977, and presents several novelties for the regional and/or Indian fauna. The book was not issued commercially, but a substantial abstract is available from the author."– (Abstractor's note: The author should be congratulated on the production of this work, which has been carried out under often hard personal circumstances and under a great financial stress).*

Although Prof. G.P. Sharma, my first PhD thesis examiner, had commented ".... a phenomenal research work" and another evaluator, Dr. P.K. Sen-Sarma wrote ".... such a massive research work must be published as a priority", nevertheless, it was admiration of Dr Kiauta which is to me the "*summa cum laude*" mark, even greater than the award of degree rendered by the Garhwal University. The following two statements will possibly elicit my evolution under the mentorship of Dr Kiauta (Fig. 21.2):

(1) Prof. T.N. Ananthakrishnan (14th March, 2009), unquestionably one of the world's most outstanding entomologists and the former Director of Zoological Survey of India, chronicled my growth in science as:

> *"... I am privileged to be able to testify to the uncanny abilities of Dr. B.K. Tyagi, Director, CRME, as a medical entomologist of high calibre and whose dedication and perseverance has earned for CRME the status of internationally acclaimed institution in the field of Medical Entomology. This is justified by the diversity of his numerous publications in journals of high standing. The solid foundation laid by the first Director, Rachel Reuben, has been further strengthened by his personal efforts, not to mention the interactions with his colleagues... . Friendly, alert, interested in diverse subjects concerned with Medical Entomology, Dr. Tyagi has been considered as an outstanding Medical Entomologist whose efforts to tide over many diseases like Dengue have been well known. The Annual Reports prepared by him reflect the diverse activities of CRME under his care and needless to say that in Dr. Tyagi we have a scientist with an uncanny ability to sustain the growth of an institution, which has adequately demonstrated the significance of the use of vector control technologies, some of which have had sufficient potentialities to end users. It is certainly a privilege to be associated with Dr. Tyagi whose sense of dedication is unparalleled ..."*

(2) Prof. Dr. Graham B. White (14th February, 2011), an international authority in medical entomology and former editor of *Medical & Veterinary Entomology*, summed up my research contribution as follows:

> "... *For the past three decades, Dr. Tyagi has contributed greatly to research and leadership on vector-borne diseases across the spectra of biological science and organizational systems. During his career with the Indian Council of Medical Research (ICMR) his effectiveness and productivity transcend the norms of ICMR responsibilities: while zealously performing his formal research management duties, Dr. Tyagi emphasizes the pastoral side of institutional development, and somehow finds the time and energy to produce inspirational books on biological themes. His personal research began with insecticidal product evaluation and development, collaborating with many agencies and industries for innovative technology, progressing to implementation research and cost-benefit analysis of components of integrated management and prevention of vector-borne diseases. One cannot exaggerate the versatile skill that Dr. Tyagi brings from PI management of grants from TDR and other agencies, for projects spanning many States of India, various disciplines and groups of TDR client countries, on wide ranging questions regarding arboviruses, malaria, filariasis, genetic control, and VBDs generally. His well balanced enthusiastic stewardship of everything he takes on, the depth and breadth of his biological knowledge, multi-disciplinary and visionary approach, administrative and international experience in fields currently acquired by STAC, make Dr. Tyagi highly suited and willing to serve TDR in his capacity.*"

Fig. 21.2. Dr B.K. Tyagi making his opening remarks at the 4th Conference of Medical Arthropodology (Madurai; Dec. 20-21, 2010).– (Source: B.K. Tyagi)

A true scientist and a great human being, Dr Kiauta possesses priceless virtues and qualities that earned him international reputation both as a scientist and a teacher. There are many events in the annals of both odonatology and the life of SIO whose brilliant execution clearly owes, to a very great extent, to his irrepressible sense of hard work and sincerity to the advancement of the science of odonatology. His odonatological knowledge and editorial skills lent him good support to elevate *Odonatologica* above the rest of the world's entomological journals dedicated to a single insect order. Its popularity could be gauged from the simple fact that the SIO soon registered a membership largest ever for a similar activity within the class Insecta. This phenomenal success was achieved only through the dint of his perseverance, far-sightedness, and focused dedication for more than four decades; all the issues of various journals have been incessantly brought out on exactly the scheduled dates of publication! *Odonatologica*, in particular, demands a greater amount of care and attention as its articles are written in either of the four prominent European languages, that is, French, Italian, English and German – languages that hardly anyone in odonatological realms knew better!

Execution of great responsibilities also demand investment of enormous amount of energy and labour. In 1985, when I went to Paris for attending the Eighth International Symposium of Odonatology on the invitation of Dr Kiauta, I was amazed to see him speaking to various delegates in different languages – all at the same time! I keenly observed that he, as a top executive of the SIO Council, participated in all meetings that continued till late hours in the night, but still he was there on the dot of time when the next day's proceedings at 9 am sharp would commence. His sense of valueing significance of time in all kinds of deeds is phenomenal. He shows great respect to the punctuality of time – a character the young scientists have to emulate in their life. With a colossal stature amongst the world's leading odonatologists, he is nevertheless a symbol of humility and camaraderie to his friends and students the world over. So, when an Indian scientist colleague of mine, Dr D.B. Tembhare from the Zoology Department of Nagpur University, once went to attend the seventh SIO Symposium held in Chur, Switzerland, he was courteously invited by Dr Kiauta to stay at his house. Dr Tembhare was simply taken aback by his huge energy invested in executing the various responsibilities with absolute precision and thoroughness. He works uninterruptedly for 14-16 hours every day without any lapse, rising early at 5 am and going to sleep not until past midnight, leaving for tomorrow no work of today! "Still he finds time to look after his garden", Tembhare told me with a clear shine of amazement in his eyes! I am reminded of the most famous and inspiring lines by Robert Frost –

The woods are lovely, dark and deep
But I have promises to keep,
And, miles to go before I sleep,
And, miles to go before I sleep.

For the young and budding odonatologists, Dr Kiauta is an epitome of hard work and application. His deep sense of sincerity for commitment is exemplary. One particular anecdote is worthwhile to be told here as an example – out of scores of such examples experienced by his friends and students – that I often tell everybody as the finest example of the custodian of science!! It was during the Paris symposium of SIO that the rare idea of penning a bio-bibliography of Col. F.C. Fraser dawned upon me. I immediately placed my thought before Dr Kiauta who, holding his favourite pipe in his right hand close to his mouth, appreciated my proposal but cautioned me saying that Dr Fraser published his papers in a large number of journals, but sometimes by himself, and there was never any authentic listing prepared on his publications nor even by himself. "But," Dr Kiauta, judging my disappointment, came to my rescue almost immediately and suggested "A rather incomplete manuscript of Dr Fraser's publications existed in Dr B.E. Montgomery's archive and perhaps you could use that as the launching pad." Then, it is a history of great learning for me, and many other students of science, to acknowledge the fact that though I was doing my own bit to complete a rather less significant part of the whole exercise, Dr Kiauta took upon himself the major burden of accomplishing the task and regularly searched the titles in the big library of Amsterdam, several kilometres away from his home in Utrecht. In spite of his very busy teaching and research schedules at the Utrecht University, he consulted libraries in Amsterdam, Utrecht and Bilthoven, and completed the daunting task of check-list of Fraser's publications numbering over four hundreds. All he did for me, and for Dr Montgomery's memory who was of course also made posthumously the author of the booklet. It was a great learning for me. He fulfils his commitments and promises (even when made in the lightest sense) – virtues of greatness in all the geniuses of the world!

●●●

22

Extraordinary Correspondence between Dr B. Kiauta and Worldwide Odonatologists

"True guidance is like a small torch in a dark forest. It does not show everything once, but gives enough light for the next step to be safe."

— Swamy Vivekananda

One of the finest characters that all scientific geniuses are said to have is to attend the correspondence promptly – on the day as they are received. Dr Kiauta's legendary prowess in this respect is something the young generation of odonatologists must emulate. He is an exceptionally gifted communicator – absolutely immaculate and meticulous in attending to a communication. Times and age have never seemed to hinder his energies and disciplined life in caring correspondence with his colleagues and students all over the world. Each of his letters is an ocean of knowledge. As the general overseer of the SIO he freely helped a large number of odonatologists across the world to exchange their requirements, in any form, such as dragonfly samples, reprints, photographs etc. Even though he has long retired from the active SIO service, nevertheless he continues to help colleagues in their matters of need, as possibly best exhibited from the correspondence between him and Dr. Gunther Theischinger over a request for a photograph of *Hemiphlebia mirabilis* from Australia by Dr Kiyoshi Inoue. In Dr Gunther's own words (28th September, 2016) this is '*An example how Bastiaan was in the middle of, and facilitating, conversation all over the world*':

"*From: Kiauta [mailto:mbkiauta@gmail.com]*
Sent: Friday, 26 February 2016 11:09 PM
To: Gunther Theischinger <Gunther.Theischinger@environment.nsw.gov.au>
Subject: Re: photo Hemiphlebia mirabilis

Dear Gunther,
Thank you so much for the photographs, which I am sending on to Kiyoshi, who is certainly going to respond and thank you as well.

With kindest regards, also on behalf of Marianne,
Yours ever,
Bastiaan
****

On Thu, Feb 25, 2016 at 2:37 AM, Gunther Theischinger <Gunther.Theischinger@environment.nsw.gov.au> wrote:

Dear Bastiaan,

I have some photos here (see attachment). They all were made by Reiner Richter. If you send them to Kiyoshi and he picks what he wants, you can let me know and I may get higher resolution photos from Reiner or Reiner can send them himself. I am on email and things can go fast.

Best wishes to you, your wife and Kiyoshi,

Yours ever,

Gunther

****

From: Kiauta [mailto:mbkiauta@gmail.com]

Sent: Thursday, 25 February 2016 10:24 AM

To: Gunther Theischinger

Subject: photo Hemiphlebia mirabilis

Dear Gunther,

Our friend, Kiyoshi Inoue (ks-inoue@mx2.nisiq.net), is working on a completely revised edition of his book "All about dragonflies". This morning I have received his request to try to get from a colleague in Australia a field photograph of adult H. mirabilis, to be reproduced in the book.

Could you please help: either with an unpublished photograph or with the permit to reproduce the one in your Australian dragonflies book of 2006?

For the courtesy and help I am thanking in advance warmly and remain with kindest regards,

Yours ever,

Bastiaan"

All of Dr Kiauta's communications carry an indelible hallmark of utmost courtesy. Dr Milen Marinov reminds me vide his email dated the 9th June, 2012 about the curious way of showing courtesy in all his letters: "*Dear Prof. Tyagi, Here is another example of Prof. Kiauta's courtesy in his relations to others. I've always been surprised by the way he finished his letters – a long sentence sending his regards, wishes and sincere thanks. These days we usually send a short "Regards", but Prof. Kiauta always spends time to express his deepest gratitude to others. Kind regards,*

Milen." He is an institution himself in developing the art of making correspondence which is a priceless treasure of odonatological knowledge for a serious student of dragonfly research.

The central theme of Dr Kiauta's correspondence was odonatology, in one or another form. Thus, all his communications bore the indelible mark of a dragonfly subject, either pertaining to the manuscripts submitted to *Odonatologica* and/or *Notulae odonatologicae*, SIO management, international symposia, SIO National Offices, dragonfly conservation or even soliciting guidance in improving the quality of scientific manuscripts etc. Dr Zhu H.-q. (Nankai University, China) described beautifully as to how much he reverently thought of Dr Kiauta in providing scientific supervision. In his communication dated 14th May, 2006, enclosing therewith the manuscript, "Chinese species of the genus *Calicnemia* Strand, 1926 (Zygoptera: Platycnemididae)" for *Odonatologica*, he very politely explained the background research experience of its authors and requested Dr Kiauta to revise the manuscript to suit its publication which, of course, later appeared in the journal following a thorough scrutiny of the manuscript [*cf.*X. Yu & W. Bu: *Odonatologica*, vol 37 (2008), nr. 3, p. 247-255.].

Dr Naoya Ishizawa (Japan) corresponded extensively with Dr Kiauta, discussing among many matters also about *Epiophlebia sinensis,* the third and latest anisozygopterous species described so far, as he wrote on 11th March, 2017, "*I owed so much to Dr. Kiauta in case of my contribution to Odonatologica. My linguistic ability is so poor that he supported me so much. I sincerely thank him.*" In the odonatological world today Dr Kiauta has become a household name where scholarly people appreciate him even without having met and/or corresponded with him. Dr Rassim Khelifa (Department of Evolutionary Biology and Environmental Studies, University of Zürich, Switzerland), vide his mail dated the 28th September, 2016, has written, "*Although I read a lot about Pr. Kiauta, but I did not have the opportunity to meet or with a clear pinch of regret for never meeting with Prof. Kiauta correspond with him*" which reminds me of the famous Indian Sufi poet Rahim's invaluable advise;

बड़े बड़ाई ना करैं बड़ो न बोले बोल

रहिमन हीरा कब कहे, लाख टकामम मोल।

Transliteration- *(Bade badai na karain, bade na bole bol*

Rahiman heera kab kahe, lakh taka mero mol.)

Translation - (*A true greatness never reveal thyself. Just like a diamond never says it's value*)

The modern age of budding odonatologists have much to learn from his life, especially the art of communication. Even in those days when e-mails did not exist, his promptness in replying to a communication, however trivial or even intricate for that matter, can be gauged from the simple fact that the inquirer was calculatingly bound to get his reply on 7th day across 6000 km from Europe to any place in the world! Today, of course, his reply is there on my computer monitor when I enter my office next morning at 9 am sharp.

COPY

S.I.O. CENTRAL OFFICE
P.O. Box 256
3720 AG BILTHOVEN
The Netherlands

Bilthoven, 20 May 1995

Dear Stephen,

Thank you so much for your letter (and for the loan of the enclosed NATURE CONSERVANCY issue!).
We are very happy, your hylas & viridulm trip was such a success!

Your visit was of course much too short. It would be nice if you could come again, stay a bit longer, and have a look at e.g. Somatochlora alpestris larvae....

In Slovenia it was OK. Somatochlora meridionalis is there very common, but we still do not know whether or not it occurrs in Switzerland at all (perhaps in the southernmost Ticino?). We shall tru to have a look this summer.

If there is anything I could do for you (e.g. xerox copies, etc.) please do not hesitate to let me know.

As "discussed" already at Flumserberg, I would be very happy to have your hylas paper for publication in OPUSCULA, when you are ready with the work.

Kindest regards and all good wishes from both of us,

Yours ever,

Bastiaan

P.S.: From the recent issue of the BDS Newsletter (page 4) I understand, you have published in 1982 a work on "The dragonflies of Shropshire and their distribution". I have never seen this publication, nor we were ever able to list it in Odonatological Abstracts. I would be very thankful if you could let me have a copy (or a xerox). - Very many thanks in advance

B

Fig. 22.1. Facsimile of a communication by Dr Kiauta to Dr Stephen (UK).

His correspondence is always like an ocean of knowledge, meaning the deeper one delves into it the better and wiser feels the reader! I was able to build the SIO National Office in India with this kind of rare training in science and the results were obvious: a series of biennial symposia in India, an international symposium of odonatology of the SIO, publication of the annual journal *Indian Odonatology*, semi-annual newsletter *Fraseria*, need-based irregularly published *Occasional Publications* and several other isolated publications of enormous value. His teachings have always been the guiding force to his innumerable students all over the world. It is this indefatigable spirit of learning that the Indian dragonfly lovers have once again mustered in enough courage to advance the science of odonatology by forming a new scientific society, the Indian Dragonfly Society (IDS), in 2017. To newcomers in dragonfly research his correspondence always served to pave the right path for exploring future subjects of investigation, as exemplified by the works of Dr Stephen (Steve) Butler (vide infra).

Dr Steve Butler was still a novial odonata biologist when he came in contact with Dr Kiauta. His life changed thereafter as he did not only receive his extremely fruitful guidance in his research work but also the full support in his endeavours to publish research findings, also in *Odonatologica* (Fig. 22.1). Dr Kiauta's 'odonatological' letter dated 20th May, 1995, to Dr Steve, is a testimony to his unflinching spirit of guidance and mentoring.

With many leading odonatologists Dr Kiauta is in touch during more than 40 years, with some of the "younger" ones since their student time and/or their first beginnings of scientific work (Fig. 22.2). With some of the great workers of the past generation (Fraser, Laidlaw, Valle, Erich Schmidt etc.) he had just some correspondence during the concluding time of their life. Assumably he had some correspondence with anything like 5-600 Odonata workers throughout the world, perhaps even more. For the sake of record it should be mentioned that over the years the guestbook at the Editorial Office was signed by more than hundred visiting odonatologists, by some on several occasions. Among the legible signatures are those of: Adamović, Ž.R.; Anselin, A.; Asahina, S; Askew, R.R.; Bechly, G.; Bedjanič, M.; Belle, J.; Bick, G.; Bick, J.; Biedermann, J.; Bischof, A.; Boon von Ochssée, T.; Butler, S.; Cannings, R.; Cannings, S.; Cempirek, J.; Corbet, P.S.; Davies, D.A.L.; De Jong, T.H.; De Marmels, J.; Deacon, K.; Dumont, H.; Dyatlova, E.S.; Erbida, N.; Falchetti, E.; Geijskes, D.C.; Geister, I.; Grossniklaus, H.-P.; Hämäläinen, M.; Haritonov, A.Yu.; Hartung, M.; Higler, L.W.G.; Hilton, D.F.J.; Holmquist, J.; Hostettler, K.; Inoue, K.; Jensen, A.; Jödicke, R.; Jurzitza, G.; Kalkman, V.; Kemp, R.G.; Komnik, H.; Kotarac, M.; Krüner, U.; Legrand, J.; Lencioni, F.; Lieftinck, M.A.; Lohmann, H.; Machado, A.B.M.; Marinov, M.; Martens, A.; Meier, C.; Michiels, N.; Mielewczyk, S.; Mill, P.; Miyakawa, K.; Mol, A.; Montgomery, B.E.; Moore, N.W.; Müller, R.; Nielsen, P.; Nimz, C.; Norma-Rashid, Y.; Parr, M.J.; Peterson, J.; Pilon, J.-G.; Pinhey, E.C.G.; Pinratana, A.; Piper, W.; Pirnat, A.; Poboljšaj, K.; Popova, O.; Pritchard, G.; Rehfeldt, G.; Robertson, H.; Rudolph, R.; Šalamun, A.; Sandhall, Å.; Schaller, F.; Schiess, H.; Schmidt, Eb.; Schorr, M.; Schröter, A.; Sherk, T.; Siegenthaler, I.; Sovinc, A.; Strub, O.; Suzuki, K.; Tembhare, D.B.; Treagust, A.; Tyagi, B.K.; Ubukata, H.; Utzeri, C.; Van Tol, J.; Velthuis, H.; Verdonk, M.; Wasscher, M.Th.; Watson, J.A.L.; Weihrauch,

F.; Westfall, M.J.; White, T.R.; Wildermuth, H.; Williams, E.C.; Winstanley, B.; Wolf, M.; Wysoki, M.

In connection with *Odonatologica,* SIO biennial symposia and/or some other associated scientific activities such as numerous odonatological missions and field work Dr Kiauta has enjoyed assistance, hospitality and/or various kinds of other help from colleagues and friends the world over, e.g., G. Alberti (Italy), S. Asahina (Japan), P. Bovey (Switzerland), J.W. Boyes (Canada), H. Burla (Switzerland), L.J.M. Butot (Netherlands), R. & S. Cannings (Canada), H. Franz (Austria/Nepal), E. Ghirardelli (Italy), A.Yu. Haritonov (Russia), U. Hellmich (Germany/Nepal), L.W.G. Higler (Netherlands), K. Inoue (Japan), G.B. Kalikote (Nepal), M. Kotarac (Slovenia), M.A. Lieftinck (Netherlands/France), J.C. Lien (Taiwan), M.F. Mörzer Bruyns (Netherlands), R.A. Müller (Switzerland), A. Piskernik (Slovenia), R. Schloeth (Switzerland), I. & O.R. Siegenthaler Strub (Switzerland), M. Titayavan (Thailand), B.K. Tyagi (India), M.J. Westfall (USA), E.C. Williams (USA) and from some dozens more persons in various parts of the world.

(a) (b)

Fig. 22.2. (a) Dr. Matjaz Bedjanič, Editor Erjavecia, exploring dragonflies in the Sri Lankan mountain rainforest (Source: M. Bedjanič); Fig, (b) Dr Kiauta and Dr Steve Butler at the Dr & Mrs Kiauta's home Casa d'Uors in Switzerland in April 1995 (Source: Stephen Butler).

Details of correspondence with a few noted odonatologists is featured below, just to mirror kind of contents of their communications:

(1) **Sajiro Makino.** – Dr Kiauta did not have any regular correspondence with him, though he did receive his book "Chromosome numbers in Animals" (1951) and the reprints of his papers relative to his interests.

(2) **Syoziro Asahina.** – Dr Kiauta was in touch with him since the early 1960s and he has met him for the first time in person in 1968 and, subsequently, at various International Symposia of Odonatology in Europe and elsewhere. Dr Asahina also brought a visit to him in their home in Holland. In Tokyo Dr Kiauta with his wife visited him. In addition to professional correspondence, during the years they also developed a personal friendship.

In Japan Dr Asahina was treated with enormous respect and he was rather formal also at the international odonatological meetings, which was in the SIO family rather unusual. At the Calgary Symposium the participants celebrated his 70th birthday and he declared this was "the most beautiful day in his life". Gordon Pritchard, the Secretary of that Symposium, tried to get him off his formal attitude and asked him how he is called by his wife. His answer was: "Shóshiro", to which Pritchard exclaimed: "oh well, then you are Shoshi!" This call-name was immediately accepted by all non-Japanese odonatologists (but not by the Japanese, who did not dare to call him like this!). Since then he was Shoshi for all of SIO family. At the Osaka Symposium, Kiyoshi Inoue was delivering a review talk on the history of Japanese odonatology and requested Dr Asahina to agree he used the name Shoshi in his talk.

Dr Kiauta has an enormous bunch of their private correspondence in his archive.

(3) **Angelo Machado.** – Dr Kiauta and Dr Machado were also in touch since early 1960s and developed subsequently an intense friendship, duly corroborated by a recent communication dated the 27th October, 2016 by Dr Machado to me which reads as follows, "*He is one of my best friends and I have been corresponding with him for about thirty years. Last month I wrote him and sent the Festschrift of the book that Zootaxa (4078, 2016) published in my honor for the 80th birthday where this friendship is mentioned. It contains a photo of me together with Bastiaan, Marianne and P. Corbet during a collecting trip in Alpes near his Casa D'hors in Switzerland. I stayed a couple of days there where I met Philip and Pinhey. After dinner Marianne invited us to go to Lichtenstein for the fest of the Prince birthday. We went by car but somebody did not allow us to cross the frontier. There is no room for more cars in Lichtenstein – he said – you will have to park in Switzerland. We entered the county by foot. This is one of the good times that I had with Bastiaan and Marianne besides meeting them in the Gainesville, Chur and Paris Symposiums. I named after him two species* Navicordulia kiautai *Machado & Costa, 1995 (Odonatologica 24-2) and* Neoneura kiautai *Machado 2007 (in your book)*". When he had visited with Dr Kiauta in his home in Switzerland,

during the Chur symposium, at that time many odonatologists were boarding and lodging in his home, so they were short of space and beds. They put Machado on a provisional bed in their Swiss laboratory and he was very happy.

Dr Machado's scientific and social interests are very broad. In a certain period he was interested in the attitude of children towards the forest. With the help of Irene Siegenthaler, Dr Kiauta had mediated for him a questionary on this subject to be circulated in Swiss primary schools. The results were very interesting and he published them in one of his books on this matter in Brazil.

Dr. Kiauta has a private library housing his entire correspondence. It amounts to some 10-15 shelf metres in his archive. The figures are hard to estimate. The *Odonatologica* editorial correspondence (with authors, reviewers, in the matters of *Odonatological Abstracts* etc., of 43 years) could consist of 50,000 letters, other scientific correspondence of the last 50 years to about 10,000 letters, the other (technical and management) correspondence to about 5,000 letters.

His dragonfly collection (Odonata and other insects) measures ca 150 drawers (the odon. specimens all papered) and numerous glasses of odonate larvae (in alcohol), mostly deposited at the Swiss home.

●●●

SECTION IV: REAL SOURCE OF STRENGTH

23. "Where is my wife?"

23

"Where is my Wife?"

"It is not how much we do,
but how much love we put in the doing.
It is not how much we give,
but how much love we put in the giving."

— Mother Teresa

Science and technology are often considered, though entirely thoughtlessly, to be the forte of men, and odonatology is no exception. Nevertheless, the contribution of women, who – in blatant contrast to men, almost always have to part their times between managing homes – *hook, line and sinker*, on one hand, and the laboratories – *with complete dedication*, on the other, and still made significant advances to the progress of diverse disciplines of dragonfly research, cannot be disregarded. There have been numerous gifted and far-famed women scientists in history who made crucial discoveries and inventions in the science of odonatology globally, such as, to mention symbolically a few only, Cynthia Evelyn Longfield (1896 – 1991), Janny Margaretha van Brink (1923 – 1993), Jill Silsby of the "Dragonflies of the World (2001)" fame etc. In the life of International Society of Odonatology (SIO) and *Odonatologica,* Marianne Kiauta's devotion as exhibited in all her energetic actions has in all times been a marked feature that furnishes the true measure of her life. Like all great men and women of science, art and literature, she cultivated meticulously the spirit of self-help, perseverance and patience converting apparently the most insuperable proved obstacles in her way to her advantage. In fact, in many instances she turned these very difficulties her best helpers, by evoking her powers of labour and endurance, and stimulating into life faculties which might otherwise have lain dormant: *Where there is a will, there is a way*!

Marianne Kiauta was born in Gouda, the Netherlands, in 1948. She studied biology, Sanskrit and Tibetology at the universities of Utrecht and Leiden and travelled extensively on scientific expeditions in the Himalaya, south-eastern Asia and Japan, together with her husband. She has published haiku, senryu, tanka and linked poetry since 1984 in literary journals and anthologies in Holland, Belgium, Germany, England, Japan, Greece, New Zealand and the USA. In 1993 a selection of her haiku, senryu and tanka poetry appeared under the title *Tule voor de berberis* (Tule for the barberry) [Dutch]. In 2000 the bilingual [Slovenian and English] *Prvižafran* (*The first crocus*) was published in Slovenia, the home country of her husband. *Poldergroen* (Poldergreen) appeared in 2002 [Dutch], with haiku that tell about the region of her childhood. In 2003 a tanka collection was published trilingual [Dutch, Slovene and English] with the title *Het binnenste van een woord* (*Notranjost besede, The interior of a word*). Next to her daily task as assistant editor of several scientific journals (on entomology, specialized in Odonata), she has a practice in (foot) reflexology.

Marianne Kiauta, the beloved wife of Dr Kiauta, a biologist (University of Utrecht) and since 1971 his assistant in the Department of Genetics, University of Utrecht, now retired, is an odonatologist and polymath. Since 2008, she is running an asylum for old and discarded animals. Her contribution to development and growth of *Odonatologica*, commenced March 1972, is yet possibly best presented in Dr Kiauta's own words: "*Since the birth of Odonatologica, two people have been crucial for its development: without the support and much more than the manifold help from the late Professor Dr Janny M. van Brink (Utrecht/Bilthoven, The Netherlands) and my wife, Marianne, it would have been unthinkable to continue in the office for so long. For some decades, Marianne has been the backbone of the SIO and Odonatologica, taking care of all the technical matters and much more.*" She actively participated in almost all of Dr Kiauta's field work. During 1972–2013 her help in the editorial matters of *Odonatologica* and in the management of SIO could not be exaggerated. Aside of her technical and other assistance in the multifaceted SIO and *Odonatologica* works, her skilful management of the logistics was indispensable particularly in the Himalayan, SE Asian and the Philippine missions, where her familiarity with some of the local languages and scripts was often a valuable asset also.

The Dutch scientific history has no dearth of great women scientists and technologists. These women dedicated their lives to science and made significant advances. In fact, superior scientific intellectuality in all walks of life, be it medicine, health, conservation, or information technology, has all continued to enrol women enrichingly much before many other countries did in the Old and the New Worlds. The instances of men and women who, by sheer dint of hard work, application and energy have raised themselves from the humblest ranks of society to eminent scientific positions of usefulness and influence, are indeed so numerous that they have long ceased to be regarded as exceptional. Looking at some of the more remarkable, it might almost be said that early encounter with difficulty and adverse circumstances was a necessary and indispensable condition of their inevitable success, just as the laws of time are inexorable. These men and women have had the benefit of no political patronage, but have invariably achieved their goals and made the country proud and prosper by sheer dint of their hard work and application under severest of unfavourable conditions. A very illustrious scientist to live in the Netherlands in recent decades is Professor Janny M. van Brink, the insect geneticist, widely remembered respectfully by the sobriquet earned for her overall care of the International Society of Odonatology as "Mother of SIO" that made all the odonatologists under its umbrella interact affectionately among themselves on all matters like a "family of friends", a philosophy innovated and nurtured by Dr Kiauta all his life! And, the greatest support to Dr Kiauta in inculcating this philosophy, in practice, is rendered by none other than Mrs Marianne Kiauta through all thick and thin (Fig. 23.1 and 23.2).

Fig. 23.1. *"This big was the dragonfly I just caught ..."* Marianne Kiauta, holding her two inseparable tools, a camera and a dragonfly net. Symposium fieldtrip Novosibirsk, 2001. (Source: M. Kiauta)

Fig. 23.2. The Kiautas at dragonfly inventarisation of a karst pond, Brje near Triest, Slovenia: summer 1997. – [Photo: Prof. Dr Matija Gogala]

She is very kind and generous, to all – human beings as well as animals! For many years now, after the emeritate of Dr Kiauta, she has been running an asylum for old and discarded animals, like rabbits, dogs, birds, horses, ponies, goats, chickens (*cf.* SITAR, 2012a). She, along with Dr Kiauta, looks after their health and life. Her home, both in The Netherlands and Switzerland, is often flocked by a number of odonatologists and the Kiautas are an excellent host to be remembered with reverence forever!

She is also a dedicated photographer and a field explorer. That is why two tools which invariably remain on her persona include a dragonfly net and a camera. She has in Dr Kiauta's archive thousands of negatives and photographs of odonatologists and the odonatological events such as the SIO International Symposia of Odonatology, and in fact they weave in a grand panoramic view of the world history of odonatology. Some of these photographs are rare as one seen for Col. F.C. Fraser [*cf.* Tyagi, B.K. (Editor), 1988. Odonatological bibliography of Frederick Charles Fraser (authored by Basil Elwood Montgomery[f]; reorganised and updated with a photograph and a biographic note by Dr. Tyagi under permission and guidance of Prof. B. Kiauta). *Occ. Publs. SIO Natl. Off. India* No.3: 1-24.] (Fig. 23.3).

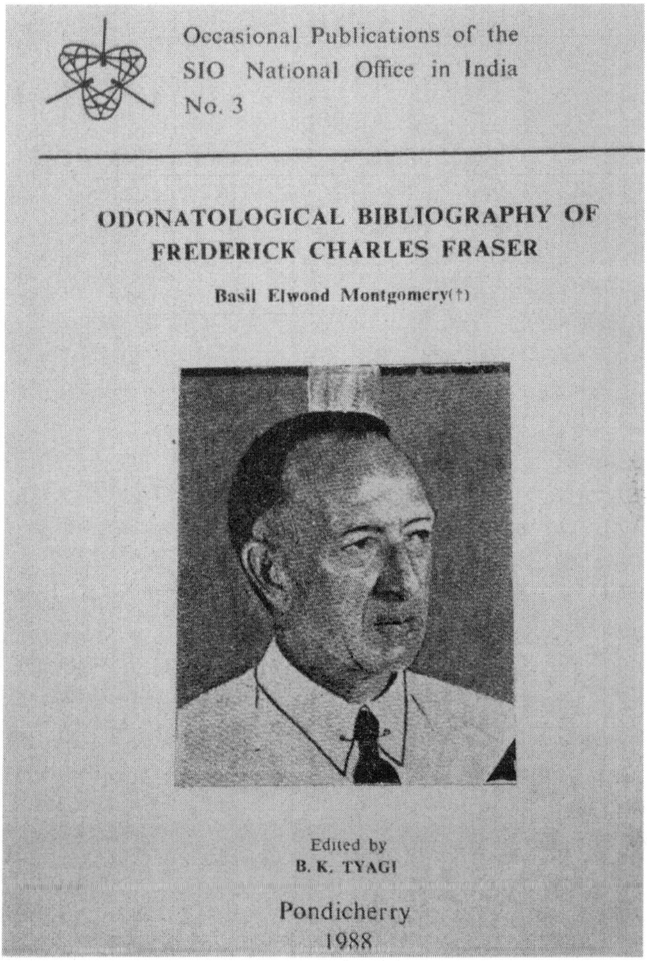

Occasional Publications of the
SIO National Office in India
No. 3

ODONATOLOGICAL BIBLIOGRAPHY OF
FREDERICK CHARLES FRASER

Basil Elwood Montgomery(†)

Edited by
B. K. TYAGI

Pondicherry
1988

Fig. 23.3. Facsimile of the title page of Tyagi (1988), showing portrait of Col. Dr F.C. Fraser (Source of supply: Marianne Kiauta and B.K. Tyagi).

During 1988, soon after 9th International Odonatological Symposium in Madurai, Dr Kiauta, Mrs Marianne and Prof. Van Brink had visited my home in Puduchery (then Pondicherry) where only four of us, myself, my wife and daughters, namely Anupama (nee Anu) aged nearly seven and Akansha (nicknamed Charu) four years, were living. A little before their visit my wife, Ajita, had an attack of facial paralysis (Bell's palsy) and she, like all of us, used to be under heavy stress always. She has tried allopathy, homeopathy, acupuncture and acupressure, in the best medical hospitals and clinics available there but the relief was, at best, minimal. Marianne, who had by then befriended Ajita, understood her concern and debonairly advised her to practice putting daily the tender palms of our younger daughter Charu on her face, and feel the warmth of love with her. We did not really know its impact on Ajita's problem but she religiously followed the instructions day and night! That wonders do happen in our life, my wife soon realized regaining much normal facial contours – thanks to the advice of Marianne whom our heartfelt gratitude will always be there.

One of the major fixture of *Odonatologica* was the regular appearance of odonatological abstracts of the literature, in any language, worldwide. In India some odonatological researches were also published in Hindi or some other native language. During early 1980s I was serving as a member of the editorial board of the *Odonatological Abstracts* section of the journal and I was communicating all the Indian dragonfly publications to Dr Kiauta. On one such occasion I came across a paper which was in Sanskrit and informed Dr Kiauta about my intention to send along with the translation as well. However, he immediately wrote to me requesting to just send the reprint or enen copy alone without taking the trouble of translating it because, he wrote, Mrs Kiauta, who is familiar with Sanskrit and some of the other oriental languages, would do that for the OA! I was wondering since Sanskrit is quite a difficult language and it needs a lot of hard work and practice for many years to command the language! Yet, Marianne Kiauta would arduously do such translations for the OA. It's admirable indeed.

Japanese language is said to be among the toughest languages to learn. Not many Europeans can boast of their potential to write haiku, which can take a lot of practice how to learn to write a haiku well even for the native Japanese. A *haiku* (俳句 high-koo), a form of poetry first made popular in Japan, is short three-line poem that uses sensory language to capture a feeling or image – in a pattern of approximately 5,7 and 5 syllables with reference to the natural world – which has become appreciated around the world. But Mrs Marianne Kiauta's interest in Oriental languages is so deep and thorough that she academically crossed all borders and delved deeply to explore more and more knowledge from these exotic communication media. She has written many haiku poems, besides four books. Some examples of her poetry are offered below (Source: Simply Haiku: A *Quarterly Journal of Haiku and Related Forms*, Spring 2005, Vol. 3, No. 1):

across the table
your glance and mine meet
in exclamation marks

such a meek look,
her hands like corduroy
search the table

a crow's feather
by the toppled tombstone –
he was a writer

almost drowned
in wavy spring grasses
the barbwire fence

just looking
at the soft-blue velvet
of the young plum

Dragonflies are fascinating animals and find a place in haiku, too. Citing a haiku of Basho (1644-1694):

crimson pepper pod
add two pairs of wings, and look
darting dragonfly

and one of herself for posterity reasons to always nurture a poet's tender heart with a scientist's robust mind.

frozen stiff
the reed that just cradled
a dragonfly

dragonfly eyes
both sides of a grass stem
watching intruders

Mrs Marianne has manuscripted some interesting odonatological research publications, both with Dr Kiauta, on odonate cytotaxonomy (e.g., KIAUTA & BRINK, 1975d; KIAUTA & KIAUTA-BRINK, 1975e) and by herself (e.g., KIAUTA-BRINK, 1976). Interestingly her last paper was possibly one among the abysmally small number of publications made on the subject and certainly a rarity from Tibet. Here she described six Tibetan expressions for "dragonfly", viz. "*sgo. na.nas.t'on.pa*" "*c'u.rkan.rten.'bu*", "*c'u.srin.rgyal.mo*", "*a.c'og.ts'al.pa*", "*p'ye. leb.sog.bzi*" and "*a.ts'ags.ts'ags.pa*", as extracted from autochthonous Tibetan informants or mentioned in various lexicons. She even ventured to correlate these expressions with the biological features of dragonflies and Tibetan demonology. She inferred that in Tibetan, like in South-East Asia, the dragonfly names appear emotionally neutral; the dragonfly is considered neither a lovable nor a frightening creature. This is at variance with the expressions in the Far East (China, Japan), where the dragonfly is generally respected and admired, while in European cultural areas its appellations often incur a superstitious dislike. This inference is of great importance and shows man's concern about the dragonfly fauna as well as their conservation in various different regions the world over!

Dr Kiyoshi Inoue, President of the Societas Internationalis Odonatologica (SIO), in his "Foreword" message dated 10th December, 2006 for Dr B.K. Tyagi's (2007) book titled "*Odonata: biology of dragonflies*" has most appropriately summarized overall scientific contributions by Marianne Kiauta, particularly in the SIO and Odonatologica-filled life of Dr Kiauta: *"When we speak of his activity, we cannot forget to mention the devoted help of his lovely wife Mrs Marianne Kiauta-Brink. She is always helping him or co-working with him in editorial and administrative matters. ... (We) thank her for all she did for us through her help to him!"* In a truly odonatologic manner even Dr Kiauta expressed his deep gratitude by naming a dragonfly in her honour, *Crocothemis servilia mariannae* Kiauta, 1983, incidentaly discovered from Japan where in native Japanese it is called Shojotombo (ショウジョウトンボ)!

Fig. 23.4. *Crocothemis servilia mariannae* Kiauta, 1983

Both, Dr Kiauta and Mrs Marianne, have in their hearts enormous affection and respect for each other. In Paris and in Madurai, during respective SIO symposia, I, almost with a katydid's remarkable acoustic prowess, overheard, during lunch or tea time, Prof. Kiauta, with his famous pipe in place, moving around very slowly and murmering almost absent-mindedly: "*Where is my wife*?" and "*Have you seen my wife*?" And, of course, she was right there, already, looking at him with a glitter of wonderment and debonair look. I can not find a better definition of their infinite love and respect for each other than to reverberate what Shri Kamlesh Patel, Head, Sri Ramachandra Mission echoed at Kanha Shantivanam, Hyderabad for 'Heartfulness', on 30th April, 2018:

> "*Love is the fuel that propels us.*
> *The ripening of the state of mergence is*
> *accelerated by love and love only.*
> *In this ocean of love, both the lover and*
> *the beloved remain only for name's sake.*
> *The trinity of lover, beloved and love*
> *all become One.*"

●●●

SECTION V: EVOLUTION OF SCIENCE OF ODONATOLOGY

24

Professor Bastiaan Kiauta— The Universal Phenomenon

अयं बन्धुरयं नेति गणना लघुचेतसां
उदारचरितानां तु वसुधैव कुटुम्बकं

Transcription - *"Ayam bandhurayam neti ganana laghuchetasam udaracharitanam tu vasudhaiva kutumbakam"*

Translation - *"Only small men discriminate saying: One is a relative; the other is a stranger.*

For those who live magnanimously, the entire world constitutes but a family."

As mentioned in the foregoing pages, Dr Kiauta had gotten in touch with dragonflies at the age of five (1942), but his first entomological interest from about that time on were the Lepidoptera. In 1953 he published his first more-or-less scientific note on these (KIAUTA, 1953a), followed by a number of papers, the last published in 1994 (SCHMIDT & KIAUTA, 1994). In 1952, he decided to concentrate his attention to dragonflies and stayed with this order ever since. His first odonatological paper was published at the age of 17 years (KIAUTA, 1954a)!

During 1960s Dr Bastiaan Kiauta was an emerging young scientist with new concepts and ideas in the world of odonatology. Being a professor of genetics, a science noticing extensive technological advancement, and editor of a few established international journals (e.g., *Genetica, Genen Phaenen* etc.) supported by his uncanny control of several European Congress and other languages (Dutch, French, German, Italian, English, Russian, Slovenian etc.), he came to be viewed by his colleagues with an eye for the freshness of fast adapting to new technologies embracing hitherto untouched frontiers and tackle new challenges in a hyperconnected world not apt to slow down any time soon. There was a certain amount of uniqueness in everything odonatological he touched upon. He soon pondered on the idea of founding an international odonatological confrerie which could bring on one platform the odonatologists, both young amateurs and seasoned, for free and frank discussions with the aim of universalizing the science of odonatology. He had hardly realized at that time that this initiative could mobilize the odonatological human resources across the world to such an extent for eventually making the International Odonatological Society (Societas Internationalis Odonatologica, S.I.O.) a robust "Family of Friends" rather than a conventionally formulated scientific association where odonatological fraternity springs up, through the Society's mainstream journal, *Odonatologica*, and the biennially organized international symposia (everytime in a different place/ country) to form new study groups, new projects of mutual understanding and, of course, new friends! Like a great leader devoted entirely to his cause or movement, he also resolved to navigate his brainchild, the International Society of Odonatology, forward without any disruption. His leadership and truly global statesmanship displays one of the most illuminating examples of how a scientific organization can mobilize, execute, and transform with agility as well as accelerate performance; within a short span of ten years, International Society of Odonatology (SIO) became by far the largest organization in the world on a single insect order with over 500 members, and its biennial symposia became landmarks for the organization in an healthy and friendly environment. The road ahead, however, was not easy and he needed to apply all his knowledge and energy to realize the dream, reminding me the words of wisdom by Dr A.P.J. Abdul Kalam:

"*Dream is not that which you see while sleeping,*
it is something that does not let you sleep."

On 7th March, 1970, Dr Kiauta has organized a Colloquium of Dutch dragonfly workers that has taken place in the Institute of Genetics, University of Utrecht. It was chaired by Dr M.A. Lieftinck and attended by some 30 persons and a few guests from Belgium and Germany. The response of the participants was overwhelming, so the 2nd Colloquium, with more international participation, was convened sooner on 21st November, 1970, followed by the 3rd one on 29th April, 1971. The response to this initiative made it clear that this kind of meetings was providing essentially in the desire of the odonatologists to get in personal touch and to learn on the current work of the colleagues.

In view of this, Dr. Kiauta, backed up by Prof. J.M. van Brink, has convened the First European Symposium of Odonatology, hosted by Dr H.J. Dumont at the University of Ghent, Belgium, during 22-23 October, 1971. In all, 39 workers had registered from Belgium, Czechoslovakia, Denmark, France, Germany, India, Italy, The Netherlands, Poland, United Kingdom and USSR, though some were unable to attend in person. At this symposium, the SIO and *Odonatologica* were born and it was decided that International Symposia are to be convened at 2 year intervals, every time in a different country. The management of SIO and editorship were entrusted to Dr Kiauta and his team. A worldwide call for membership was circulated in November, 1971, *Odonatologica* commenced publication in March 1972 and by mid-1972 several hundred copies were circulated in all continents.

From the onset, *Odonatologica* became the centre of SIO, and the Society a warm family of friends, worldwide. Along with the symposia, these initiatives triggered much impetus in the worldwide dragonfly research. In this connection are noteworthy numerous statements made on various occasions by different members. These are too many and too diverse to be quoted, but a few are mentioned, just to substantiate the positive impact of SIO on the lives of various odonatologists, e.g.:

- At the celebration of Dr M.A. Lieftinck's 80th birthday, organized by SIO at his place of residence (Rhenen), Prof. P.S. Corbet stated at the banquet: "*The SIO has entirely changed my life and gives a great impetus to my work*".
- At the 6th International Symposium of Odonatology (Chur, Switzerland, 1981) Dr R.W. Garrison, at that time a young and much promising odonatologist from California (USA) said: "*I would attend these Symposia even if I had to swim across the ocean*".
- Dr S. Asahina, the doyen of Japanese odonatology, said at the 7th Symposium (Calgary, Canada, 1983): "*These are the finest days of my life*".
- Prof. Dr Norman Moore (2007), Ex-Chairman, Odonata Specialist Group, Species Survival Commission (IUCN), was expressing the feelings of many by writing: "*I did not know personally any of those working on dragonflies outside the British Isles... The initiatives of Bastiaan and the colleagues opened up a new world... I for one will always be grateful to Bastiaan's leadership, so ably supported by Marianne and tirelessly assisted by Janny van Brink, for the enrichment of my life. Thank you Bastiaan for your pioneering work.....*".

Fig. 24.1. Professor Dr Bastiaan Kiauta. – [Source: Published previously in Erjavecia (2007) 22 and Trdoživ (2015) 6(2)]

- Dr Kiyoshi Inoue, a celebrated Japanese odonatologist and the Past-President of SIO, with a keen eye for wild photography, summed up the achievements of Dr Kiauta in a communication of 10th December, 2006: "*Dragonflies occur worldwide, and the odonatologist need to know their own odonate fauna as well as those of other areas. On this account the international communication and mutual help are indispensable. What Dr Kiauta has done and continues to do is indispensable for current and forthcoming research in odonatology alike... Dr Kiauta always calls his colleagues in odonatology a 'family'... He is so to speak the 'Father of the large family of odonatologists worldwide'.*"

No contemporary dragonfly researcher has made a greater impact on the advancement of the science of odonatology in recent decades than Dr Kiauta. He, in my honest opinion, is the greatest odonatologist today who may be referred, with due reverence for his lifelong and outstanding scientific contributions to the science of odonatology, to belong to the same extraordinary league of scientists to which the all-time greats like Dr Edmond de Selys-Longchamps, Dr W.F. Kirby, Dr R.J. Tillyard and Dr Philip S. Corbet have been associated to be decorated! Dr Kiauta, now healthily 82, is the ever shining star of our world's odonatological constellation.

Professor Dr Philip S. Corbet, undeniably one of the world's greatest ecologists on dragonflies ever, who produced masterpieces like, "A biology of dragonflies" (1962) and "Dragonflies: Ecology and behaviour" (1999), was in association with Dr Kiauta in an odonatologically highly productive camaraderie both in the matters of SIO and dragonfly research. On the initiative of Dr Kiauta, Dr Corbet wrote an assay on the history of odonatology, critically reviewing the various strands of its development through centuries! He recounted the contribution of many outstanding odonatologists in the pre-SIO era, especially about Syoziro Asahina, who was producing the dragonfly journal *Tombo* since 1958, and about an important move of B. Elwood Montgomery, who convened a Colloquium on Odonata at Purdue University, Indiana, USA in 1963, yet at the international level he eulogised Dr B. Kiauta's initiative to found in 1971 the Societas Intemationalis Odonatologica (S.I.O.) – a herculean task to bring all the world's odonatologists at one platform for the first time in the history of odonatology for which Professor Dr Kiauta would ever be opulently remembered. Corbet (1991) has summed up Dr Kiauta's endeavours as:

> "*The prime mover and current President of S.I.O. is Bastiaan Kiauta, the magnitude of whose contribution to this conspicuously successful venture cannot be exaggerated: it includes editing* Odonatologica — *a journal of international standing now in its twentieth year of publication — compiling research abstracts, arranging biennial international symposia and generally facilitating and overseeing initiatives in the Society. As one*

of the many fortunate users of the services offered by S.I.O., I may mention here a provision that is of inestimable value: the section of the journal Odonatologica *that is devoted to abstracts of articles published elsewhere on dragonflies. These Odonatological Abstracts, now numbering more than 7,000, constitute an essential source of reference for the serious student of Odonata."*

Dr Kiauta is a rare scientist, and a rare odonatologist; simple and yet so extraordinary! He enunciated anew the underlying core principle of the SIO, as a "family of friends' which is something unheard of earlier in the realms of science anywhere in the world. Always a man of few words, he means every bit of it; believing strictly in doing what he says and vice versa. For nearly a decade, until I finally met with him in 1985 in Paris during the SIO's 8th International Symposium of Odonatology, I knew him only through correspondence. He influenced my life much ever since by his deep knowledge, constant guidance and parent-like mentoring. My life was destined to change for the better and I knew I would make a better tomorrow - as a teacher, scientist and philosopher - just like scores of his other students spread out in various different parts of the world!

●●●

25

S.I.O. Dichotomy or 'Speciation' of a New Confrerie, the 'WDA'?

The basis on which the International Society of Odonatology (= Societas Internationalis Odonatologica, S.I.O.) was established, was not only very strong but was also in safe hands of its general overseer, Professor Dr Bastiaan Kiauta and his highly efficient team of advisors and learned colleagues that included some of the world's most eminent and influential odonatologists such as Prof. Minter J. Westfall Jr. (USA), Prof. F. Schaller (France), Prof. Philip S. Corbet (UK), Dr B.F. Belyshev (USSR, now Russia), Dr S. Asahina (Japan), Dr J.A.L. Watson (Australia), Prof. Janny M. van Brink and Dr M.A. Lieftinck (The Netherlands), just to name a few. Guided by a meticulously drafted "Constitutition and By-Laws" of the SIO, together they fortified each and every decision taken collectively and unanimously that included, among most important, the following:

- Publication of *Odonatologica*, the mainstream journal of the society, on quarterly basis.
- Publication of up-dated world odonatological literature's reviews in each issue of *Odonatologica*, as absracts under a permanent fixture, *Odonatological Abstracts.*
- Organization of the biennial organization of the International Symposia of Odonatology, in different countries all over the world each time.
- Establishment of the International Odonatological Research Institute (IORI) in the United States of America.
- Establishment of the National/Regional Offices of SIO, in different regions.
- SIO affiliation to various different entomological associations worldwide.
- Representation of SIO in the IUCN/SSC's Odonata Specialist Group.
- Assistance to odonatologists in their research works by providing literature and other materials.
- Financial help to budding and promising odonatologists.
- Developing and financing research projects in different countries.

By mid-1980s the SIO, groomed on the philosophy of "Family of Friends" by Dr Kiauta, evolved at a blitzkrieg pace to become the only entomological society for a single order to have enrolled more than 500 members, and backed by so many remarkable scientific activities spearheaded prominently by its regularly organized biennial symposia, quarterly *Odonatologica*, and the establishment of nearly a dozen National/Regional Offices across the world! The uniquely ambitious project of IORI also got underway for an early establishment in the USA under the charge of Prof. Minter J. Westfall, as its first Director.

The SIO was becoming a highly powerful nucleus of odonatological activities at a time when the world of science was itself moving toward a great deal of transformation, due largely to the availability of personal computer and knowledge on huge data retrieval with better systematic arrangements or moieties. In a few minds with different thinking, the powerhouse of the SIO was assuming the form of a 'world tree', or *Kalpavriksha* (Devanagari: कल्पवृक्ष, also known as kalpataru, kalpadruma or kalpapādapa, is a wish-fulfilling divine tree in Hindu mythology. It is mentioned in Sanskrit literature from the earliest sources), and the first sign of "speciation" appeared during the Tenth International Symposium of Odonatology (1989) when in the framework of General Body Meeting, chaired by Dr Bastiaan Kiauta, the then SIO President, a voting exercise was enacted, only to assert sowing of the seeds for an inevitable 'dichotomy' of the society in near future! The incessant events that followed made the writing on the wall crystal clear – the birth of a new odonatological association, Worldwide Dragonfly Association, in 1997. The mother society of SIO, too, was reorganized as the International Odonatological Foundation, Societas Internationalis Odonatologica (FSIO). To Dr Kiauta, who has always regarded and maintained the SIO as an association of friends, and for which he has selflessly devoted all his prime period of life, often at the cost of social and family life, leave aside many other worldly matters, this was the most painful day hardly endurable by ordinary mortals in any way. But, Dr Kiauta braved his way out of three heart attacks, two of them severe leading, at one time, even to paralysis of one-half of his body! Nothing new under the sun, not even his constantly failing sight and a hugely worrisome major surgery on Mrs Marianne Kiauta, could shake Dr Kiauta's resolve to bring multifaceted academic and scientific prosperity to the SIO, publishing *Odonatologica* and *Notulae odonatologicae* as scheduled and organizing biennial symposia in different parts of the world! The world of odonatological fraternity would possibly never know about similar endless stories of his braveheartness since what mattered to Dr Kiauta was only dragonflies – the SIO!

Dr Kiauta kept me updated with his long letters, often with several enclosures, on all the developments within and outside the SIO. So did Professor Philip S. Corbet (UK) and also Dr Gordon Pritchard (Canada). I knew that Dr Kiauta held both of them in high respects, and I also then infallibly understood from the communications from both Drs Corbet and Pritchard as to how lofty reverence they showered on him. The inevitable split or dichotomy of the SIO, therefore, was at best faux pas which nobody wanted, but it had to be, just as the laws of time with respect to biological speciation from the parental stock are inexorable! More challenges awaited ahead for both the SIO(F) and the WDA!

During 2008 when I met with him in his home in Bilthoven, I saw into his eyes a clear gloom of worry related to future organization of international odonatological symposia. While, the SIO's symposia were being regularly organized biennially on odd years since 1971, the WDA had also programmed to hold their series of biennial conferences which often fell close to the dates of the former. Under the circumstances, Dr Kiauta was constantly receiving sentiments and feelings of a large number of odonatologists toward a difficult situation of fund raising to enable participants to attend the symposia. He was really very much worried about the 'waste" of valuable resources, both money and time, and was contemplating to make an arrangement with the WDA for holding joint conferences so that all odonatologists could best participate with little efforts. With a glimmer of hope in his half-opened, pensive eyes he would say, "Dragonfly researchers need a pleasant atmosphere", reminding me of the soulful composition '*Early Spring*' by William Wordsworth:

Through primrose tufts, in that green bower,
The periwinkle trailed its wreaths;
And 'tis my faith that every flower
Enjoys the air it breathes.

●●●

26

The Kiauta-Corbet Era

In modern times of dragonfly research, i.e., Corbet's (1991) "The Intercommunicating Strand" or, for convenience of better understanding, post World War II period, the world has witnessed emergence of a large number of odonatologists who attracted international attention for their indispensable contribution to the science of Odonatology, such as, Dr S. Asahina, Dr B. Elwood Montgomery, Dr Minter J. Westfall, Dr F.C. Fraser, Dr R.M. Gambles, Dr E. Pinhey, Dr. B.F. Belyshev, Dr A.B.M. Machado, Dr M.A. Lieftinck, and Dr N.W. Moore, just to name a few for example. Yet, two of the geniuses who, in my opinion, influenced the world dragonfly community to an inestimable extent are Prof. Dr B. Kiauta and Professor Dr Philip S. Corbet.

Like Dr Kiauta, the life of Dr Corbet (21 May 1929 -13 Feb. 2008), too, has been enormously illustrious and successful in making a specific niche for him in the annals of odonatology as well as becoming the world's foremost odonatological authority. The range and depth of his knowledge was astonishing and his recall of scientific details and events was unparalleled. He remembered almost everything he ever saw, heard, or read, and never ever forgot the sequence of events in biology and behaviour of the vast number of dragonfly species he encountered during his frequent travels across the world. He was an excellent and all-round naturalist and biologist who studied various different animal groups, such as the non-cichlid fishes and mosquitoes, besides of course the dragonflies - a subject which absorbed him for the whole of his working life whenever circumstances allowed it, producing some of the most valuable literature, including books, on dragonflies; the last book he wrote on "Dragonflies", together with Stephen Brooks, was published in 2008! While his prowess as a biologist of international standing is widely recognised and respected, his professional working life too always seemed to be exceptionally well ordered and planned and he meticulously executed all his research work. He had an immense sense of right and wrong, and he considered one needed to observe the official guidelines and rules governing the organisation of scientific associations. An abundantly prolific writer, immaculate speaker and discussant, Dr Corbet was truly an odonatologist *par excellence*. In a way both Dr Kiauta and Dr Corbet possessed certain similarities; both of them had an early grooming as a naturalist-entomologist, dragonflies became their life-long object of intensive and extensive investigations, they were both utterly dedicated to the growth and development of the International Odonatological Society (SIO) as a 'family of friends' and invariably always discussed every minute detail associated with any activity under the Society, such as, for example, the establishment of International Odonata Research Institute (IORI), and, above all, helping each others in editing manuscripts submitted to SIO journals and improving unequivocally the quality of SIO conferences (Fig. 26.1).

(a)

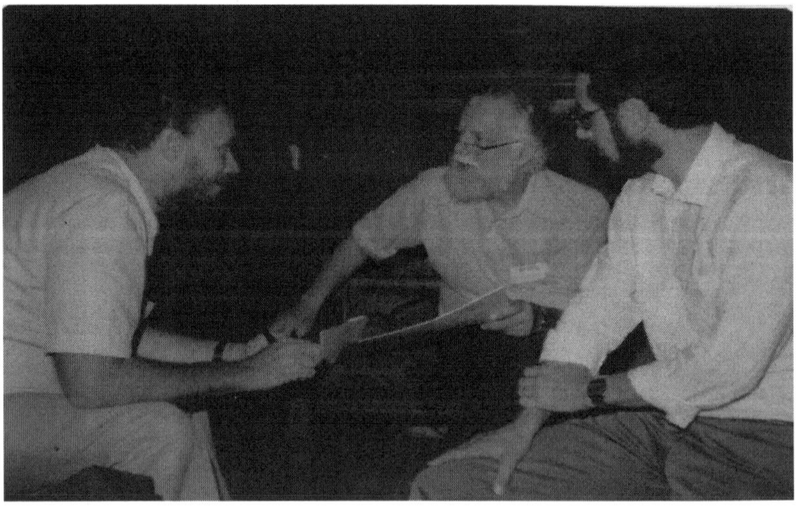

(b)

Fig. 26.1. (a) Dr B. Kiauta and Dr P.S. Corbet, among other distinguished guest-participants of the 9th SIO Symposium (1988), being greeted by Dr. S. Krishnaswamy, Vice Chancellor of Madurai Kamaraj University, Madurai, TN, India, and (b) Prof. Kiauta in discussion with Prof. Philip S. Corbet, as another researcher looks on. (Source: B.K. Tyagi)

I have had great pleasure in meeting with both, Dr Kiauta and Dr Corbet, in Paris (1985), Madurai (1988) and Tennessee (1989). I found both of them being always very pleasant, composed and friendly – blessed with lofty human qualities. They were always happy to share their expert knowledge, treating young and older odonatologists as equals, so that numerous odonatologists of all ages and levels of expertise and experience would be inspired by their enthusiasm and encouragement. Best of all, their patient listening capacities have been legendary and they always appreciated one's own knowledge and research work. I could often spot both of them sitting together and having some kind of a dragonfly discourse, in low voice and smilingly. Obviously Dr Kiauta regarded Dr Corbet beyond comprehension, and *vice versa*. I recollect here under a few episodes down the memory lane which show how much respect and understanding they had for each other in their heart. In his highly laudable seminal essay on "A brief history of odonatology" (1991), Dr Corbet has opulently praised Dr Kiauta and remarked, "*The explosive increase of scientific publication since the Second World War has demanded measures to come to terms with it. Odonatologists are exceptionally fortunate that intitiatives have been generated within their science to develop effective systems for communicating and disseminating information. At the national level, the first step was taken in Japan, by Syoziro Asahina, with the establishment in 1958 of Tombo, a journal devoted exclusively to reports about Odonata. At the international level, an important move was made by B. Elwood Montgomery when he convened a Colloquiumon Odonata at Purdue University, Indiana in 1963 (Montgomery, 1963), the year in which he started Selysia, the first international newsletter of odonatology. But by far the most significant development for the subject as a whole has been the formation in 1971 of the Societas Internationalis Odonatologica (S.I.O.) (Kiauta & van Brink, 1972). The prime mover and current President of S.I.O. is Bastiaan Kiauta, the magnitude of whose contribution to this conspicuously successful venture cannot be exaggerated....*".

In Paris Dr Corbet had arrived on the evening of the pre-symposium get-together, at a time when some of us, the participants, were enjoying light discussions over a cup of tea. I joined the table with Dr Kiauta, Mrs Marrianne Kiauta and Dr Schmidt. Suddenly I saw a pure smile coming on the face of Dr Kiauta as he murmered about the entry of Dr Corbet, "Philip has arrived now". Dr Corbet came straight to Dr Kiauta who stood up to greet him huggingly. Both exchanged briefly some pleasantries and, before Dr Corbet could think to move elsewhere, Dr Kiauta asked him to join on the table and additionaly arranged a chair for him then and there! Dr Corbet greeted every one of us in his inimitable manner, giving me a moment to wonder on his similarity to the great zoologist, Charles Darwin. While I was awe-struck and seemed to be lost in some thoughts, Dr Kiauta brought in another pleasant news to me, "Philip is the nephew of the famous naturalist, Jim Corbet, who founded the famous Jim Corbet National Park in northern India". This information was new to me and I felt a great feeling of pride inside me. Above all, their friendship had put an enormous influence on me.

After the Paris Symposium I began corresponding with Dr Corbet and on request he even gifted me a copy of his *magnum opus*,"A biology of dragonflies" (1962), with a highly inspiring signed dedication. During this symposium I, on behalf of Dr S. Mathavan, had accepted the 'the SIO Insignia' as an indication of taking responsibility to organize the next SIO symposium in India, in 1988. I requested and welcomed all the participants to the holy city of Madurai, on the side of the famous Western Ghats, and to the picturesque venue of the historical Madurai Kamaraj University in the southern Indian state of Tamil Nadu. As conventional, the SIO used to organize within the framework of each of its symposia a special seminar on a specific topical issue and this time at the Madurai symposium the topic was about migration of *Pantala flavescence.* Dr Kiauta suggested to Dr Corbet to coordinate the seminar and the latter was all too pleased to oblige. Later, Dr Kiauta published the outcome of this interactive meeting in the form of a report in the SIO-IUCN series. This kind of camaraderie between the two senior odonatologists was a great example of efficient teamwork for many young scientists who will be the torch-bearer tomorrow, as the poet and essayist Samuel Taylor Coleridge wrote, "*The dwarf sees farther than the giant, when he has the giant's shoulder to mount on*". Contributions of both these giants have been disproportionately great and influential over many decades.

The history of odonatology is very rich with enormous knowledge of fossil Odonata as well as every conceivable discipline commencing from extensive work on taxonomy, systematics, evolution, biogeography, biology, physiology, anatomy, neurology, cytogenetics, biocontrol tools, and behaviour etc. by several distinguished researchers. However, before the International Society of Odonatology was founded in 1971, most of the dragonfly workers were unconnected, at least on regular basis, nor were they having a platform dedicated to publishing their research results. The SIO aptly filled in this gap, on one hand, and provided them a number of means to grow scientifically in their respective fields. The main anchor of this continuing process was its mainstream journal, *Odonatologica*, which, together with its side stream periodical, *Notulae odonatologicae*, will publish more than 60-80 manuscripts annually. Subsequently some more subject specific periodicals also joined these, such as, for example *Advances in Odonatology* and *Rapid Communications* etc. SIO's biennial symposia were always eagerly awaited since they provided an opportunity to meet and discuss with many giants of the subject, on one hand, and participants of all ages could freely exchange their ideas about dragonfly research so as to identify experts in their chosen fields or disciplines, on the other. At the same time systematic compilation of Odonatological Abstracts (OA) from around world in each issue of *Odonatologica* amassed an immeasurable support of the world dragonfly community about which Dr Corbet too spoke in great veneration of Dr Kiauta's achievement,"*These Odonatological Abstracts, now numbering more than 7,000, constitute an essential source of reference for the serious student of Odonata*". Thousands of dragonfly researchers, both professional and amateur, were fortunate users of the OA services of unfathomable value offered by S.I.O.

Odonatologists and the science of odonatology were never so strongly integrated, both personally and professionally, in past as possibly during the SIO. While the SIO's backbone, *Odonatologica,* successfully disseminated the results of scientific works of odonatologists from all over the world, in four languages, i.e., English, French, Italian and German, with an inexorable provision of *Odonatological Absetracts* services with each of its issues, the SIO biennial symposia provided the world's dragonfly enthusiasts as well as serious students of odonata a unique platform for free, frank and friendly exchange of scientific views under the umbrella of SIO being a 'family of' friends' rather than a hierarchically structured society. Odonata were always the most fascinating subject to their investigators in past, but, in my opinion, both the science of odonatology and odonatologists have taken a giant leap forward only during this truly 'integrating and intercommunicating' era – the Kiauta-Corbet Era!

●●●

27

To India, with Love!

"I saw her dance so comelily,
Carol and sing so sweetly,
And laugh and play so womanly,
And look so debonairly,
That, certes, I trow that nevermore,
Was seen so blissful a treasure."

[Though the above lines, originally composed in French about 500 years ago, were made in a different context, yet, in my studied opinion, may certainly also be seen as most commensurably appreciating the ever mesmerizing nuptial maneuvers of a female dragonfly, such as any of the four *Rhinocypha* species common in India.]

The instances of men and women scientists in the annals of odonatology, who by sheer dint of hard work, application and energy emerged champions in their respective disciplines, are indeed so numerous that they have long ceased to be regarded as exceptional. Looking at some of the more remarkable, it might almost be said that they all rose to the ranks of usefulness and influence by their irrepressible love for dragonflies, on one hand, and respect for the science of odonatology, on the other. One of the most illustrious geniuses in the field is Professor Dr B. Kiauta, widely remembered for, other than the achievements in odonate cytotaxonomy and cytophylogeny, his concept of regarding SIO as a 'family of friends' which invariably transcended the ever beatifying spirit of camaraderie among hundreds of dragonfly enthusiasts across the world and bringing them together periodically, every alternate year during the SIO biennial symposia, for free and frank discussions – all for the evolution of odonatology! A tireless crusader of the SIO spirit and dragonfly biodiversity conservation, Dr Kiauta is truly a global statesman with a great vision for the growth and dissemination of dragonfly science worldwide. During the time of his N.C.A.B.R. assignment in Nepal he had flown over India, by only touching Delhi, as many as nineteen times, and always nurtured a strong wish to visit this great country some day in near future. Therefore, while working on the Nepalese fauna, he also thoroughly studied and inventorized simultaneously all the existing data on cytogenetics and cytotaxonomy of the Indian dragonflies. His two monographs as well as several other publications are a testimony to this aspiration (*cf.* KIAUTA, 1974, 1975a). India, for its rich dragonfly fauna and a great odonatological history, thanks largely due to Col. Dr F.C. Fraser's extensive works culminating into his *magnum opuses* – the three Odonata volumes published in the world famous *Fauna of British India* series (*cf.* FRASER, 1933, 1934, 1936), attracted his attention and it naturally became the apple of his eye.

Dr Kiauta's one of the first contacts in India was with Dr S.K. Sangal, who was serving as a faculty at the Department of Zoology, in Dayanand Anglo Vedic College, Dehra Dun and he reverently associated the remarkable rise in odonatological explorations and transformation of the Dehra Dun Valley into a stronghold of dragonfly research tradition as a result of his personal influence. Having been impressed by the rare dragonfly collection at the museum of the Forest Research Institute & Colleges, located in Dehra Dun where some of the most distinguished entomologists such as Dr E.P. Stebbing, Dr C.F.C. Beeson, Dr. J. C. M. Gardner, Dr A.D. Imms and Dr M.L. Roonwal had worked, he chose to research on *Crocothemis servilia* morphology and biology for his PhD thesis, later awarded by the Agra University in 1961. A kind hearted and generous human being, Dr Sangal was as enthusiastic and energetic in the field as the dragonflies themselves. He had a great networking with almost all famous institutions in the north-western India, in particular.

During late 1960s Dr Sangal's first PhD student was the famous odonatologist Dr Arun Kumar with whom Dr Kiauta corresponded for many years and also supplied much literature on dragonflies, besides publishing his couple of research papers in the inaugural volume of *Odonatologica,* Nos. 2 and 4. Dr Arun Kumar was awarded PhD degree on his thesis titled "*Taxonomic Studies on the last Instar Larvae of Odonata from the Dehra Dun Valley, India*", under Meerut University in 1971. Later Dr Arun Kumar made great strides in course of his professional evolution at the famous Zoological Survey of India, retiring as its Additional Director and an expert animal ecologist. Almost simultaneously, four more ZSI researchers, i.e., Dr Mahabir Prasad, Dr Asket Singh, Dr A.R. Lahiri and Mr T.R. Mitra, came in contact with Dr Kiauta and were carefully groomed to become specialist in their respective disciplines. All of them have brought out many significant regional monographs on dragonflies and published papers in both *Odonatologica* and *Notulae odonatologicae,* but also occasionally in *Opuscula zoologica Fluminensia.* Among them they have described as many as a dozen new species of Odonata from different parts of the country.

One of the most strongly-marked features of Dr Kiauta's personailty is his spirit of industry, standing out prominent and distinct in his past history, and as strikingly characteristic of him now as at any former period. It is this sense of humility, camaraderie and yeomanship, displayed in all his deeds accomplished, which has laid the foundations and built up the scientific greatness of discipline, odonatology, in India. This vigorous growth of the science of odonatology has been mainly the result of the free and frank discussions among all dragonfly workers, old and young as well as seasoned and amateur alike. Thus, when Dr Kiauta organized the 1st International Symposium of Odonatology in Ghent, Belgium, it was attended (in absentia) by at least two Indian scientists, with one of whom, i.e., Dr K. Chatterjee, he had later collaborated to publish a paper (CHATTERJEE, K. & B. KIAUTA, 1973). His name figures as an expert in some of the major publications such as Zoological Survey of India's (1991) "Animal Resources of India – Protozoa to Mammalia: State of the Art". During the period of his editorship of *Genetica,* some of the prominent Indian coleopteran cytogeneticists (e.g., Dr J.S. Yadav, Kurukshetra University, Rohtak, Haryana State) and arachnid karyosystematists (e.g., Dr. O.P. Mittal, Panjab Uniuversity, Chandigarh) were in extensive correspondence with him. At least two of his student-like odonatologists went on to join a rather unusual line of profession and yet appreciably lofty ranks; Dr. R. Sandhu, an odonate cytogeneticist and a PhD from Punjabi University, Patiala, is currently a senior Deputy Superintendent of Police in Punjab State and Dr B. Suri Babu, a biologist and a PhD from Sagar University, Madhya Pradesh, is a Joint Director in Forensic Cell, Police Department, Madhya Pradesh. Several universities have empaneled Dr Kiauta for the evaluation of doctoral dissertations.

In mid-1975, Dr. S.K. Sangal, under whom I had enrolled for PhD in his department, got me introduced to Dr Kiauta through two manuscripts intended to be published in *Odonatologica.* Within a fortnight Dr Kiauta sent Dr Sangal his reply, with elaborate comments on each manuscript, copy of which was also marked to me by name. Thus began a seemingly endless journey of correspondence between Dr Kiauta and me, the quantum of which touching a thousand mark by now and dealing with only odonatological subjects, viz., manuscripts, administration and management of SIO National Office in India, production of the annual journal *Indian Odonatology* and six-monthly newsletter *Fraseria,* organization of biennial Indian Symposium of Odonatology. *Fraseria* (Editor: Dr B.K. Tyagi) was to be produced soon after the establishment of the SIO National Office in India, located at my official address of posting in a tribal region hyperendemic for malaria in southern Gujarat State. As such my resources fell short of expectation to produce a newsletter – a new experiment for the science of odonatology in India! I explained the situation to Dr Kiauta who immediately came to my relief, assuring production of at least first few issues till my office became self-sufficient in the matter. I learnt my first lrssons in producing and editing a periodical during this long span of *Fraseria* production under his watchful eyes since the initiation of the mainstay journal, *Indian Odonatology* (Executive Editor: Dr B.K. Tyagi), had still to wait for mnay years to come. Constitution of editorial board for *Indian Odonatology* was a huge challenge since the journal, an official organ of the SIO National Office in India covering formally, besides the host nation, also Sri Lanka, Nepal and Bangladesh, was presented rather as a regional carrier. Here at this moment, once again, Dr Kiauta came to my rescue and suggested inclusion of names of some of the odonatologists from anywhere in the world who had great practical experience with the south Asean regional odonata (e.g., S. Brooks of BMNH, UK and M. Hämäläinen from Helsinki University, Finland). As a policy all manuscripts submitted to the journal were peer reviewed. In order to sustain the journal he even arranged a sizeable subscription from overseas odonatologists. Besides, in order to assist in matters of editorial excellence he benevolently considered supply of SIO publications, especially *Odonatologica* and *Notulae odonatologicae*, on regular basis which proved vital.

Dr Kiauta invariably shared with me all his actions taken at the level of SIO Central Office, as his transparent style of functioning. This infused a lot of confidence in my own management of the SIO National Office in India. Accordingly literature archived in the SIO National Office/Regional Office in South Asia was open to all and, in the footprint of Dr Kiauta, assistance with xerox copies of reprints etc. were provided to anyone who sought them and dragonfly samples collected by an odonatologist were got identified by experts and outcome conveyed to the concerned. Management of SIO National Office finances, including membership, sale proceeds etc., though small, was nevertheless a big challenge albeit a bank pass-book and a proper day-to-day closure of ledger register of all expenditure! To tide ove this I followed the advice from Dr J.M. van Brink which she offered to me during her visit to India in 1988: "*Complete daily the ledger book of expenditure and receipts, and tally the total, for if it was not done then it would be like'No expenditure, no receipts' that day*."I emulated her advice ever since in all my financial transactions!

Just as Dr Kiauta would help anyone else anywhere in the world, he has also helped me enormously during the days of my doctoral research and, in fact, thereafter ever since. More than the manifold materialistic assistance of diverse nature, he most providentially offered his invaluable guidance throughout my professional development and truly mentored in all strives for excelling in scientific research. In essence, he sowed in the mind of his Indian odonate associates the seeds of ingenious research spirit – through ceaseless perseverance, focused application and singleminded devotion to the science of odonatology, which reminds me of the famous saying by Samuel Smiles in his classic "Happy Homes and the Hearts That Make Them":

"*Sow a thought, and you reap an act;*
Sow an act, and you reap a habit;
Sow a habit, and you reap a character;
Sow a character, and you reap a destiny."

Dr Kiauta initiated four major activities in India for which the entire odonatological community continue to reap richly benefits of his perseverance, farsightedness and patronage:

(i) Established the SIO National Office in India

(ii) Organized the 9th and 18th International Symposia of Odonatology

(iii) Appointed a Member in-Charge in India of Odonata Specialist Group (IUCN/SSC)

(iv) Encouraged establishment of Indian Dragonfly Society (IDS) (2017)

(a)

(b)

(c)

(d)

(e)

(f)

(g)

(h)

(i)

(j)

(k)

(l)

(m)

(n)

(o)

(p)

(q)

(r)

Fig. 27.1 (a-r). Some of the Indian odonatologists who have associated and/or corresponded with Dr B. Kiauta:- (a) Dr S.K. sangal, (b) Dr Arun Kumar, (c) Dr B.K. Tyagi, (d) Dr D.B. Tembhare, (e) Dr V.K. Srivastava, (f) Dr S. Mathavan, (g) Dr B.K. Srivastava, (h) Dr T.R.

Mitra, (i) Dr A.R. Lahiri, (j) Dr B. Suri Babu, (k) Dr Gurinder Walia, (l) Dr Mahabir Prasad, (m) Dr Francis K. Kakkasserry, (n,o) Mr C.G. Kiran, and his new species, *P. ponmudiensis* (p) Dr R.J. Andrew, (q) Mr Kumaran Sathasivam, (r) Dr K.A. Subramaniam.
(Source: B.K. Tyagi)

Compared to a near standstill or, at best, snail-paced research scenario of any concerted research activity till 1970, in India Dr Kiauta invariably infused life and helped resuscitate the science of odonatology in a massive way particularly by mentoring young and budding odonatologists during 1970s onward! A remarkable illustration of his energetic promptitude was the manner in which he solidly provided, on a great emergency, literature on various different aspects of dragonfly biology in other countries of the world to Drs Sangal, Arun Kumar, B.K. Tyagi, Mahabir Prasad, S. Mathavan, D.B. Tembhare, R.J. Andrew, T.R. Mitra, A.R. Lahiri, B.K. Srivastava, V.K. Srivastava, B. Suri Babu, R. Sandhu, Gurinder K. Walia etc., just to name a few. This kind of mentoring gesture, or 'early schooling' in the science of odonatology, apparently helped sow the seeds of a great learning and a never ending process of developing a healthy tradition of dragonfly research in India.

With the enrichment of indigenous research, Indian odonatology saw a spurt in publication of a large number of highly useful regional reports, inventories and field guides, notable among which are LAHIRI (1987), MITRA (2002), TYAGI (2008), SUBRAMANIAN (2009), KIRAN & RAJU (2013), to name a few. While Mr Mitra, Dr M. Prasad, Dr Tyagi and Dr Subramanian have been associated with Dr Kiauta through editorial boards of *Odonatological Abstracts* and *Notulae odonatologicae,* Dr Lahiri extensively corresponded with him about his manuscripts. Mr Kiran, discoverer of *Protosticta ponmudiensis* Kiran et al, 2015, from the Western Ghats region and a down-to-earth naturalist, knew about Dr Kiauta mostly through my discussions and the initiative to establish Indian Dragonfly Society, with dragonfly habitat conservation as a major objective.

Today, besides the traditional Indian Dragonfly Society and the SIO South Asian Council of Odonatology. the dragonfly research in India is tremendously promoted by several web-based portals/Facebook groups among which most prominent sites are (i) *dragonflyindia* by Parag Ajit Rangnekar/Pankaj Koparde, who have been annually organizing a hands-on field-based training since 2014, sometimes in collaboration with the SIO SACO, also the latter's Symposium of Indian Odonatology (9th in Goa), (ii) *International Odonates* and *Indian Biodiversity Conservation Network* by Mr Dheerendra Singh, who also authored (along with Brijendra Singh and Jan T. Hermans) the book "*Dragonflies and Damselflies (Odonata: Insecta) of Keoladeo National Park, Rajasthan, India*". Dr Hermans, a Dutch odonatologist teaching in a college there, is a close acquaintance of Dr Kiauta and, led by Mr Dheerendra, is currently extensively mapping the Indian dragonflies' distribution, and (iii) *indiabiodiversity* which extensively deals with the dragonflies of India, as a specialized page.

From a mere 110 odonatologists in India during the 1970s (TYAGI & TYAGI, 1986) the quantum of dragonfly enthusiasts on various different electronic media has risen manifold probably crossing the 3000 mark, thanks to fruitful and encouraging initiatives of a large number of mentors. However, the time is witness that of all the senior odonatologists who fruitfully guided Indian workers during past five decades, Dr Kiauta's contribution as a mentor is particularly noteworthy. He actually reached out on his own to all even on a mildest whisper of need. He invited manuscripts and published them in *Odonatologica* and *Notulae odonatologicae*, sought abstracts and published them in *Odonatological Abstracts*, invited active and promising dragonfly enthusiasts and arranged funding to help them participate in the SIO International Symposia of Odonatology, arranged for budding odonatologists university scholarships in different countries, and even helped many research scholars secure employment. His contribution to the science of odonatology and odonatologists in India has been seen as a benchmark for generations to follow. Much of this he loves to do even now regardless of an advancing age and many other imminent preoccupations!

In India, Dr Kiauta is looked upon with great veneration, almost like a father-figure, whose guidance and mentoring have always shown a right path to tread. His life — inevitably indispensably singularly dedicated to dragonfly research and a family-like environment within the SIO, has been tremendously influential in questing our own objectives in India. To many in India his life and research works are lessons to comprehend that it is not the result in any case that is to be regarded so much as the aim and the effort, the patience, the courage, and the endeavour with which desirable and worthy objects are pursued with diligent self-culture, self-discipline and self control – and above all … that honest and upright performance of individual duty which is the true glory of character:

"Success is a journey, not a destination.
The doing is often more important than the outcome."

– Arthur Ashe.

●●●

SECTION VI: IN THE SERVICE OF GOD'S OWN CREATIONS

28

Engagement Post-Retirement

"... I did what I could, but there were many imperfections and shortcomings in everything."

—Dr. B. Kiauta

On January 20th, 2017, Emeritus Professor Dr B. Kiauta celebrated the 80th anniversary of his life that was and, although having taken retirement from the position of Executive Editor of *Odonatologica* in December 2013, continues to be largely devoted to the science of Odonatology, as abundantly exhibited by over a dozen-and-a-half research publications on a variety of subjects, interviews to journals and scientific societies, scientific expeditions, involvement in editing for several journals etc. etc. With the initiative on the foundation of Societas Internationalis Odonatologica, SIO (1971), the journal *Odonatologica*, including its section of *Odonatological Abstracts* (covering 19 871 works, published between 1971–2013), with the initiation of various other odonatological periodicals and regional newsletters, with the establishment of International Odonata Research Institute and a string of National/Regional Offices, and with the stimulation of biennial International Symposia of Odonatology, he has given an impetus to the worldwide Odonata research and greatly facilitated the communication within the odonatological community of the globe during the past half of the century. His life and works have been earlier described by various notable odonatologists and others (e.g., CORBET, 1991; BEDJANIČ, 2007, 2015; GOGALA, 2007; MOORE, 2007; TYAGI, 2007a, 2007b; INOUE, 2007; 2017). From 2014 onwards Dr Kiauta is continuing his work on the SIO Board of Trustees and on various odonate research projects, reminding me of the most inspiring lines from the famous poem, *Agneepath*, by Shri Harivansh Rai Bachchan, dedicated to those whose life is a tough struggle all the way, but whose resolve is superhuman:

तू न थकेगा कभी,
तू न थमेगा कभी,
तू न मुड़ेगा कभी,
कर शपथ! कर शपथ! कर शपथ!
अग्निपथ! अग्निपथ! अग्निपथ!

Transliteration: *Tu na thakega kabhi,*
Tu na thamega kabhi,
Tu na mudega kabhi,

Kar shapath! Kar shapath! Kar shapath!
Agneepath! Agneepath! Agneepath!
Translation (adopted from Riku Sayuj):
You shall never tire,
You shall never slow down,
You shall never turn back,
This oath you will take today!
This oath you will fulfill in your life!
Take this oath!
And walk the Path of Fire, every single day!
The oath of fire! Yes, That Path of Fire!

A scientist and teacher *extraordinaire*, he has students and colleagues in all continents of the world, some of them leading odonatologists (e.g., Dr Milen Marinov, Dr. Lillian M. Mola). Dr Kiauta is immortalized in the annals of odonatology through the names of 14 odonate spp. by 18 authors (1986-2013). For the list of these see Annex 4. A keen taxonomist himself, Dr Kiauta has also described *Croatocordulia platyptera* (Charpentier) gen. nov. in 1969 and *Crocothemis servilia mariannae* ssp. nov. in 1983.

With several honours and decorations won (1981- Societas Internationalis Odonatologica (S.I.O.): *Membership of Honour*; 1993 - Slovene entomological Society: *Golden Medal*; 2002 - Queen of the Netherlands: *Knighthood in the order Oranje-Nassau;* and 2005 - The Netherlands Association of Odonatologists: *Golden Dragonfly Medal*, Dr Kiauta is always extremely polite, soft-spoken and generous to all his friends, colleagues and students. In the final issue of *Odonatologica* under his charge, he went all out to express his genuine thankfulness for the initiative of Drs M.Th. Wasscher (Utrecht) and Prof. Dr M. Gogala (Ljubljana) that resulted in a "national acknowledgement" of his work in, respectively, The Netherlands (2002) and Slovenia (2007). He was elected an Associate (2007) and subsequently a Full Member (2015) of the Slovenian Academy of Sciences and Arts in Ljubljana. This is a rather prestigious position, rendering him lifelong also an additional monthly income, which he is spending entirely in his odonate research. The number of Members in the Academy is strictly limited and includes fewer than 20 outstanding biologists from all over the world. *Odonatologica* was not only a lifeline for the SIO but to Dr Kiauta, too; the journal was life personified, along with its authors and contents for which his heartfelt thanks reverberated for all who were/are supporting *Odonatologica*, either by submitting or mediating the submission of manuscripts, offering suggestions and/or constructive criticism, helping in editorial details, promoting circulation of the journal and/or assisting in any other way, not the least by their encouragement, expressed in correspondence or otherwise. He was ever prompt to specifically thank individually, and though an exhaustive list of the persons involved in rendering help to him would be too long to mention here, he even thanked Dr B.K. Tyagi (Madurai, India), for his very humble and sincere contribution to *Odonatologica* via *Odonatological Abstracts* during the years he served as the Head of the SIO National /Regional Office in India and SE Asia but most particularly also for his initiative and editorial preparation of the monumental Festschrift "*Odonata: biology of dragonflies*" (2007, Scientific Publishers, Jodhpur). The book took him completely by surprise on his 70th birthday and, in view of Dr Kiauta's opinion, continues rendering manifold encouragement ever since. Not to forget in the least the real strength to this book, the authors, he warmly thanked also to the 38 authors, from 17 countries that had contributed papers published in that volume. All these laurels and honours bestowed on the great Dr Kiauta owe to a very dear and respected person, his beloved wife - Marianne Kiauta, herself a highly dedicated odonatologist, who is the greatest source of inspiration and support for Dr Kiauta.

Since his retirement from the responsibilities for SIO (1stDecember 2013), Dr Kiauta's days continue to be filled up with odonatological works (Fig. 28.1). This includes the completion of various research papers, for which he had no time during the all too busy service years. He is meticulously following up the current odonatological literature and helping the younger colleagues. Among his current research projects is also the history of odonatology in Slovenia. His odonatological correspondence continues to remain appreciable, helping with suggestions, advice and otherwise, whenever requested. His appreciable odonatological library of some 60,000 reprints and books remains open for consultation to all who need it. One of his younger colleagues, Dr Hermans, also from the Netherlands, who has been recently visiting India regularly for dragonfly exploration, informed that Dr Kiauta's library is so rich and diverse that it is always an uphill task to arrange the literature systematically, and yet he does it – absolutely meticulously!

Fig. 28.1. Professor Dr B. Kiauta at the desk in his study (Photo: Jure Javoršek, May 2012).

Dr Kiauta, along with his wife, Marianne, lives (since 2008) in the southeastern province of the Netherlands. There they have realized a centre for animals that are old or otherwise not anymore taken care of, all are welcome. They live now in company with 16 rabbits (in a special room in their house). During my visit to their earlier house in Bilthoven, I have of course seen some of them already. Outside, they have a meadow with 5 goats, another meadow with 8 Shetlander ponys and again another meadow with 2 Arabian horses. Around the house a flock of 20 chickens are roaming free, they were saved from the battery cages. And there are always birds to feed, not only the small ones, but also pheasants, pigeons, jays etc. For the hedgehogs, and the wild rabbits they also serve apples and grains. This puts them both to quite a daily task, but very much satisfying (Fig. 28.2).

(a) (b) (c) (d)

Fig. 28.2 (a-d). Dr Kiauta and Mrs Marianne at the asylum for the discarded animals.

Dr Kiauta is only 82 years young now and may the Almighty God shower on him as many years of youthful good health, love and affection.

REFERENCES CITED

ASAHINA, S., 1964. Nepalese Odonata taken by the botanical expedition of Tokyo University in 1963. *Akitu* 12(1): 11-12.

ASAHINA, S., 1965. The dragonflies taken by the Rikkyo University Himalayan expedition, 1964. *Akitu* 12(3-4): 33-34.

ASAHINA, S., 1974. Nepalese Odonata taken by Dr. J. Martens in 1969/70 and 1973. *Senckenbergiana biol.* 55 (4/6): 281-291.

ASAHINA, S., 1993. A list of the Odonata from Thailand: Parts 1-221. Brothers of St. Gabriel in Thailand, Bangkok, 448pp..

BEDJANIČ, M., 2007. Prof. Dr Boštjan Kiauta – the father of Slovenian odonatology. *Erjavecia* 22: 6-10. – [Slovenian]

BEDJANIČ, M., 2015. Interview with Boštjan Kiauta. *Trdoživ* 4(2): 23-28, with portrait. – [Slovenian]

CORBET, P.S., 1991. A brief history of odonatology. *Advances in Odonatology* 5: 21-44.

CORBET, P.S., 1999. *Dragonflies: behaviour and ecology of Odonata.* Harley Books, Martins. xxxix+830 pp.

DARWIN, C.R., 1859. *On the origin of species by means of natural selection, or the preservation of favoured races in the struggle for life.* John Murray.

DASGUPTA, J. (1957). Cytological studies on the Indian dragonflies. 2. A study of the chromosomes during meiosis in thirty species of Indian Odonata (Insecta). *Proc. zool. Soc.* Calcutta 10: 1–66.

DAEMS, G. & H.J. DUMONT, 1974. Rotifers from Nepal, with the description of a new species of Scaridium and a discussion of the Nepalese representatives of the genus Hexarthra. *Biol. Jb. Dodonaea* 42: 61-81.

FRASER, F.C., 1933. *Fauna of British India, including Ceylon and Burma, Odonata,* Vol. 1, Taylor & Francis, London. xiv+ 423 pp., 1 map.

FRASER, F.C., 1934. *Fauna of British India, including Ceylon and Burma, Odonata,* Vol. 2. Taylor & Francis, London. xxiv+ 398 pp., 4 pls.

FRASER, F.C. 1936. *Fauna of British India, including Ceylon and Burma. Odonata,* Vol. 3, Taylor & Francis, London. xii + 461 pp., 2 pls.

GAMA, V., C. DA CRUZ LANDIM & A. FERREIRA, 1981. Estudos morfológicos da estruturados cinetocores de Enallagma cheliferum (Odonata). *Naturalia,* Sao Paulo 6: 75-87.

GOGALA, M., 2007. On the 70th birthday of Prof. Boštjan Kiauta. *Erjavecia* 22: 1-5, with portrait. – [Slovenian]

GOGALA, M., 2013. *On the trail of mountain cicadas.* Slovenian Academy of Sciences and Arts, Ljubljana & Royal Entomological Society, London 175 pp.

HAMALALNEN, M., 1989. Odonata from the Dehra Dun Valley (Uttar Pradesh,India) with notes on synonymy of some west Himalayan species. *Odonatologica* 18(1): 13-20.

HÄMÄLÄINEN, M. & A. PINRATANA, 1999. *Atlas of the dragonflies of Thailand: distribution maps by provinces*. Brothers of St. Gabriel in Thailand, Bangkok. vi + 176 pp.

INOUE, K., 2007. Foreword. *In*: B.K. Tyagi,[Ed.], *Odonata: biology of dragonflies*, pp. v-vi, Scientific Publishers (India), Jodhpur.

INOUE, K., 2017. Bastiaan Kiauta – octogenarian (1937–2017). *Odonatologica* 46(1/2): 1-24.

KIAUTA-BRINK, M.A.J.E.,1976. Some Tibetan expressions for "dragonfly",with special reference to the biological featuresand demonology. *Odonatologica* 5 (2): 143-152.

KIAUTA-BRINK, M.A.J.E., 1977. Further annotations on the Tibetan expressions for "dragonfly". *Odonatologica* 6(2): 69-76.

KIRAN, C.G. & DAVID V. RAJU, 2013. *Dragonflies and damselflies of Kerala*. Tropical Institute of Ecological Studies, Thiruvananthapuram.

LAHIRI, A.R., 1987. Studies on the Odonate fauna of Meghalaya. *Rec. zool. Surv. India, Misc. Publ., Occ. Pap.* No. 99. 402 pp.

LIEFTINCK, M.A., 1984. Further notes on the specific characters of Calicnemia Strand, with akey to the males and remarks on some larval forms (Zygoptera: Platycnemididae). *Odonatologica* 13: 351-375.

LINNAEUS, C., 1758. *Systema naturae*, Vol. 1. 10th ed. Laurentius Salvius, Holmiae.

MATHAVAN, S. & P.L. MILLER, 1989. A collection of dragonflies (Odonata) made in the Periyar National Park, Kerala, southern India, in January 1988. *Soc. Int. Odonatol. rapid Comm.* (Suppl.), 10: 1-10.

MAKINO, S. 1956. *A review of chromosome numbers in animals*. (revised ed.). Hokuryukan, Tokyo.

MITRA, T.R., 1982. Epiophlebia laidlawi Tillyard in Schedule I. *Selysia* 11(2): 28-29.

MITRA,T.R., 2002. Geographical distribution of Odonata (Insecta) of Eastern India. *Mem. zool. Sur. India* 19(9):1-208.

MOLA, L.M., 2007. Cytogenetics of American Odonata. (*In*: B.K. Tyagi [Ed.], Odonata: biology of dragonflies). Scientific Publishers (India), Jodhpur pp.153-173.

MOORE, N., 2007. Before and after 1971: a personal tribute to Bastiaan Kiauta. *In*: B.K. Tyagi, [Ed.], *Odonata: biology of dragonflies*, Scientific Publishers (India), Jodhpur p.ii..

PINRATANA, A., B. KIAUTA & M. HÄMÄLÄINEN (1988). List of the Odonata of Thailand and annotated bibliography. Brothers of St. Gabriel in Thailand, Bangkok.

POLUNIN.O., 1950. Plant hunting in the Nepal Himalayas. *Geogr. Mag.* 23(4): 132-147.

RAY-CHAUDHURI, S.P. & J. DASGUPTA, 1949. Cytological studies on the Indian dragonflies(Odonata). *Proc. Zool. Soc. Calcutta* 2: 81-93.

SANGAL, S.K. & B.K. TYAGI, 1984. Calicnemia miniata doonensis, a new subspecies from the Dehradun Valley (Dehradun), India (Zygoptera: Platycnemididae). *Abstr. Pap. 1st Indian Symp. Odonatol.*, Madurai, p.2.

SANGAL, S.K. & B.K. TYAGI, 1985. Calicnemia miniata doonensis, a new subspecies from the Dhradun Valley (Dehradun, India), (Zygoptera: Platyenemididae). *Proc. 1st Indian Symp. Odonatol.*, Madurai, pp. 193-198.

SANGAL, S.K., 1961. *Morphology and biology of Crocothemis servilia servilia (Drury) (Odonata: Libellulidae*). PhD thesis, University of Agra, India. 197 pp.

SITAR, S., 2012a. The Kiauta horse asylum. *Revija o Konjih*, Ljubljana 20(7): 54-55. – [Slovenian]

SITAR, S., 2012b. The Knight and the dragonflies: Boštjan Kiauta (*1937). *In*: S. Sitar, *Slovenian citizens of the World [Slovenci, državljani sveta]*, pp. 174-181. Kmečki Glas, Ljubljana (ISBN 978-961-203-414-0). – [Slovenian]

ST. QUENTIN, D., 1970. Odonata aus Nepal. *Khumbu Himal* 3(3):389-411.

SUBRAMANIAN, K.A., 2009. *Dragonflies of India: a field guide*. Vigyan Prasar, Department of Science and Technology. 168 pp.

TILMAN, H.W., 1952. *Nepal Himalaya*. Cambridge Univ. Press. xii + 272 pp., 39 pls.

TYAGI, B.K., 1977. A note on the karyotypes of Burmagomphus pyramidalis Laidlaw and Onychogomphus saundersi duaricus Fraser (Anisoptera: Gomphidae). *Odonatologica* 6:(4): 277-282.

TYAGI, B.K., 1978. *Studies on the chromosomes of Odonata of Dun Valley* (*Dehra Dun, India*). Ph.D. Thesis, Garhwal Univ., Srinagar, Garhwal. viii +285 pp. (17 tabs, 335 figs.excl.).

TYAGI, B.K., 1985. Dragonfly conservation in India – present status and future prospects. *Rep. Odon. Specialist Group Int. Un. Conserv*. 6:1-5.

TYAGI, B.K., 1988a. An introduction to the activities of the S.I.O. National Office in India. *SIO National Office in India,* Pondicherry, 37, pp.

TYAGI, B.K., 1988b. Odonatological bibliography of Frederick Charles Fraser [originally manuscripted by Basil Elwood Montgomery; re-organized, updated and edited by Dr B.K. Tyagi with the help and guidance of Prof. B. Kiauta, Editor, Odonatologica]. *Occ. Publs. SIO Natl. Off. India* 3: 1-24.

TYAGI, B.K., 2007a, [Ed.], *Odonata: biology of dragonflies. Festschrift [...] dedicated to Professor Dr Bastiaan Kiauta*. Scientific Publishers (India), Jodhpur. Frontispiece (portrait), xix + 366 pp.

TYAGI, B.K., 2007b. Glimpses into Prof. Dr Bastiaan Kiauta's life and odonatological achievements. *In*: B.K. Tyagi, [Ed.], *Odonata: biology of dragonflies*, pp. 2-8, Scientific Publishers (India), Jodhpur.

TYAGI, B.K. & A. TYAGI, 1986. Directory of odonatologists from the Indian subcontinent. *Occ. Publs S.I.O. Natl. Office India*, 1: 1-20.

VENKATESH, A. & B.K. TYAGI, 2013. Predatory potential of Bradinopyga geminata and Ceriagrion coromandelianum larvae on dengue vector Aedes aegypti under controlled conditions (Anisoptera: Libellulidae; Zygoptera: Coenagrionidae; Diptera: Culicidae). *Odonatologica* 42(2): 139-149.

WALKER, E.M. and P.S. Corbet. 1975. The Odonata of Canada and Alaska. Volume 3. Univ. Toronto Press, Toronto. 307 pp.

WHITE, M.J.D. 1973. *Animal cytology and evolution*. (3rd ed.). Cambridge University Press. 961 pp.

●●●

SECTION VII: SCIENTOMETRY OF PROFESSOR DR B. KIAUTA'S PUBLICATIONS AND OTHER ODONATOLOGICAL AWARDS, DISTINCTIONS AND HONOURS ETC.

EPILOGUE: SKY IS THE LIMIT TO AN ILLUSTRIOUS CAREER

Dr Kiauta's bibliography encompassing both Odonata and other invertebrates, but also certain vertebrates, has all the ingredients of making a monumental epic. To comprehend the scenario of his publications it is considered worthwhile to analyse his bibliographic scientometry. Ironically, though prepared singlehandedly by himself, those approximately 20,000 odonatological abstracts published in *Odonatologica* are not a part of his bibliography, nevertheless the Herculian effort of bringing out in light so many articles, often originating in different languages, nondescript periodicals, and in lesser known countries or regions, cannot be disregarded a bit at all!

"*Extraordinary men do not do extraordinary things, they do things extraordinarily*", goes the famous maxim thus which applies fully to Dr Kiauta, too, whose odonatological bibliography (1954-2018) has anything like 400 bibliographic units, including 168 of these in the *Odonatologica* section of *Odonatological Abstracts* (abstract Nos 1-19,871). Though he mostly published solitarily, yet several papers were published jointly with various colleagues. The non-odonatological papers (probably ca 150, of which 64 selected titles are listed here) constitute about 27.27%, albeit covering a wide range of disciplines, invertebrate and vertebrate groups and countries, such as, e.g., Slovenia, where Dr Kiauta's systematic research works on various animal groups, both vertebrates and invertebrates, including Odonata, are considered a landmark for professionals and beginners for making in-depth investigations in these groups. As single author (75%) he has published scientific research papers, notes, abstractions etc., as twin authors 86 (0.42%), three-authors 9 (0.044%), and finally four authors 2 (0.0099%) – which speak volumes of his scientific and academic prowess in documenting varied research undertakings. Notwithstanding 95% of his research manuscripts having been written in English, yet a sizable quantum of papers are also published in Slovenian, Dutch, German and French. It is noteworthy here that on many occasions, either in international conferences or universities across the world, he has formally delivered presidential, inaugural and/or keynote lectures in Russian, Italian, German etc. which have not been considered in present analysis for reason of clarity in documentation and analysis.

Most of Dr Kiauta's publications are published in international journals of high repute such as, e.g., *Odonatologica, Genetica, Chromosome Information Service, Tombo* etc. Odonatologically, nearly all of his research works are published in English, save for a few exceptions in French, German and in Slovenian. The latter during the initial stage of his career and, once again, after formal retirement from the SIO/ *Odonatologica* services in 2013, at the age of 76 years!

Dr Kiauta's dragonfly research publications, though mostly revolving around taxonomy through cytotaxonomy, faunistic surveys, inventorization of taxa and chorogeographic populational characteristics, nevertheless, deal extensively with a variety of subjects such as biology, ecology, conservation, description of new taxa, cytogenetics, karyosystematics, karyophylogeny, ethology, speleology, alpine odonatology etc. Several of his research papers have appeared as series (e.g., KIAUTA, 1972k, 1972l), while others as monographs (e.g., KIAUTA 1974, 1975a). Even though he seems to be a lone crusader in accomplishing scientific tasks for publication, yet, on a close scrutiny, he has actually established himself as a team contributor, having associated with several reputed odonatologists of late 20th and early 21st centuries. Several papers were published jointly with various colleagues, namely, Drs G. Jurzitza, J.M. van Brink, P.S. Corbet, H.J. Dumont, J.W. Boyes, K. Chatterjee, A.Yu. Haritonov, M.A.J.E. Kiauta, B.F. Belyshev, A. Ferreira, A. Zaha, G.A. Boon von Ochssée, M. Scrimgeour, J.E. Holmquist, A. Pinratana, M. Hämäläinen, D.C. Geijskes, C. Soler, D.J. Thompson, M. Kotarac, K. Inoue, O.N. Popova, S.N. Borisov, O.E. Kosterin, M. Bedjanič, F. Leben,L. Lankhorst, G. van Griethuysen, L.J.M. Butot, J. Dvořak, , P. Brandmayr, J.H. De Haas, H.J.P. Lambeck, A.W.M. Mol, R.L. Shrestha, R. Müller, B.K. Tyagi, and Vijay Veer. In over 30 publications his consort was the joint author. Although he exchanged material and information with a large number of odonatologists across the world, he never published anything on such secondary data. There are, however, instances where on his own initiative some scientific publications were produced with his central role in their preparation but he only preferred to appear as a co-author (e.g., CORBET, P.S., C.M., SCRIMGEOUR & B. KIAUTA, 1984e; CORBET, P.S., C.M., SCRIMGEOUR, J.E. HOLMQUIST & B. KIAUTA, 1984f).

Dr Kiauta has contributed also in non-odonate groups, both vertebrates (Chiroptera) and invertebrates, such as, e.g, Collembola, Ephemeroptera, Lepidoptera, Orthoptera, Coleoptera, Homoptera, Heteroptera, Trichoptera, Diptera, Neuroptera etc. as well as other Invertebrates like phylum Mollusca's Gastropoda (Succineidae, Hydrobiidae, Assimineidae, Rissoacea) and Bivalvia (Unionidae, Helicidae, Hygromiidae Zonitacea). In the process he covered many disciplines such as, e.g., ecology, biology, distribution, speleology, speleomorphology, biospeleology, natural history of various faunae, conservation, cytology, cytotaxonomy etc. He has also compiled biographies and bibliographies of several eminent scientists.

Annex 1. Odonatological Bibliography of Professor Dr B. Kiauta (1954-2018, *partim*)

Note: Titles of papers written in non-congress languages are given in the language of the summary or in English translation. Book reviews, minor notes, forewords in various odonatological books etc. are mostly omitted.

Dr. Bastiaan Kiauta's odonatological bibliography (1954-2018 *in partim*) has anything like 400 bibliographic units, including 168 of these in the *Odonatologica* section of Odonatological Abstracts (Abstract Nos 1-19, 871). Several papers were published jointly with various colleagues; in over 30 publications his consort was the joint author. In the below appended Bibliography, the minor notes, most book prefaces, addresses in opening sessions of various odonatological symposia and similar published texts are not listed. These amount to some 40 bibliographic units, most in English, some in Dutch.

1954a The Odonata of the surroundings of Ljubljana. *Proteus*, Ljubljana 16(8): 220-222, fig. 1. – [Slovenian]

1954b Biogeographically alien elements in the odonate fauna of the Ljubljana Moor. *Proteus*, Ljubljana 17(2): 44-48, figs 1-4. – [Slovenian]

1954c Biologist's impressions from the Mala Paklenica Gorge. *Proteus*, Ljubljana 17(4/5): 115-118, figs 1-2. – [Slovenian; includes notes on the Odon. of the Velebit Range, northern Dalmatia and the first record of Sympetrum flaveolum for Croatia]

1959 On the Pleistocene dragonfly relicts. *Proteus*, Ljubljana 21(10): 260-263, fig. 1. – [Slovenian]

1960 La faune des libellules (Odonata Fabr.) dans le Parc national du Triglav. *Varstvo Spom.* 7: 376-377.

1961a Beitrag zur Kenntnis der Odonaten-Fauna Sloweniens. *Biol. Vest.* 8: 31-40.

1961b Beitrag zur Kenntnis der Libellen-Fauna (Odonata Fabr.) in der Umgebung von Škofja Loka. *Loški Razgledi* 8: 174-182, figs 1-6.

1962 Dragonfly fauna of the Triglav National Park and its surroundings (Odonata Fabr.). *Varstvo Narave* 1: 99-117, figs 1-7.

1963a Copper engravings in Scopoli's Entomologia carniolica. *Kronika*, Ljubljana 11(1): 57- 60, fig. 1. – [Slovenian; the origin and fate of the plates are traced, the odon. figs are analyzed in depth, and the odon. inventory is critically discussed]

1963b Lindenia tetraphylla v.d.Lind. und Somatochlora metallica meridionalis Nielsen aus Nordwestistrien (Jugoslawien) (Odonata: Gomphidae, Corduliidae). *Beitr. naturk. Forsch. SüdwDtl.* 22(1): 65-66.

1963c Additions to the Cyprus list of Odonata, with some notes on the annotated list given by K.J. Valle. *Commentat. biol.* 26(8): 1-6.

1963d A note on an unusual habitat of Calopteryx virgo (L.) (Calopterygidae). *Tombo*, Tokyo 6(3/4): 25-26, fig. 1.

1963e The dragonfly fauna (Odonata Fabr.) of the Kamniška Bistrica river basin (Slovenia, Yugoslavia). *Kamn. Zbornik* 9: 137-153.

1964a A note on the discovery of Somatochlora alpestris in the Slovenian Alps. *Plan. Vest.* 20(8): 359-360. – [Slovenian]

1964b Notes on some field observations on the behaviour of Leucorrhinia pectoralis Charp. (Odonata: Libellulidae). *Ent. Ber., Amst.* 24(5): 82-86, figs 1-2.

1964c On the occurrence of Somatochlora arctica (Zetterstedt, 1840) in the Netherlands (Odonata: Corduliidae). *Ent. Ber., Amst.* 24(11): 235-238, fig. 1.

1964d (VAN BRINK, J.M. & –). Notes on the chromosome behaviour in the spermatogenesis of the damselfly Enallagma cyathigerum (Charp.) (Odonata: Coenagrionide). *Genetica* 35: 171-174, 2 pls (figs 1-9) excl.

1964e On the dragonfly fauna of the Peel peatbog area, the Netherlands. *Levende Nat.* 67(1): 12-17, figs 1-3. – [Dutch]

1964f On a migration of Libellula quadrimaculata L. (Odonata: Libellulidae). *Levende Nat.* 67(5): 59-63, figs 1-3. – [Dutch]

1964g Beobachtungen betreffs der Biologie, der Ökologie und der Ethologie der Quelljungfern im Bergland von Škofja Loka (Odonata: Cordulegastridae). *Loški Razgledi* 11: 183-192.

1965a Notes on the odonate fauna of some brackish waters of Walchern island. *Ent. Ber., Amst.* 25(3): 54-58, fig. 1.

1965b Notes sur le dépouillement de Cordulia aenea (L.) (Odonata: Corduliidae). *Ent. Ber., Amst.* 25(6): 111-113, fig. 1.

1965c On the odonate fauna of Warmbad Villach in southern Carinthia. *Beaufortia* 13(152): 35-46, figs 1-4.

1965d The chromosome behaviour in spermatogenetic meiosis of Anax imperator Leach (Odonata: Aeshnidae). *Tombo*, Tokyo 7(3/4): 18-21, figs 1-5.

1965e The behaviour of the ultimate instar larva of the dragonfly Anax imperator Leach (Odonata: Aeshnidae). *Levende Nat.* 65(2): 35-40, figs 1-4.

1965f Libellula quadrimaculata migration in the Netherlands, in 1964. *Levende Nat.* 65(2): 35-40, figs 1-4.

1965g Dragonflies – victims of traffic and modern civilisation. *Proteus*, Ljubljana 27(7): 196-197. – [Slovenian]

1966a [Descriptive review] "On a probable centre of origin and on primary directions of the dispersal of holarctic Odonata", by B.F. Belyshev. *Ent. Ber., Amst.* 26(7): 152.

1966b Some nature conservation aspects of the odonatological program of the RIVON, *Ann. Rep. St. Inst. Nat. Conserv. Res.*, Zeist 1965: 35-36, 62, 1 col. pl. excl. – [Dutch]

1966c The chromosome behaviour in spermatogenetic meiosis of the dragonfly Sympetrum striolatum (Charp.) (Odonata: Libellulidae) from Luxembourg. *Bull. Soc. Nat. Luxemb.* 69: 54-60, figs 1-12.

1967a Considerations on the evolution of the chromosome complement in Odonata. *Genetica* 38(4): 430-446, textfigs 1-4, 1 pl. (figs 1-3) excl.

1967b Meiotic chromosome behaviour in the male damselfly, Calopteryx virgo (Linnaeus), with a discussion on the value of chromosome numbers and karyotype morphology in odonate systematics. *Dt. ent. Z.* (N.F.) 14(3/4): 339-348, pls 1-2 (11 figs) excl.

1967c A new hypothesis on the karyotypic evolution in Odonata. *Tombo*, Tokyo 10(1/4): 29-33, figs 1-4.

1967d Dragonfly motifs on postal stamps. *Proteus*, Ljubljana 30(1): 28-29, figs 1-3. – [Slovenian]

1968a Variation in size of the m-chromosome of the dragonfly Calopteryx virgo (L.) and its significance for the chorogeography of the Calopteryx virgo superspecies. *Genen Phaenen* 12(1): 11-16, fig. 1.

1968b Additions to the list of Odonata of the Dutch Wadden islands, with an account of the water quality data of larval habitats, and a review of the dragonfly fauna of the Dutch and German Northsea islands. *Biol. Jaarb. Dodonaea* 36: 88-114, textfig. 1, pls 1-2 (figs 2-6) incl., 2 fold. tabs excl.

1968c Morphology and kinetic behaviour of the odonate sex chromosomes, with a review of the distribution of sex determining mechanisms in the order. *Genen Phaenen* 12(2): 21-24, figs 1-3.

1968d Annotations on the Dutch Odonata, 1. Crocothemis erythraea (Brullé, 1832) in the Netherlands (Libellulidae, Sympetrinae). *Ent. Ber., Amst.* 28(8): 156-160, fig. 1.

1968e (– & J.M. VAN BRINK). Sex chromosomes and sex determining mechanism in the order Odonata. *Proc. 12th int. Congr. Genet.*, Tokyo 1: 206.

1968f Variation in size of the dragonfly m-chromosome, with considerations on its significance for the chorogeography and taxonomy of the order Odonata, and notes on the validity of the rule of Reinig. *Genetica* 39(1): 64-74, textfig. 1, 1 pl. (figs 2-5) excl.

1968g The chromosome numbers of eight Old World dragonflies (Odonata). *Chrom. Inf. Serv.*, Tokyo 9: 3-4, figs 1-8.

1968h The chromosomes of the male dragonfly Cordulegaster boltonii (Donovan, 1807) (Odonata: Cordulegastridae). *Biol. Vest.* 16: 87-94, figs 1-18.

1968i Dragonfly migration. *Levende Nat.* 71: 95. – [Dutch]

1969a Sex chromosomes and sex determining mechanisms in Odonata, with a review of the cytological conditions in the family Gomphidae, and reference to the karyotypic evolution in the Order. *Genetica* 40(2): 127-157, textfigs 1-6, pls 1-8 (figs 1-51) excl.

1969b Autosomal fragmentations and fusions in Odonata and their evolutionary implications. *Genetica* 40(2): 158-180, pls 1-5 (figs 1-37) excl.

1969c *Methods for study of dragonfly chromosomes*. Inst. Genet., Univ Utrecht. 4 pp.

1969d The chromosomes of the Hawaiian endemic dragonflies, Megalagrion oahuense (Blackburn) (Coenagrionidae: Pseudoagrioninae) and Nesogobia blackburni (McLachlan) (Libellulidae: Sympetrinae), with a note on the cytotaxonomic affinities between the genera Nesogonia Kirby and Sympetrum Newman (Odonata). *Proc.Hawaii. ent. Soc.* 20(2): 429-433, figs 1-6.

1969e The chromosomes of eight dragonfly species from continental Africa and Madagscar (Odonata). *Arnoldia, Rhod.* 4(15): 1-8, figs 1-7.

1969f *Studies on karyotypic evolution in Odonata*. Bronder, Rotterdam. 112 pp., 105 figs.

1969g Survey of the odonate fauna of the Autonomous Region Friuli-Venezia Giulia (northern Italy). *Atti Mus. civ. Stor. nat. Trieste* 26(6): 177-247, figs 1-18.

1969h The odonate collection from the karst of Inner Carniola and the Slovene Littoral in the Natural History Museum of Triest, with a review and zoogeographical considerations on the dragonfly fauna of this and the adjacent territories. *Biol. Vest.* 17: 101-111, fig. 1.

1969i A proposal for the protection of some Slovene dragonfly species that are rare or threatened with extinction (Odonata). *Varstvo Narave* 6: 121-130, figs 1-2.

1969j Croatocordulia platyptera (Charpentier, 1843) gen. nov., from the Lower Miocene of Radoboj in Croatia (Odonata, Anisoptera, Corduliidae, Coduliinae). *Geologija* 12: 85-90, figs 1 (excl.)-2.

1970a The chromosomes of four neotropical dragonflies from Mexico. *Chrom. Inf. Serv.*, Tokyo 11: 8-9, figs 1-4.

1970b Some remarks on the evolution of the centromere, based upon the distribution of centromere types in insects. *Ent. Ber., Amst.* 30(4): 71-75. – [A preprint was published in *Genen Phaenen* 13(1969): 1-6]

1970c Colloquium of Dutch dragonfly workers. *Selysia* 5(1): 3. – [Report on the proceedings of the first of the two Colloquia which were forerunners of the International Symposia of Odonatology, and at which the idea of founding the Societas Internationalis Odonatologica (SIO) was born]

1970d The karyotype of the damselfly Epallage fatime (Charpentier, 1840), with a note on the cytotaxonomic affinities in the superfamily Calopterygoidea. *Genetica* 41(4): 525-531, pl. 1 (figs 1-4) excl.

1970e Bibliography on the Swiss odonate fauna. *ContactBr. ned. LibellenOnderz.* 2(Suppl.): 1-6.

1971a Studies on the germ cell chromosome cytology of some cytotaxonomically interesting or hitherto not studied Odonata from the Autonomous Region Friuli-Venezia Giulia (northern Italy). *Atti Mus. civ. Stor. nat. Trieste* 27(2): 65-127, fig. 1, pls 1-12 (figs 1- 84) incl.

1971b Predation by ants, Formica fusca L. and F. rufa polyctena Bondr., on the emerging dragonfly, Aeshna juncea (L.), and its teratological consequences. *Tombo*, Tokyo 14(1/2): 2-5, figs 1-4.

1971c An unusual case of precocious segregation and chromosome fragmentation in the primary spermatocytes of the damselfly Calopteryx virgo meridionalis Selys, 1873 as evidence for a possible hybrid character of some populations of the Calopteryx-virgo-complex (Odonata, Zygoptera: Calopterygidae). *Genen Phaenen* 14(2): 32-40, figs 1-10.

1971d Cytotaxonomic peculiarities in the neotropical odonate genera Leptagrion Selys, Orthemis Hagen and Macrothemis Hagen. *Abstr. Pap. 1st Europ. Symp. Odonatol.*, Gent. pp. 27-28.

1972a Scientific results of the Yugoslav 1969 Himalaya Expedition. Odonata. *Biol. Vest.* 20: 109-119, figs 1-9.

1972b The First European Symposium of Odonatology. *Ent. Ber., Amst.* 32(2): 23-24.

1972c (– & G. JURZITZA) Erstes Europäisches Symposium für Odonatologie und Gründung der "Societas Internationalis Odonatologica". *Beitr. naturk. Forsch. SüdwDtl.* 31: 185.

1972d (– & J.M. VAN BRINK) Societas Internationalis Odonatologica. *Vakbl. Biol.* 52(1): 14. – [Dutch]

1972e (– & J.M. VAN BRINK) Editorial. *Odonatologica* 1(1): 1.

1972f The karyotype of the damselfly Leptagrion macrurum (Burmeister, 1839) and its possible origin, with a note on the cytotaxonomic affinities of the genus (Zygoptera: Coenagrionidae). *Odonatologica* 1(1): 31-35, figs 1-2.

1972g Synopsis of the main cytotaxonomic data in the order Odonata. *Odonatologica* 1(2): 73-102, figs 1-2.

1972h (DUMONT, H.J. & –) [Obituary]. Dr Filimon Cirdei. *Odonatologica* 1(3): 165-166.

1972i Mechanisms of karyotypic evolution in dragonflies. *Abstr. 14th Int. Congr. Ent.*, Canberra, p. 107.

1972j (– & J.W. BOYES). Cytology of ten South American Libellulidae, with cytophylogenetic considerations on the genera Orthemis Hagen and Erythrodiplax Brauer (Odonata, Anisoptera). *Genetica* 43(3): 407-421, textfig. 1, pl. 1 (figs 1-19).

1972k Notes on new or little known dragonfly karyotypes, 1. The germ cell chromosomes of three Latin American species: Argia funebris (Hagen), Megapodagrion contortum (Selys) (Zygoptera: Coenagrionidae, Megapodagrionidae) and Castoraeschna castor (Brauer) (Anisoptera: Aeshnidae). *Genen Phaenen* 15(1): 23-26, figs 1-10.

1972l Notes on new or little known dragonfly karyotypes, 2. Male germ cell chromosomes of four East Mediterranean species: Lestes barbarus (Fabricius), Calopteryx splendens amasina Bartenev (Zygoptera: Lestidae, Calopterygidae), Caliaeschna microstigma (Schneider) and Orthetrum taeniolatum (Schneider) (Anisoptera: Aeshnidae, Libellulidae). *Genen Phaenen* 15(2/3): 95-98, figs 1 -11.

1972-2013 Odonatological Abstracts, Nos 1-19871. *Odonatologica* 1[1972]: 53-62, 119-130, 167-179, 263-278; 2[1973]: 49-64, 125-137, 175-189, 335-347; 3[1974]: 79-92, 143-151, 191-201, 273-292; 4[1975]: 35-53, 101-124, 191-206, 277-297; 5[1976]: 79-95, 181-192, 291-307, 391-404; 6[1977]: 35-54, 119-141, 211-226, 297-312; 7[1978]: 91-114, 167-186, 287-307, 391-414; 8[1979]: 71-89, 135-154, 227-244, 333-350; 9[1980]: 107-124, 197-215, 259-278, 329-352; 10[1981]: 65-83, 159-177, 237-257, 333-354; 11[1982]: 63-87, 167-187, 251-270, 347-368; 12[1983]: 93-113, 179-200, 297-313, 401-421; 13[1984]: 169-185, 317-334, 481-500, 603-623; 14[1985]: 65-80, 163-175, 261-278, 369-380; 15[1986]: 137-155, 223-243, 353-371, 469-491; 16[1987]: 103-118, 209-223, 315-334, 407-428; 17[1988]: 65-83, 147-177, 289-311, 435-458; 18[1989]: 107-132, 219-236, 297-324, 397-420; 19[1990]: 91-115, 195-221, 309-332, 403-426; 20[1991]: 127-149, 253-279, 375-399, 483-509; 21[1992]: 119-140, 253-273, 385-408, 515-542; 22[1993]: 101-127, 243-271, 379-397, 511-535; 23[1994]: 77-101, 193-214, 303-327, 431-454; 24[1995]: 129-148, 249-267, 361-382, 485-505; 25[1996]: 87-107, 197-220, 313-333, 405-425; 26[1997]: 95-120, 227-247, 347-373, 483-506; 27[1998]: 125-147, 267-286, 391-414, 495-512; 28[1999]: 93-115, 187-208, 293-315, 439-455; 29[2000]: 75-94, 167-194, 273-294, 369-392; 30[2001]: 123-144, 233-254, 351-373, 457-474; 31[2002]: 95-115, 211-228, 323-343, 415-434; 32[2003]: 99-118, 183-204, 303-320, 393-411; 33[2004]: 105-127, 217-236, 333-352, 443-462; 34[2005]: 83-102, 189-209, 313-334, 415-434; 35[2006]: 89-116, 205-224, 305-327; 36[2007]: 95-116, 219-241, 315-338, 421-440; 37[2008]: 79-99, 172-192, 283-304, 381-399; 38[2009]: 73-92, 179-202, 273-292, 375-394; 39[2010]: 75-96, 163-184, 265-286, 367-386; 40[2011]: 67-88, 149-167, 255-276, 347-368; 41[2012]: 61-88, 177-200, 293-308, 361-384; 42[2013]: 73-96, 163-187, 263-284, 427-448.

1973a First European Symposium of Odonatology. *Selysia* 6(1): 7-8.

1973b Notes on new or little known dragonfly karyotypes, 3. Spermatocyte chromosomes of four nearctic anisopterans: Aeshna californica Calvert (Aeshnidae), Cordulia shurtleffi Scudder (Corduliidae), Sympetrum internum Montgomery, and S. madidum (Hagen) (Libellulidae). *Genen Phaenen* 16(1): 7-12, figs 1-12.

1973c Notes on new or little known dragonfly karyotypes, 4. Spermatocyte chromosomes of Calopteryx s. splendens (Harris) (Zygoptera: Calopterygidae), Gomphus pulchellus Selys and Libellula depressa Linnaeus (Anisoptera: Gomphidae, Libellulidae) from southern France. *Genen Phaenen* 16(2): 55-60, figs 1-19.

1973d Infraspecific karyotypic variability in some oriental dragonflies. *Abstr. Pap. 2nd int. Symp. Odonatol.*, Karlsruhe, p. 16.

1973e A note on the dragonfly folk names in Nepal. *Odonatologica* 2(1): 29-32.

1973f (CHATTERJEE, K. & –) Male germ cell chromosomes of two Calopterygoidea from the Darjeeling Himalaya (Zygoptera: Chlorocyphidae, Euphaeidae). *Odonatologica* 2(2): 105-108, figs 1-2.

1974 *Introduction to insect cytotaxonomy*. Lectures delivered at the Tribhuvan University, Kathmandu, Vol. 1. Nepal Research Centre, Kathmandu. xii+81 pp., fig. 1 – [Reprint edition published in 1978]

1975a *Cytotaxonomy of dragonflies, with special reference to the Nepalese fauna*. Lectures delivered at the Tribhuvan University, Kathmandu, Vol. 2. Nepal Research Centre, Kathmandu. xiv+78 pp., textfigs 1-5, pls 1-3 (figs 1-42).

1975b On a small dragonfly collection from Nepal. *Kailash*, Kathmandu 3(3): 289-294.

1975c (HARITONOV, A.Yu. & –). To Dr Boris Feodorovich Belyshev on his 65th birthday. *Odonatologica* 4(4): 207-218, figs 1-3, portraits. – [Biography, appreciation of odonatological work, and annotated bibliography]

1975d (– & M.A.J.E. BRINK). Cytotaxonomic notes on the Sympetrum pedemontanum complex (Anisoptera: Libellulidae). *Odonatologica* 4(4): 249-254, figs 1-3.

1975e (– & M.A.J.E. KIAUTA-BRINK). The chromosomes of the dragonfly Sympecma annulata braueri (Yakobson & Bianki, 1905) from the Netherlands, with a note on the classification of the family Lestidae (Odonata, Zygoptera). *Genen Phaenen* 18(2/3): 39-48, figs 1-4.

1976a (BELYSHEV, B.F. & –). To Nikolai Nikolaevich Akramowski on his 70th birthday. *Odonatologica* 5(4): 309-312, fig. 1. – [Biography, appreciation of odonatological work and annotated bibliography]

1976b (– & M.A.J.E. KIAUTA). The chromosomes of some dragonflies from the Langtang Valley, central Nepal. *Odonatologica* 5(4): 347-354, figs 1-12.

1977a Notes on new or little known dragonfly karyotypes, 5. The male germ cell chromosomes of Macromia moorei Selys from Nepal (Anisoptera: Corduliidae). *Genen Phaenen* 19(2/3): 49-51, figs 1-8.

1977b (VAN BRINK, J.M. & –). Annotations on the chromosome complements of some dragonflies from southern Africa. *Abstr. Pap. 4th Int. Symp. Odonatol.*, Gainesville/FL, pp. 18-19.

1977c (VAN BRINK, J.M. & –). To Mrs Leonora K. Gloyd on her 75th birthday. *Odonatologica* 6(3): 143-149, figs 1-5, portrait. – [Biography, appreciation of odonatological work and bibliography]

1978a (– & J.M. VAN BRINK). Male chromosome complements of some Florida dragonflies, United States. *Odonatologica* 7(1): 15-25, figs 1-40.

1978b An outline of the history of odonatology in Switzerland, with an annotated bibliography on the Swiss odonate fauna. *Odonatologica* 7(3): 191-222, figs 1-10, 3 pls incl.

1978c Two cytotaxonomically interesting cases of irreversible autosome fusion in dragonflies: Argia moesta (Hagen) (Zygoptera: Coenagrionidae) and Anaciaeschna isoceles (Müller) (Anisoptera: Aeshnidae). *Notul. odonatol.* 1(1): 7-9, figs 1-4.

1979a (FERREIRA, A., – & A. ZAHA). Male germ cell chromosomes of thirty-two Brazilian dragonflies. *Odonatologica* 8(1): 5-22, figs 1-54.

1979b (– & G.A. BOON VON OCHSSÉE) Some dragonfly karyotypes from the Voltaic Republic (Haute Volta). *Odonatologica* 8(1): 47-54, figs 1-12.

1979c The karyotypes of some Anisoptera from Surinam. *Odonatologica* 8(4): 267-283, figs 1-56.

1979d The karyotype of Ischnura pumilio (Charp.) (Zygoptera: Coenagrionidae). *Notul. odonatol.* 1(3): 47-48, fig. 1.

1979e (– & M.A.J.E. KIAUTA) The karyotype of Libellula fulva Müll. from Switzerland (Anisoptera: Libellulidae). *Notul. odonatol.* 1(4): 73-74, figs 1-2.

1979f (– & M.A.J.E. KIAUTA) Cytotaxonomy of the genus Argia Rambur (Zygoptera: Coenagrionidae). *Abstr. Pap. 5th Int. Symp. Odonatol.*, Montreal, pp. 17-19.

1979g (VAN LEEUWENHOEK, A. & –) Letter No. 140 (85), dated Nov. 30, 1694, addressed to Petrus Rabus. *In*: L.C. Palm, [Ed], *The collected letters of Antoni van Leeuwenhoek. A complete edition of the works of Van Leeuwenhoek, annotated and considered in the perspective of the history of science and medicine, with special reference to general and biological microscopy*, Vol. 10 (*1694-1695*), pp. 152-165. Swets & Zeitlinger, Lisse. – [Annotations and comments]

1980a (– & M.A.J.E. KIAUTA) Introduction to the cytotaxonomy of the odonate genus Argia Rambur (Zygoptera: Coenagrionidae). *Odonatologica* 9(1): 35-56, figs 1-38, 1 col. pl. incl.

1980b (BOYES, J.W., J.M. VAN BRINK & –) Sixteen dragonfly karyotypes from the Republic of South Africa and Swaziland, with evidence on the possible hybrid nature of Orthetrum julia falsum Longfield (Anisoptera: Libellulidae). *Odonatologica* 9(2): 131-145, figs 1-37.

1980c (– & M.A.J.E. KIAUTA). On a small collection of dragonfly karyotypes from the Philippines. *Odonatologica* 9(3): 237-245, figs 1-19.

1980d (– & M.A.J.E. KIAUTA) Considerations on cytotaxonomy and speciation in the dragonfly genus Rhinocypha (Odonata: Chlorocyphidae). *Abstr. Pap. 16th Int. Congr. Ent.*, Kyoto, p. 19.

1980e (– & M.A.J.E. KIAUTA) The karyotypes of Aeshna subarctica elisabethae Djak. and Somatochlora alpestris (Sel.) from Switzerland (Anisoptera: Aeshnidae, Corduliidae). *Notul. odonatol.* 1(6): 104-105, figs 1-3.

1981a *Annotated catalogue and bibliography of taxa introduced in Odonata from 1971 to 1980*. Soc. Int. Odonatol., Utrecht. 64 pp.

1981b A review of the main achievements in odonate taxonomy and systematics from 1971-1980. *Abstr. Pap. 6th Int. Symp. Odonatol.*, Chur, pp. 22-25.

1981c Ten years of the International Odonatological Society: history, development and perspectives. *Abstr. Pap. 6th Int. Symp. Odonatol.*, Chur, p. 72.

1981d (– & M.A.J.E. KIAUTA) The karyotype of Risiocnemis incisa Kimmins, 1936 from Luzon, the Philippines (Zygoptera: Platycnemididae). *Odonatologica* 10(2): 151-154, figs 1-5.

1981e (– & M.A.J.E. KIAUTA) The Engadine Post-Symposium Tour. *In*: *Field Trips in the framework of the Sixth International Symposium of Odonatology*, Chur, pp. 13-20. Soc. Int. Odonatol., Utrecht. – [Includes a checklist of the Engadine Odonata]

1981f (KIAUTA, M. & –) Aspirator for collecting of small zygopterans close above water surface and among dense vegetation. *Notul. odonatol.* 1(8): 133-134, fig. 1.

1981g In memoriam Luitenant Kolonel B.D. Dr Cornelis de Jong. *ContactBr. ned. LibellenOnderz.* 2: 6. – [Dutch]

1982a (– & M. KIAUTA) The chromosome numbers of sixteen dragonfly species from the Arun Valley, eastern Nepal. *Notul. odonatol.* 1(9): 143-145, figs 1-5.

1982b (– & M. KIAUTA) The chromosome numbers of eleven dragonfly species from Singapore. *Notul. odonatol.* 1(10): 164-165, figs 1-7.

1982c (– & M. KIAUTA) *List of species, with chromosome numbers and preliminary notes on the karyotypes of the Odonata, collected in May 1979 and August 1980 by the members of the Kansai Research Group of Odonatology and examined by B. and M. Kiauta*. Report for the Kansai Research Group of Odonatology, Osaka, 8 pp.

1983a In memoriam Professor Basil Elwood Montgomery. *Odonatologica* 12(1): 1-4, 14 figs.

1983b (– & M. KIAUTA) Further notes on the Philippine odonate karyotypes. *Notul. odonatol.* 2(1): 14-15, fig. 1.

1983c (– & M. KIAUTA) The chromosome numbers of some Odonata from Thailand. *Notul. odonatol.* 2(2): 27-28, figs 1-6.

1983d The status of the Japanese Crocothemis servilia (Drury) as revealed by karyotypic morphology (Anisoptera: Libellulidae). *Odonatologica* 12(4): 381-388, figs 1-15. – [C. s. mariannae ssp. n.]

1983e Über das Vorkommen der Südlichen Heidelibelle, Sympetrum meridionale (Selys), in dem Engadin. *Jber. naturf. Ges. Graubünden* 100: 151-156, fig. 1.

1984a (– & D.C. GEIJSKES) To Dr h.c. Maurits Anne Lieftinck on his 80th birthday. *Odonatologica* 13(1): 1-4, portrait incl.

1984b (GEIJSKES, D.C. & –) Annotated catalogue of taxa introduced in Odonata by M.A. Lieftinck, with his complete bibliography (1919-1983). *Odonatologica* 13(1): 21-50.

1984c Aktuelle Probleme der Zytotaxonomie, erläutert bei südasiatischen Prachtlibellen (Odonata: Chlorocyphidae) und schweizerischen Köcherfliegen (Trichoptera: Limnephilidae), mit Bemerkungen über die Bedeutung der Zytotaxonomie für die Umweltforschung. *Opusc. zool. flumin.* 1: 1-20, figs 1-6.

1984d Breeding habitat of Planaeschna risi in Taiwan. *In*: P.S. Corbet, Current topics in dragonfly biology, pt 1, p. 12. *Soc. int. odonatol. rapid Comm.* (Suppl.), Vol. 2, Utrecht.

1984e (CORBET, P.S., C.M. SCRIMGEOUR & –) Author index for Odonatological Abstracts 1-4225 from Odonatologica volumes 1-12. *Soc. int. rapid Comm.* (Suppl.) 4(A), viii + 63 pp.

1984f (CORBET, P.S., C.M. SCRIMGEOUR, J.E. HOLMQUIST & –) A topic index for Odonatological Abstracts 1-4225 from Odonatologica volumes 1-12. *Soc. int. odonatol. rapid Comm.* (Suppl.) 4(B), vii + 41 pp.

1984g (– & M. KIAUTA) Crocothemis erythraea (Brullé) recorded from canton St Gallen, Switzerland (Anisoptera: Libellulidae). *Notul. odonatol.* 2(4): 65-66.

1984h (– & M. KIAUTA) The true chromosome number of Aka-tombo, Sympetrum frequens (Sel.) of Japan, with a note on the karyotype of S. depressiusculum (Sel.) (Anisoptera: Libellulidae). *Notul. odonatol.* 2(4): 66-67, figs 1-2.

1985a Foreword. *Proc. 1st Indian Symp. Odonatol.*, Madurai, p. ix.

1985b The identity of F. Ris's correspondent and odonate collector "H. Spoerry jun. in Flums", Switzerland. *Notul. odonatol.* 2(6): 103.

1986a (– & M. KIAUTA) The dragonfly fauna of the Flumserberg region, canton St. Gallen, eastern Switzerland (Odonata). *Opusc. zool. flumin.* 3: 1-14.

1986b To Professor Dr Minter Jackson Westfall, Jr on his 70th birthday. *Odonatologica* 15(1): 1-4, portrait.

1986c (– & M. KIAUTA) Beitrag zur Kenntnis der Libellenfauna des St. Galler Rheintales zwischen Altenrhein (Bodensee) und Bad Ragaz, Schweiz (Odonata). *Ber. bot.-zool. Ges. Liechtenstein-Sargans-Werdenberg* 15: 127-136, fig. 1.

1986d [Breeding habitat of Somatochlora alpestris at Amdenerhöhe, canton St Gallen, Switzerland]. *In*: P.S. Corbet, Current topics in dragonfly biology, pt 2, p. 3. *Soc. int. odonatol. rapid Comm.* (Suppl.), Vol. 6, Bilthoven.

1986e [A comment on dragonfly predation on insects associated with coniferous forests in the United States and the possibility of its beneficial bearing on forest conservation in the air-polluted areas]. *In*: P.S. Corbet, Current topics in dragonfly biology, pt 2, p. 6. *Soc. int. odonatol. rapid Comm.* (Suppl.), Vol. 6, Bilthoven.

1986f [A note on the problems of the field assessment of the odonate component in the structure of biotic communities]. *In*: P.S. Corbet, Current topics in dragonfly biology, pt 2, p. 7. *Soc. int. odonatol. rapid Comm.* (Suppl.), Vol. 6, Bilthoven.

1986g [A note on the odonate community structure in a Luzon fishpond, the Philippines, with references to the role of dragonflies as human food in Indonesia and as vectors of human diseases in Thailand]. *In*: P.S. Corbet, Current topics in dragonfly biology, pt 2, pp. 10-12. *Soc. int. odonatol. rapid Comm.* (Suppl.), Vol. 6, Bilthoven.

1986h [Effects of "acid rain" on the Leucorrhinia larval populations in Fennoscandia]. *In*: P.S. Corbet, Current topics in dragonfly biology, pt 2, p. 13. *Soc. int. odonatol. rapid Comm.* (Suppl.), Vol. 6, Bilthoven.

1986i [The modified life history of Lestes macrostigma in Dalmatia, as conditioned by the early summer dry season]. *In*: P.S. Corbet, Current topics in dragonfly biology, pt 2, p. 15. *Soc. int. odonatol. rapid Comm.* (Suppl.), Vol. 6, Bilthoven.

1986j [A statement on the variation in the numbers of dimorphic females in geographically distant populations of Boyeria irene]. *In*: P.S. Corbet, Current topics in dragonfly biology, pt 2, p. 16. *Soc. int. odonatol. rapid Comm.* (Suppl.), Vol. 6, Bilthoven.

1986k [A comment on the geographic variation in the number of larval instars in Ceriagrion coromandelianum, Cordulegaster and Epiophlebia]. *In*: P.S. Corbet, Current topics in dragonfly biology, pt 2, p. 24. *Soc. int. odonatol. rapid Comm.* (Suppl.), Vol. 6, Bilthoven.

1986l Aeshna caerulea (Ström, 1783) vs Aeshna coerulea (Anisoptera: Aeshnidae). *Notul. odonatol.* 2(8): 133.

1987 The first All-Union Symposium on Odonatology in the USSR. *Selysia* 16(1): 5-6.

1988a (PINRATANA, A., – & M. HÄMÄLÄINEN) *List of the Odonata of Thailand and annotated bibliography*. Viratham Press, Bangkok. iv+20 pp., 20 col. pls excl.

1988b [Migratory habits and vertical distribution of Pantala flavescens in the Himalayan region]. *In*: P.S. Corbet Current topics in dragonfly biology, pt 3, p. 5-6. *Soc. int. odonatol. rapid Comm.* (Suppl.), Vol. 8, Bilthoven.

1988c (– & M. KIAUTA) The unusual recombination potential and its ecological implications in Coenagrion mercuriale (Charp.) from Liechtenstein (Zygoptera: Coenagrionidae). *Notul. odonatol.* 3(2): 34-35, fig. 1.

1988d The path towards the International Odonata Research Institute. *Anu. Rep. Int. Odonata Res. Inst.* 1: 1-5, fig. 1.

1990a (– & M. KIAUTA) Early summer dragonflies of the Ryder Lake area, Chilliwack, British Columbia, Canada (Odonata). *Opusc. zool. flumin.* 58: 1-10, fig. 1.

1990b (– & M. KIAUTA) Some dragonflies collected around Ryder Lake, Chilliwack district, in 1990. *Boreus* 10(2): 7.

1991 (– & M. KIAUTA) Biogeographic considerations on Coenagrion hylas freyi (Bilek, 1954), based mainly on the karyotypic features of a population from North Tyrol, Austria (Zygoptera: Coenagrionidae). *Odonatologica* 20(4): 417-431, figs 1-2.

1992 (– & M. KIAUTA) Aeshna caerulea (Ström), a new dragonfly for the fauna of Slovenia (Anisoptera: Aeshnidae). *Notul. odonatol.* 3(10): 168.

1993a (– & M. KIAUTA) In memory of Janny Margaretha van Brink (5 May 1923 – 28 March 1993). *Odonatologica* 22(2): 129-131, portrait. – [Appreciation of her work in the SIO]

1993b (– & C. SOLER) Die rätoromanischen Namen für die Libellen Graubündens: ein Diskussionsvorschlag (Odonata). *Opusc. zool. flumin.* 110: 1-8.

1994a Two photographs relative to the history of odonatology in Slovenia. *Notul. odonatol.* 4(4): 69-70, figs 1-2.

1994b (THOMPSON, D.J. & –) Odonatospeleology: dragonflies in caves, with a checklist of the known records (Odonata). *Opusc. zool. flumin.* 118: 1-10.

1994c (– & M. KIAUTA) Dragonfly records from the districts of Ribnica and Kočevje, Lower Carniola, Slovenia (Odonata). *Opusc. zool. flumin.* 124: 1-8.

1994d (– & M. KIAUTA) A note on the dragonfly assemblage at the intermittent pond of Ledine, Julian Alps, NW Slovenia (Odonata). *Opusc. zool. flumin.* 124: 9-12, fig. 1.

1994e (– & M. KIAUTA) On a small autumnal dragonfly collection from the Cariboo and southwestern British Columbia, Canada (Odonata). *Opusc. zool. flumin.* 126: 1-8.

1994f A contribution towards the odonatological bibliography of Slovenia. *Exuviae* 1(1): 9-15.

1995a (– & M. KIAUTA) Odonatological periodicals in-print. *Notul. odonatol.* 4(5): 77-80.

1995b (– & M. KOTARAC) Two dragonfly records from karst caves in Bosnia-Herzegovina and Slovenia (Anisoptera: Aeshnidae, Corduliidae). *Notul. odonatol.* 4(6): 106-107.

1995c (– & M. KIAUTA) The karyotypes of Somatochlora meridionalis Nielsen from Slovenia and S. metallica (Vander L.) from Switzerland, with a tentative note on the origin of the central European S. metallica (Odonata: Corduliidae). *Opusc. zool. flumin.* 137: 1-5, figs 1-6.

1998 An interesting old observation on dragonfly behaviour at the seaside, Indonesia. *Notul. odonatol.* 5(2): 21.

1999 (– & M. KIAUTA) A note on dragonfly response during the 93.8% solar eclipse of 11 August 1999 in the Netherlands (Odonata). *Opusc. zool. flumin.* 172: 1-6.

2000 Ferdinand J. Schmidt (1791-1878). *Erjavecia* 9: 1-9, figs 1 -5, portrait. – [Slovenian; biography and appreciation of his odonatological work]

2001 Snake associations in the European "dragonfly" folk appellations: distributional patterns reflecting the Urnfield culture expansion during the first millenium B.C.? *Proc. 1st int. topical Conf. 'The Veneti within the Ethnogenesis of the central-European Population'*, Ljubljana, pp. 199-200.

2002a A visit of Professor František Klapálek, the founder of Czech scientific entomology, to Slovenia – a century ago. *Erjavecia* 14: 1-4, portrait. – [Slovenian; biography and appreciation of odonatological work]

2002b Davy Humphry. *Enciklopedija Slovenije* 16: 37. Mladinska Knjiga, Ljubljana. – [Slovenian]

2003a (–, M. KIAUTA & K. INOUE) To Professor Dr Norman Winfrid Moore for his 80th birthday. *Odonatologica* 32(1): 1-4, portrait.

2003b A note on the Odonata collection of the Rev. Father Gabriel Strobl (1846-1925) in the Benedictine Abbey at Admont, Austria. *Notul. odonatol.* 6(2): 14-19, figs 1-3, portrait.

2003c Rev. Father Gabriel Strobl (1846-1925) and the dragonflies of Slovenia. *Erjavecia* 15: 1-5, portrait. – [Slovenian]

2003d Dr Roman Puschnig, the father of Carinthian odonatology (1875-1962). *Erjavecia* 16: 1-6, fig. 1, portrait. – [Slovenian; biography, appreciation of odonatological work and bibliography]

2004a On the 1834 assemblage of the odonate fauna of Ljubljana listed by Franz Wilhelm Lippich (1799-1847). *Erjavecia* 17: 1-7, figs 1-2, portrait. – [Slovenian]

2004b Odonate faunistics at the southwestern reaches of the Slovenian ethnic territory: incidental collectors of the Alessandro Carlo principe Della Torre e Tasso circle (1881-1937). *Erjavecia* 18: 1-8, figs 1-3, portrait. – [Slovenian]

2005 Professor Dr Ciril Ažman, insect collector from Novo mesto (1881-1952). *Erjavecia* 20: 1-7, portrait. – [Slovenian; biography and appreciation of odonatological work]

2008 Dragonflies (Odonata) in the works of I.A. Scopoli revisited. *Idrijski Razgledi* 53(1/2): 59-62.

2012 Alfredo Lazzarini (1871-1945) and his odonatological work on the western outskirts of the Slovenian territory. *Erjavecia* 27: 1-4, figs 1-2, portrait. – [Slovenian]

2013a Franc Anton von Breckerfeld (1740-1806), the first collector of Slovenian dragonfly folk names. *Erjavecia* 28: 1-5, figs 1-2. – [Slovenian]

2013b [Obituary] Gordon Pritchard (1939-2012). *Odonatologica* 42: 257-264, figs 1-4, portrait.

2013c Editorial at the conclusion of four decades of editorship [of Odonatologica] (1971-2013). *Odonatologica* 42(4): 285-293, figs 1-13.

2013d Editorial at the conclusion of the editorship [of Notulae odonatologicae] (1978-2013). *Notul. odonatol.* 8(2): 17-18.

2013e (POPOVA, O.N., S.N. BORISOV, O.E. KOSTERIN & –). [Obituary] Anatoly Yuryevich Haritonov (1949-2013). *Odonatologica* 42(4): 409-425, figs 1-7, portrait.

2014a Odonatological profile of Mr Kiyoshi Inoue. *Tombo/Acta odonatologica japonica* 56: 18.

2014b Sketches for the dragonfly fauna (Insecta: Odonata) of the metropolis of Ljubljana, Slovenia. *Natura Sloveniae* 16(1): 15-40.

2014c Message to the participants of the 8th Indian Symposium of Odonatology and tropical Biodiversity. *Souvenir 8th Indian Symposium of Odonatology and tropical Biodiversity*, Nagpur, pp. xi-xii.

2014d I.A. Scopoli, author of generic name Aeschna (nomen emendatum), 1777. *Erjavecia* 29: 1-4, figs 1-2. – [Slovenian]

2014e Garden pond at the former Classical College, the first wetland object for teaching purposes in Ljubljana, Slovenia. *Erjavecia* 29: 32-35, fig. 1. – [Slovenian]

2014f A note on the book W.E. Holzinger & B. Kamposch, 2012: "Die Libellen Kärntens". *Erjavecia* 29: 35-38, fig. 1. – [Slovenian]

2014g Glimpses into the dragonflies of Ljubljana. *Trdoživ* 3(2): 16-18, figs 1-3. – [Slovenian]

2014h In memoriam Gerhard Jurzitza. *Odonatologica* 43(3/4): 137-142, figs 1-2.

2015a The key-word "kačji pastir" [= "dragonfly"] of the old Slovenian dictionary authors, with a review of Slovenian dragonfly appellations used in the literature until the end of the 20th century. *Erjavecia* 30: 1, 3-12, figs 1-3. – [Slovenian]

2015b Some recollections of my childhood "entomologing" in Ljubljana during the Second World War and in the years immediately thereafter (1941-1953). *Erjavecia* 30: 16-32, figs 1-3. – [Slovenian; unabridged Engl. translation is available]

2015c On the appearance of the work, W.E. Holzinger, A. Chovanec & J.A. Waringer, 2015: "Checklisten der Fauna Oesterreichs, No. 8: Odonata (Insecta)". *Erjavecia* 30: 71-72. – [Slovenian]

2015d [BEDJANIČ, M.] Interview with Boštjan Kiauta. *Trdoživ* 4(2): 23-28, 11 figs, portrait. – [Slovenian]. – [Date of publication 31 December 2015, published February 2016]

2016a (– & K. INOUE). Obituary Professor N.W. Moore, the pioneer of Odonata conservation. *Tombo/Acta odonatologica japonica* 58-74-78, figs 1-3, portrait.

2016b In memoriam Norman Winfrid Moore (1923-2015). *Odonatologica* 45(1/2): 1-6, figs 1-4.

2016c Dragonflies in the works of Fran Erjavec (1834-1887), with a note on his view concerning the occurrence of a pupal stage in odonate metamorphosis in the light of the two major hypotheses on the evolution of hemimetaboly. *Erjavecia* 31: 1-11. – [Slovenian]

2016d Critical comments on the terminology of insect larvae as proposed by S.M. Bybee and co-workers (2015), with special reference to Odonata. *Erjavecia* 31: 56-58. – [Slovenian]

2016e [Book review]. G. Galliani, R. Scherini & A. Piglia, 2015. Odonati d'Italia. Guide de riconoscimento allo studio di libellule e damigelle. *Erjavecia* 31: 60. – [Slovenian]

2016f A dragonfly folksaga from the Posočje region, Slovenia. *Erjavecia* 31: 62-64. – [Slovenian]

2017a Message to the participants of the 9th Indian Symposium of Odonatology. *Introduction and Abstracts of the 9th Indian Symposium of Odonatology, Goa, September 15-19*, pp. iii-iv.

2017b Review of the hitherto recorded odonate fauna of the historical and current Slovenian ethnic territories outside the Republic of Slovenia. *Erjavecia* 32: 1-13. – [Slovenian]

2017c Dragonflies in academic works of Biology students at the University of Ljubljana, with a complete bibliography (1959-2017). *Erjavecia* 32: 13-26. – [Slovenian]

2018a Message to the participants at the 1st IDS Conference of Odonatology. Abstracts [of the] 1st Indian Conference of Odonatology on "Dragonflies (Insecta: Odonata) bioecology & distribution dynamics v/s environment amelioration, with special reference to Global warming and Climate change", Department of Zoology, Dolphin (PG) Institute of Biomedical and Natural Sciences, 21-23 March 2018, Dehra Dun, p. 6.

2018b Kiauta, B., B.K. Tyagi & M. Marinov. Concept and realization of the World Odonata Manual. Abstracts [of the] 1st Indian Conference of Odonatology [...], Dolphin PG) Institute of Biomedical and Natural Sciences, 21-23 March 2018, Dehra Dun, p. 12.

2018c Odonates used in materia medica in the traditional Eurasian pharmacology. Abstracts [of the] 1st Indian Conference of Odonatology [...], Dolphin PG) Institute of Biomedical and Natural Sciences, 21-23 March 2018, Dehra Dun, p. 23.

2018d Some personal recollections of the late Dr Shigeru Tsuda (1922-2016). – Ko Tsuda Shigeru Hahushi (1922-2016) no Tsuioku. Gracile, Osaka 77. – [Bilingual: English & Japanese; in press]

2018e Dragonflies in the Prekmurje schoolbook of Imre Agustich "Prirodopis s kepami za národne šole", Lástno Drüžbe Svétoga Števana, Buda-Pešt, ii + 3-58 pp.; 1878. *Erjavecia* 33. – [Slovenian; in press]

2018f Notulae ad faunam odonatorum Sloveniae pertinentes, 1: Some additions and armchair thoughts on the book M.E. Siesa, 2017, Le Libellule delle Alpi: come riconoscerle, dove e quando osservarle, Blu Edizioni, Torino. *Erjavecia* 33. – [Slovenian; in press]

Annex 2. Dr B. Kiauta's Editorial Work on Odonatological Periodicals, 1970-2018

1970-1974 *CONTACTBRIEF NEDERLANDSE LIBELLENONDERZOEKERS / BULLETIN OF THE NETHERLANDS ODONATA WORKERS*, Association of the Netherlands Odonata workers, Utrecht. Irregular intervals. – [Dutch & English]. – **Editor**, Pts 1-13 and Supplements

1971 *ABSTRACTS OF PAPERS READ AT THE FIRST EUROPEAN SYMPOSIUM OF ODONATOLOGY*, Gent/Belgium. – [Western congress languages and Russian]. – **Editor**

1972-2013 *ODONATOLOGICA*. Journal of the Societas Internationalis Odonatologica (SIO), Utrecht. Quarterly. International Editorial and Referee Boards. Indexed in *Current Contents, Science Citation Index* and *Research Alert*. – [Western congress languages]. – **Executive Editor**

1973 *ABSTRACTS OF PAPERS READ AT THE SECOND INTERNATIONAL SYMPOSIUM OF ODONATOLOGY*, Karlsruhe/Germany. – [Western congress languages]. – **Joint Editor**

1978-2013 *NOTULAE ODONATOLOGICAE*. Bulletin of the Societas Internationalis Odonatologica (SIO), Utrecht. Semiannual. International Editorial Board. [Western congress languages and Spanish]. – **Editor**

1980-2013 *SOCIETAS INTERNATIONALIS ODONATOLOGICA RAPID COMMUNICATIONS* and *SUPPLEMENTS*. Irregular intervals. – [Western congress languages]. – **Editor**

1994 → *EXUVIAE*. Journal of the Slovene Odonatological Society, Ljubljana. Irregular intervals. – [Slovenian & English]. – **Member of the Editorial Board**

2014 → *ODONATOLOGICA*. – **Member of the Editorial Board**

Annex 3. New Odonate Taxa Described by Professor Dr B. Kiauta

Croatocordulia platyptera gen.n., 1969j
Crocothemis servilia mariannae ssp.n., 1983d

Annex 4. List of New Odonate Taxa Described in Honour of Professor Dr B. Kiauta (1986-2013)

A total of 14 odonate taxa, described during 1986–2013, were named in the honour of Dr B. Kiauta.

1. *Drepanosticta kiautai* Zhou, 1986
2. *Elasmothemis kiautai* (De Marmels, 1989) [orig. *Dythemis kiautai*]
3. *Risiocnemis kiautai* Hämäläinen, 1991
4. *Paragomphus kiautai* Legrand, 1992
5. *Macromia kiautai* Zhou, Wang, Shuai & Liu, 1994
6. *Navicordulia kiautai* Machado & Costa, 1995
7. *Telebasis bastiaani* Bick & Bick, 1996
8. *Argia kiautai* Steinmann, 1997 [syn. of *Rhipidolestes okinawanus* Asahina, 1951]
9. *Rhipidolestes bastiaani* Zhu & Yang, 1998
10. *Priscagrion kiautai* Zhou & Wilson, 2001
11. *Austroagrion kiautai* Theischinger & Richards, 2007
12. *Neoneura kiautai* Machado, 2007
13. *Teinobasis kiautai* Theischinger & Richards, 2007
14. *Pericnemis kiautarum* Orr & Hämäläinen, 2013

Annex 5. Some Selected Non-Odonatological Publications by Professor Dr B. Kiauta (1953-2018)

1953a Our alpine Lepidoptera. *Proteus*, Ljubljana 15(6): 160-161, figs 1-2. – [Slovenian; notes on some endemic or otherwise notable taxa]

1953b Some ecological parameters controlling butterfly coloration and size. *Proteus*, Ljubljana 16(2): 52-53. – [Slovenian; a report on some experimental work]

1955a Ecological conditions and fauna of the North Dalmatian salinae. *Proteus*, Ljubljana 17(7): 189-191. – [Slovenian]

1955b Insect collection of the Naturalist's Club at the 2nd High School in Ljubljana. *Proteus*, Ljubljana 18(2): 50. – [Slovenian]

1957 From the Plenary Business Meeting of the Academic Alpine Society, "Univerza", in Ljubljana. *Proteus*, Ljubljana 19: 228-229. – [Slovenian; outline of current research program, by the Secretary of the Alpine Research Section of the Society]

1960a The writer Ivan Tušek as a speleologist. *Proteus*, Ljubljana 22(7): 184-185. – [Slovenian; biographic note and appreciation of work of a member of the 19th century biospeleology school of Ferd. J. Schmidt]

1960b A new bat species for the cave fauna of Slovenia. *Proteus*, Ljubljana 23(1): 27. – [Slovenian]

1960c Notes on the ecology of troglopylous Orthoptera. *Proteus*, Ljubljana 23: 45-47. – [Slovenian; mostly on the Balkan members of the genera Dolichopoda and Troglophilus]

1960d (– & F. LEBEN). Systematische Beschreibung der Höhlen in der Umgebung von Škofja Loka, Slowenien. *Loški Razgledi* 7: 157-178, figs 1-18, topogr. map. – [Speleomorphology]

1960e Les chauves-souris dans les grottes des environs de Škofja Loka, Slovenië. *Loški Razgledi* 7: 179-183.

1961a History of the speleological explorations in the districts of Moravče and Domžale, Slovenia. *Proteus*, Ljubljana 23(8): 213-216, figs 1-3. – [Slovenian]

1961b Autobiography of Ferdinand J. Schmidt. *Proteus*, Ljubljana 23(9/10): 269-270. – [Slovenian]

1961c Eptesicus serotinus Schreiber recorded from a cave in Slovenia. *Naše Jame / Our Caves* 2(1/2): 80. – [Slovenian]

1961d Les lieux de découverte originaux des coléoptères cavernicoles des environs de Škofja Loka, Slovenië. *Loški Razgledi* 8: 183-184, 1 fig.

1961e The abyss of the source of the Soča river in the Trenta valley, Slovenia. *Speleolog*, Zagreb 9: 21-22, fig. 1. – [Speleomorphology]

1962a Velo polje considered in the light of nature conservation. *Delo*, Ljubljana 4(34): 8, 1 fig. – [Slovenian; a contribution to the polemics relative to the construction of a ski centre at the said locality, Julian Alps]

1962b Velo polje, the test case of ethical maturity. *Naši Razgledi*, Ljubljana 11(3): 58. - [Slovenian]

1962c Ferdinand Schmidt, the founder of biospeleology. *Kronika*, Ljubljana 10(1): 54-59, 1 map incl. – [Slovenian; biography, assessment of work, list of taxa and caves known to him, based on unpublished archive material and museum collections]

1962d Rhinolophus hyposideros. *Proteus*, Ljubljana 24(7): 187. – [Slovenian; notes on its ecology and distribution in the caves of Slovenia]

1962e The chasm in the Gabrovški Lazi near Škofja Loka, Slovenia. *Proteus*, Ljubljana 23(8): 218-219, fig. 1 – [Slovenian; speleomorphology]

1962f Slovenian karsology at the crossroads. *Naši Razgledi*, Ljubljana 11(12): 228-229. – [Slovenian; critical considerations on karst research policies in Slovenia]

1962g Contribution à la connaissance de la faune des cicadidés (Homoptera: Cicadinea) dans les environs de Škofja Loka, Slovenië. *Loški Razgledi* 9: 59-65, figs 1-6.

1963a A note at the bicentenary of the first insect inventory of the Slovenian Alps (1763). *Plan Vest.* 19(2): 78. – [Slovenian]

1963b The bicentenary of Scopoli's Entomologia carniolica. *Proteus*, Ljubljana 25(4/5) 122-123. – [Slovenian]

1963c Beitrag zur Kenntnis der Heteropterenfauna in der Umgebung von Škofja Loka, Slowenien (Hemiptera-Heteroptera). *Loški Razgledi* 10: 67-73.

1964a International coordination in nature conservation. *Naši Razgledi*, Ljubljana 13(5): 87. – [Slovenian; Essay on the development of nature conservation ideas and policies from 1873 to present, with an account of international institutions and research progress in this field, and critical considerations on the role of Yugoslavia in these]

1964b (BUTOT, L.J.M. & –). Applications of chromosome cytology in modern molluscan systematics. *Corrbl. ned. malacol. Ver.* 111: 1161-1162. – [Dutch]

1964c (BUTOT, L.J.M. & –). Preliminary report on the chromosomes of four species of the family Succineidae (Gastropoda: Pulmonata). *Corrbl. ned. malacol. Ver.* 111: 1163-1164, figs 1-3.

1966 (BUTOT, L.J.M. & –). Notes on the cytology of Rissoacea, 1. Cytotaxonomical conditions in some Hydrobiidae and Assimineidae (Gastropoda, Streptoneura). *Basteria* 30(2/3): 21-34, pl. 1 (figs 1-6) excl.

1967a (BUTOT, L.J.M. & –). Cytology and nature conservation. *Almanac St. Inst. Nature Cons. Res.*, Zeist, pp. 90-91, figs 1-2. – [Dutch]

1967b (BUTOT, L.J.M. & –). The chromosomes of Catinella arenaria (Bouchard-Chantereaux, 1837), with a review of cytological conditions within the genus Catinella and considerations on the phylogenetic position of the Succineoidea ord. nov. (Gastropoda: Euthyneura). *Beaufortia* 14(174): 157-164, pls 1-2 (8 figs) incl.

1967c (– & L.J.M. BUTOT). Notes on the cytology of Rissoacea, 2. The chromosomes of Assiminea grayana Fleming, 1828 (Gastropoda: Streptoneura). *Basteria* 31(4/5): 64-69, pls 1-2 (11 figs) excl.

1968a (– & L.J.M. BUTOT). The chromosomes of Succinea (Succinella) oblonga Draparnaud, 1801, with considerations on the status of the subgenus Succinella Mabille, 1870 (Gastropoda, Euthyneura: Succineoidea). *Beaufortia* 16(210): 1-11, pls 1-3 (figs 1-11) incl.

1968b Distribution of the chromosome numbers in Trichoptera in the light of phylogenetic evidence. *Genen Phaenen* 12(4): 110-113, fig. 1.

1969a (– & L. LANKHORST). The chromosomes of the caddis-fly Glyphotaelius pellucidus (Retzius, 1783) (Trichoptera: Limnephilidae, Limnephilinae). *Genetica* 40(1): 1-6, figs 1-8.

1969b (VAN GRIETHUYSEN, – & L.J.M. BUTOT). The chromosomes of Anadonta anatina (Linnaeus, 1758) and Unio pictorum (Linnaeus, 1758) (Mollusca, Bivalvia: Unionidae). *Basteria* 33(1/4): 51-56, figs 1-5.

1969c (BUTOT, L.J.M. & –). Cytotaxonomic observations in the stylommatophoran family Helicidae, with considerations on the affinities within the family. *Malacologia* 9(1): 261-262.

1969d (– & L.J.M. BUTOT). Contribution to the knowledge of the cytotaxonomic conditions in the stylommatophoran superfamily Zonitacea. *Malacologia* 9(1): 269-270.

1970a (BUTOT, L.J.M. & –). *Bibliography of the molluscan cytotaxonomy*. Inst. Genet., Univ. Utrecht. iv+119 pp.

1970b (BUTOT, L.J.M. & –). New chromosome numbers in some European Campylaeninae, Helicinae, Helicellinae and Hygromiinae (Gastropoda, Stylommatophora: Helicidae). *Chrom. Inf. Serv.*, Tokyo 11: 9-11, figs 1-20.

1970c Review of the germ cell chromosome cytology of Collembola, with a list of chromosome numbers and data on two species new to cytology. *Genen Phaenen* 13(4): 89-99, figs 1-.

1970d (DVOŘAK, J., J.M. VAN BRINK & –). A note on the germ cell chromosomes of three allied species of the genus Glyptotendipes Kieffer, 1913 (Diptera, Nematocera: Chironomidae, Chironominae). *Genen Phaenen* 14(1): 5-8, figs 1-10.

1971a The chromosome conditions in the spermatogenesis of the caddis-fly Goera pilosa (Fabricius, 1775) (Trichoptera, Integripalpia, Limnephiloidea: Goeridae) from the God dal Fuorn in the Swiss National Park, with considerations on the phylogenetic position of the family. *Ergebn. wiss. Unters. schweiz. Natnparks* 9(64): 174-185, figs 1-4, pl. 1 (figs 1-9) incl.

1971b (– & P. BRANDMAYR). Cytotaxonomic notes on the endemic ground-beetles (Coleoptera, Adephaga: Carabidae) of the Southeastern Alps and the Dinaric Karst. 1. Carabus (Megodontus) caelatus schreiberi Kraatz, 1877. *Genen Phaenen* 14(3): 61-64, figs 1-9.

1972a (VAN BRINK, J.M. & –). Editorial. *Genetica* 43(1): 1.

1972b (– & J.M. VAN BRINK). Association of Alpine Biological Research. *Vakbl. Biol.* 52(2): 3. – [Dutch; on foundation of the Netherlands Centre for Alpine Biological Research]

1972c (VAN BRINK, J.M. & –). Editorial. *Genen Phaenen* 15(1): 1-2. – [Dutch]

1973a (BRANDMAYR, P. & –). Cytotaxonomic notes on the endemic ground-beetles (Coleoptera, Adephaga, Carabidae) of Southeastern Alps and the Dinaric Karst. 2. Carabus (Platycarabus) creutzeri rinaldoi Meggiolaro, 1958 and C. (Megodontus) croaticus croaticus Dejean, 1826. *Genen Phaenen* 16(3): 73-79, figs 1-8.

1973b Osservazioni cariologiche sugli spermatociti di Troglophilus cavicola (Kollar, 1833) della grotta Cianevate presso Cavazzo Carnico, Friuli, Italia (Orthoptera, Grullacridoidea: Rhaphidophoridae). *Mondo sotteraneo*, Udine 1972/1973: 29-36, pls 1-3 (incl.).

1973c (DE HAAS, J.H. & –). Biological studies at the foot of Mt Bernina Switzerland: field-work of the Netherlands Centre for Alpine Biological Research on Isla Persa. *Levende Nat.* 76(1): 7-12, figs 1-3.

1973d (– & L.J.M. BUTOT). The chromosomes of Macrochlamys tugurium (Benson, 1852) from Nepal (Stylommatophora, Sigmurethra: Ariphantidae). *Basteria* 37(5/6): 145-148, fig. 1.

1973e (LAMBECK, H.J.P. & –). On a small collection of syrphid flies (Diptera: Syrphidae) from the Kathmandu Valley and the Khumbu Himal region (Nepal). *Ent. Ber., Amst.* 33(4): 70-78, figs 1-3.

1977a (–, M.A.J.E. KIAUTA & R.L. SHRESTHA). The karyotype of the Common Green Lacewing, Anisochrysa (Chrysoperla) carnea (Stephens, 1836) from Nepal, with a synopsis of the male cytotaxonomic features of the family Chrysopidae (Neuroptera, Planipennia). *Genen Phaenen* 18(2/3): 19-29, figs 1-10.

1977b (– & A.W.M. MOL). Behaviour of the spermatocyte chromosomes of the mayfly, Cloeon dipterum (Linnaeus, 1761) s.l. (Ephemeroptera: Baetidae), with a note on the cytology of the order. *Genen Phaenen* 19(2/3): 31-39, figs 1-18.

1977c (VAN BRINK, J.M. & –). On the current Master of Science projects of the Department of Animal Cytogenetics and Cytotaxonomy, University of Utrecht. *Utrecht UniversBl.* 9(17): 12. – [Dutch]

1979 (– & M.A.J.E. KIAUTA). Ecology, case structure, larval morphology and chromosomes of the caddisfly, Allogamus auricollis (Pictet, 1834), with discussion on the variation in recombination indices in the Stenophylacini (Trichoptera: Integripalpia: Limnephilidae). *Genetica* 50(2): 119-126, figs 1-10.

1982 Invertebrates above the timber-line and in mountain nival zone. *In*: J.G. van Gelderen, J. van der Kan & C. Laban, [Eds], *Encyclopedie van de natuur van Europa*, Vol. 4, pp. 1000-1001, Spectrum, Utrecht-Antwerpen. – [Dutch]

1984 Die Schwebefliegenfauna der Gletscherinsel Isla Persa, Bernina Massiv, Oberengadin (Diptera: Syrphidae). – The hoverfly fauna of the glacier island Isla Persa, Bernina Massiv, Upper Engadine, Swizerland (Diptera: Syrphidae). *Jahresbericht naturforschende Gesellschaft Graubünden* 101: 179-188, fig. 1.

1987 (–, M. KIAUTA & R. MÜLLER). Beiträge zur Kenntnis der Schmetterlingsfauna von Flums und des Flumserberges, Kanton St. Gallen, Ostschweiz. 1. Die Tagfalter (Lepidoptera: Hesperioidea, Papilionidea). *Opusc. zool. flumin.* 9: 1-25, figs 1-5.

1994 (SCHMIDT, J. & –). Beiträge zur Kenntnis der Schmetterlingsfauna von Flums und des Flumserberges, Kanton St. Gallen, Ostschweiz. 3. Spinner und Schwärmer (Lepidoptera: Hepialidae, Cossidae, Limacodidae, Psychidae, Thyrididae, Lasiocampidae, Attacidae, Drepanidae, Thyatiridae, Sphingidae, Notodontidae, Lymantriidae, Arctiidae). *Opusc. zool. flumin.* 120: 1-10.

1998 (– & M. KIAUTA). Der Nashornkäfer, Oryctes nasicornis (L.): Erstmeldung für Kanton St. Gallen, Ostschweiz (Coleoptera, Scarabeidae: Dynastinae). *Opusc. zool. flumin.* 137: 7-8.

2015 Natural History Museum of Slovenia – in Europe unique in its structure and operation. Some considerations on the required change of its housing and on the pending increase of scientific and technical staff. *Delo* (Znanost/Science), Ljubljana 57(222): 14. – [Slovenian]

2016 Foreword. *In*: B.K. TYAGI & V. VEER, *Entomology in the Doon Valley (Garhwal Himalaya): a stronghold for insect research*, pp. v-vi, Scientific Publishers (India), Jodhpur-New Delhi.

Annex 6. Editorial Work on Selected Non-Odonatological Periodicals (1962-2018)

1962 *Varstvo narave / Nature Protection*, State Department of Nature Conservation, Ljubljana, Vol. 1. – **Editor-in-Chief**

1971-1990 *Genetica. International Journal of Genetics*, Kluwer Academic Publishers, Dordrecht-Boston-London, Vols 42-83. – **Joint Editor**, including special volumes/issues, such as:

1980, *Animal genetics and evolution* – (Vol. 52/53)

1984, *Studies on avian karyology* – (Vol. 65/1)

1987, *Genetic research with non-human primates* – (Vol. 73/1-2)

1972-1977 *Genen en Phaenen. Journal of the Netherlands Genetic Society*, Utrecht, Vols 15-19 – **Joint Editor**

1973 *Field-Reports of the Netherland Centre for Alpine Biological Research*, Utrecht, Vol. 1 – **Editor**

1984→ *Opuscula zoological fluminensia. International Serial on Invertebrate Zoology*, Ursus Scientific Publishers, Flums/Switzerland, Pts 1→ [so far 224 pts published] – **Editor**

Annex 7. Handwritten (and Never Published) Text of a Talk by Dr B. Kiauta

This is a facsimile of the text of an oral presentation titled "History of dragonfly exploration in Slovenia" made by Dr B. Kiauta in the Entomology Section of the Society of Biologists of Slovenia (Ljubljana) during October 1955. This manuscript has never been published. Even though the outline of history is greatly incomplete, since numerous related data were found not until many years later (for these, see: KIAUTA, 1963a, 1994a, 1994f, 2000, 2001, 2002a, 2002b, 2003b, 2003c, 2003d, 2004a, 2004b, 2005, 2008, 2012, 2013a, 2014b, 2014d, 2014e, 2015a, 2015b, 2016c, 2017b, 2018c), nevertheless for posterity in the science of odonatology it is but a rare of the rarest piece of scientific treasure that showcases the strength of background research in a country that had been the cradle of so many animal explorations, involving also dragonflies. Yet, before Dr Kiauta set his foot on a long odonatological peregrination and began to systematically investigate into many aspects of odonate biology, particularly faunal inventory, taxonomy, cytogenetics, evolution and adaptation to different ecological niches (e.g., high altitudes, caves etc.), with special consideration to the fauna of Slovenia, the interest in Slovenian Odonata was less focused and intensive. For his massive contribution to the Slovenian dragonfly research Dr B. Kiauta can be most appropriately regarded as the "Father of Slovenian Odonatology".

Since early 1970s I have been archiving handwritten letters and/or other textual materials, sometimes represented only by signature and/or edits on a manuscript or even a dedicatory note on a book copy gifted etc., from a large number of entomologists and/or odonatologists across the world – for the sheer purpose of comprehanding the scientific prowess! I have been impressed by the writings of many contemporarians, but the three scientists or odonatologists who impacted me most are: Dr M.A. Lieftinck, Dr P.S. Corbet and Dr B. Kiauta. Handwriting of Dr Lieftinck, from whom I have quite a few typed and handwritten letters, including a post-card, was like putting an ocean into a pot, still maintaining utmost legibility and elegance. Dr Corbet's style of writing, like Dr Lieftinck's, was extremely neat and clean, and systematic. He was a highly skilled and prolific writer. Dr Kiauta and I must have exchanged several hundreds of letters; almost all in typed form, but some, like the copies of OA typescripts and other manuscripts, also bore his hand-written annotations, besides of course an inevitable 'blitzkrieg' signature. Dr Kiauta has influenced me most by his correspondence which, like that of both Dr Lieftinck and Dr Corbet, were tidy, well-ordered and always carried a crystal clear meaning. They were like 'a teacher standing before me and giving lessons to learn'.

Why is it so inevitable to bring this handwritten piece of scientific material, almost unknown to the modern world for several decades, on record through this book? Scholarly written manuscripts are treasures that store and disseminate information to absorb and update knowledge – the very essence of life every great scientist keep questing for. Spoken words in scientific colloquia may benefit at best some scores of attendees, but the printed version of speech certainly communicate with millions, and for a much longer time! In a way, these manuscripts are even more irreplaceable than jewels. New deposits of diamond and rubies might one day yield another Koh-I-Noor; there will never be another *Systema Naturae* (LINNAEUS, 1758), *Origin of Species* (DARWIN, 1859), *Dragonflies* (CORBET, 1999) or *Cytotaxonomy of Dragonflies* (KIAUTA, 1975a). Written painstakingly by hand such a manuscript is in fact a work of incandescent and gorgeous beauty that sometimes radiate rays of wisdom leading to Ariadne's thread for unraveling an enigmatic phenomenon out of its labyrinth.

Thus, in view of these facts, I am inclined to produce this manuscript here with the singleness of aim to ignite minds of young odonatologists to delve deeper with equal enthusiasm and determination into the science of odonatology, through all possibilities of media available, i.e., speech, manuscript, print and electronic, and leaving none whatsoever!

For an inquisitive and exploratory eye the manuscript – a kind of 'The Bastiaan Kiauta Code' – encompasses within its enigmatic folds a message to the budding odonatologists and a key to building early in life a great scientist, through only dint of hard work and focused application.

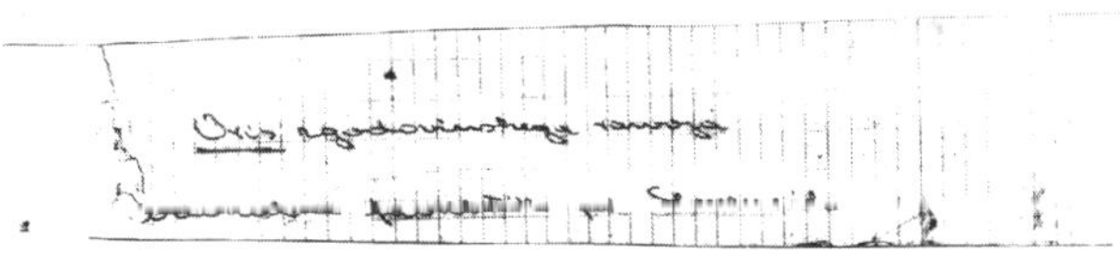

(History of dragonfly exploration in Slovenia)

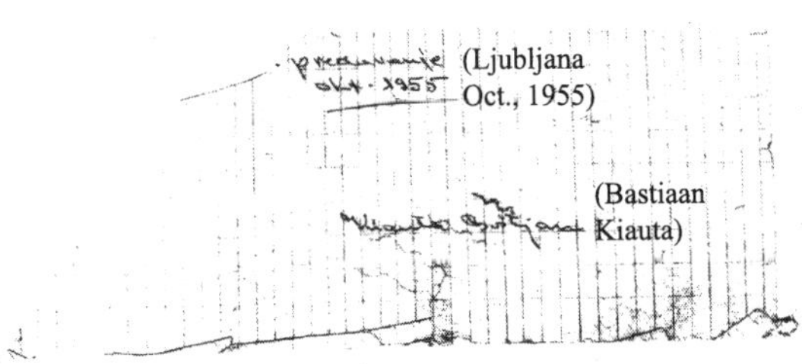

Uvod

Čeprav je minilo že več kot 200 let, odkar je Scopoli zaoral v ledino na področju slovenske entomologije, še do danes niso bili inventarizirani in favnistično obdelani vsi oddelki domače entomofavne. V tem času se je nabrala že skoro nepregledna vrsta tujih in domačih entomologov, ki so na kakršenkoli način sodelovali pri odkrivanju in študiju domačih insektov, bodisi kot znanstveniki-biologi, bodisi kot ljubitelji entomologije. Njihova imena in dela tonejo vedno globlje v pozabo. Velika škoda bi bila za domačo kulturno zgodovino in še posebej za zgodovino razvoja naravoslovja pri Slovencih, ako bi ne poizkušali ohraniti, mnogokrat pa šele odkriti, kolikor nam je še mogoče, njihovo znanstveno zapuščino.

Zgodovinski material in biološka problematika sta preobsežna, da bi bilo mogoče obdelati podatke za vse skupine insektov hkrati. Potrebno bi bilo, da bi strokovnjaki za posamezne rede skušali zbrati literaturo in arhivalno zapuščino entomologov, ki so se pri nas ukvarjali z njimi. Tako bi bilo mogoče istočasno z obdelavo starih zbirk, v kolikor še obstoje, rekonstruirati sliko dejavnosti nekdanjih domačih entomologov in spoznati kriterije po katerih so usmerjali svoje delo. Na taki podlagi bo mogoče sestaviti kompilacijo in pregled celotne favnistične dejavnosti pri Slovencih na področju entomologije in pomeje zoologije sploh, od prvih začetkov do današnjega časa.

Pričujoča študija je poizkus takega prispevka k ~~poznavanju historičnega razvoja domače entomo:~~

[illegible]

logije. Zbral sem vse znane in dostopne mi podatke o razvoji domače odonatne favnistike iz literature, arhivalnih virov in še ohranjenih zbirk. Gotovo se bo dalo to delo pomnožiti z novimi tovrstnimi, zlasti arhivalnimi najdbami kot so n.pr. stari rokopisni seznami še izgubljenih zbirk, korespondenca entomologov, dnevniki itd. še dopolniti.

Zaradi popolnosti navajam ob koncu seznam Schmidtove in dr. Staudacherjeve odonatne zbirke. Nobena od njiju do danes še ni bila obdelana v celoti, pač pa so bili objavljeni iz njih že nekateri podatki, nanašajoči se na posamezne vrste (Kiauta, 1953/54, str. 221; 1954/55, str. 46). Material Schmidtove zbirke je žal zaradi starosti že močno uničen, predstavlja pa brez dvoma enega najdragocenejših spomenikov amaterske favnistične dejavnosti v dobi slovenskih klasičnih entomoloških raziskav.

V seznamu literature navajam le najvažnejša dela. Marsikaj, kar bi osvetlilo splošne kulturne prilike dob, v katerih so delovali naši entomologi in so bile za njihov znanstveni razvoj in delovno metodo [illegible], nisem mogel upoštevati, ker bi s tem seznam preveč narasel. Sploh bi obdelava ničejne bibliografije za zgodovino domače entomologije zahtevala posebno delo.

Prijetna dolžnost mi je zahvaliti se g. nadzorniku Alfonzu Gspanu, starosti današnjih slovenskih entomologov, ki mi je v prijetnih pogovorih odkril marsikatero črto iz življenja nekdanjih domačih entomofavnistov, s katerimi je bil bodisi še sam v stikih ali je vedel o njih iz pripovedovanji svojih starejših entomoloških tovarišev, ter [illegible] del iz rokopisnega [illegible] vsakega entomološkega [illegible]

Oris zgodovinskega razvoja odonatne favnistike pri Slovencih

V skladu z duhom časa v naravoslovnih deželah pa je tudi pri nas zelo zgodaj pojavilo zanimanje za naravoslovne probleme in pojave v deželi. Od Valvasorja preko enciklopedistov konec sedemnajstega in v začetku osemnajstega stoletja do Zoisa so se tedanji kulturni krogi živo zanimali za prirodoznanska opazovanja v deželi. Vendar je do začetka prvih resnejših zooloških, predvsem entomoloških znanstvenih zapiskov pri nas doba evropske zoografije že ugašala. Tako tiho in neopazno kakor mimo Slovencev je šla tudi mimo ostalih narodov Habsburške monarhije (Grošelj, 1939, str. XIII). Obdobje sistematike to je zbiranje živalskih in rastlinskih vrst po posameznih favnističnih skupinah in sistematskih oddelkih, ki je z Linnejevim sistemom leta 1735 dobilo svojo najtrdnejšo oporo, se je pričelo že s pričetkom osemnajstega stoletja. V slovenski entomologiji je trajala ta doba vse od Scopolijevega prihoda v Idrijo nekako do Schmidtove smrti (1754–1878) in še danes prevladuje sistematska smer v študiju marsikatere skupine domačih insektov

Od vseh jugoslovanskih pokrajin je bila Slovenija prva pri proučevanju domače favne žuželk. Geografska lega in ~~politična~~ pripadnost sta ji v okviru nekdanje ~~Avstrije~~ habsburške monarhije nudili še posebno ugodne pogoje za razvoj favnistike, zlasti ker je s svojo splošno manj pestro favno vzbujala interese strokovnjakov in ljubiteljev tudi daleč izven svojih meja.

K temu sta v veliki meri brez dvoma pripomogli tudi ambicioznost nekaterih vodilnih slovenskih racionalistov – najvidnejši med njimi je bil Žiga baron Zois Edelsteinski (Kidrič, 1938, str 212; 220; 284, 285, 485; Gspan, 1956, str 379) – in marljivost mnogih tujih pri nas delujočih naravoslovcev. Za entomologijo je iz Zoisovega kroga pomemben Franc Anton pl. Breckerfeld, zbiralec gradiva za baronovo dopolnjevanje Pohlinovega slovarja. Poleg seznama slovenskih rastlinskih imen iz leta 1784 je zbral namreč tudi slovenske nazive za insekte, ribe in ptiče, ki jih je poslal Zoisu v začetku 1785 leta (Kidrič, 1938, str 235). Značilna za vse favnistično kot tudi drugo kulturno delovanje v Sloveniji do konca prve svetovne vojne, je razkosanost delovnih področij na posamezne politično-upravne enote: Kranjsko, Primorsko, Štajersko in Koroško.

Od vseh favnističnih disciplin je bila entomologija prva, ki je vzbudila zanimanje za biologijo Slovenije. Speleobiologija se je pojavila namreč kot samostojna dejavnost mnogo kasneje: šele z nastopom mentorja slovenskih naravoslovcev Ferdinanda Schmidta (1791–1878), čeprav so bile posebnosti slovenskega Krasa znane že iz antike od Posidonijevih in Strabonovih časov (Pretner, 1952, str 135), in sta od domačinov opozarjala na nekatere zanimivosti Kranjskega podzemeljskega ~~sveta~~ konec sedemnajstega stoletja že J.L. Schönleben (1618–16 (Schönleben, 1681, str. 123) in zlasti še J. V. Valvasor (1641–1693) (Valvasor, 1689, I. str 465–696)

Prvi začetki entomofavnističnega dela na področju etnografske Slovenije segajo v leto 1754, ko je bil imenovan za fizika v Idriji Joannes Antonius Scopoli (1723–1788), po rodu italijanski Tirolec iz Fleimsthala, po zanimanju pa prirodoslovec širokega obsega in raznovrstne delavnosti. Pred njegovih številnih potovanj po Notranjskem in Gorenjskem je leta 1763 izšla Entomologia carniolica, v kateri opisuje 1153 vrst artropodov iz ozemlja nekdanje Kranjske. Delo predstavlja monumentalni in pionirski prispevek k poznavanju slovenske favne živali. Mnoge od obravnavanih vrst so nove in prvič opisane. Zbirka, na podlagi katere je delo nastalo in ki je bila zbrana, kot navaja avtor v uvodu, v letih 1759–1762, žal ni ohranjena.

Za Scopolievo Entomologia carniolica je prispeval nekaj podatkov Franciscus Xaverius Wulfen (1728–1805), švedski plemič, rojen v Beogradu. V letih 1762 in 1763 je predaval na ljubljanskem jezuitskem kolegiju logiko, matematiko, metafiziko in fiziko. Podobno kot Scopoli se je tudi on udejstvoval na področju botanike, mineralogije in zoologije.

V delu avtor navaja za Slovenijo 6 vrst z 22 „varietetami". Pri njihovi determinaciji je uporabljal sledečo literaturo: Linne: Systema naturae, Roesel: Insecta aquatica in P. Poda: Museum graecum (Scopoli, 1763, str 259). Po nomenklaturi, ki jo je uporabljal so to: Libellula *grandis*; *L. depressa* (s 4 „varietetami"), *L. quadrimaculata* (z 2 „varietetama), *L. vulgata* (z 2 „varietetama), *L. virgo* (s 6 „varietetami") in *L. puella* (z 8 „varietetami)

Glede metode pri obdelavi materiala pravi avtor sam v uvodu: „Ordines et genera a celeberrimo Linnaeo stabilita secutus sum, reformatis tamen characteribus, fixisque limitibus, quibus unum ab alio rectius distingui possit .... Nomina trivialia rebus a Linnaeo data retinui, differentias vero mutavi quoties dubias notas pro certis adsumptas, et in earum nomina descriptionibus collocatas deprehendi, ..." To je vzrok, da je opisal dve novi vrsti *Cordulegaster annulatus* Latr. in *Orthetrum coerulescens* F. pod imeni po Linnéju opisanih vrst *Libellula grandis* in *L. vulgata*. Iz opisov t. im. variacij za katere daje Scopoli vedno le morfološke karakteristike, nikdar pa ne tudi imena, je razvidno, da avtor ni upošteval spolnega dimorfizma, prav tako pa tudi ne razlik med juvenilnimi in adultnimi stadiji. Zaradi pomanjkljivih opisov t. im. variacij in neohranjenosti zbirke iz katere so opisani primerki, žal ne moremo ugotoviti zanesljivo v vseh primerih katere vrste je Scopoli opisal kot prvi in so bile šele pozneje od drugih avtorjev ponovno opisane in opremljene s samostojnimi imeni. S popolno gotovostjo lahko spoznamo iz opisa „variacij" le redeče, šele pozneje od drugih avtorjev imenovane in opisane vrste: *Pyrrhosoma nymphula* Sulz (navaja kot dve variaciji: posebej obliko samca in samice pri vrsti *Libellula puella*), *Calopteryx splen-*

senz Harr. (kot tri variacije pri Libellula virgo - ena oblika samca in dvoje samic) in končno Libellula fulva Müll. (kot dve variaciji Libellula depressa - enkrat samec pred kopulacijo, drugič po kopuli z značilno temno obarvanim abdomnom).

Glede larvalnih stadijev pravi avtor sam: „... pau-cas metamorphoses, neque eas semper debita at-tentione rimatus sum, ..." Kljub temu navaja glavne morfološke tipe odonatnih larv: Libellula puella, L. grandis in L. depressa, torej oblike z listastimi in rektalnimi škrgami.

Točnih lokalitet ne navaja nikjer, razen pri vrsti Libellula quadrimaculata: „in fossis circa La-bacum" (Scopoli 1763 str. 261) ter v uvodu je pove-dano, da je bil material zbran na področju vojvo-dine Kranjske. Prav tako ne navaja nikjer feno-loških podatkov, pač pa dosledno za vse vrste krat-ko ekološko karakteristiko za imago. Po pomanj-kljivosti nekoliko presenečajo, kar ugotavlja avtor sam: „... raro enim perfecta est differentia illa specifica, quae ab uno aliove specimine in Museo viso, aut ab Amicis communicato desumitur, cum eadem in aliis regionibus alio modo se habeat, ..." Tako bi torej upravičeno smeli pričakovati nekoliko večjo pozornost do ožjih lokalitetnih podatkov.

Zanimive so ~~risarske~~ barvne(!) skice vseh odonatnih vrst, ki jih v delu omenja in jih dodaja kot prilogo na koncu knjige. V splošnem so risane precej realistično - z edino izjemo Libellula puella, ki ima popolnoma ne-naravne telesne proporcije - vendar so tudi ostale risbe brez podrobnosti: nervatura je docela zanemarjena, seg-mentacija pa močno shematizirana.

Tretji član pionirske entomološke trojice pri Slovencih je <u>Baltazar Hacquet</u> de la <u>Motte</u>, doma iz Le Conquet-a v Bretagni — „mož misterioznega porekla, morda nezakonski sin neke tuje visoke osebe, ki je v življenju vedno moral računati na pomoč iz neznanega vira" (Kidrič, 1938, str. 687). Na Kranjsko, v Idrijo je prišel l. 1766 na častno željo in proti volji kranjske dežele (Wester, 1954, str 12). Kot zdravnik in naravoslovec v najširšem smislu: biolog, geolog, paleontolog, petrograf, etnolog itd je kmalu dosegel visok ugled in leta 1773 službo v Ljubljani. Na mnogih številnih popotovanjih je poleg drugega zbral tudi bogato entomološko zbirko, ki pa je žal ostala neobdelana in je danes izgubljena. Bil je prijatelj in sodelavec baronov Žige in Karla Zoisa. ter Scopolia.

Pionirski koraki na področju naše odonatne in splošne entomološke favnistike v drugi polovici osemnajstega in na prehodu v devetnajsto stoletje, ki ji dajejo obeležje imena Scopoli, Wulfen, Hacquet, je sledilo kreativno obdobje v večini devetnajstega stoletja to je doba favnistične dejavnosti Ferdinanda Schmidta

Veliko vlogo v tem obdobju je odigrala ustanovitev ljubljanskega muzeja do katere je prišlo šele leta 1826, čeprav je že leta 1821 ob priliki ljubljanskega Kongresa Sv. alianse uradno načel vprašanje ustanovitve Kranjskega deželnega muzeja po vzoru drugih avstrijskih deželnih muzejev tedanji guverner Ilirije grof <u>Sweerts-Spork</u> in je cesar Franc s to namero popolnoma soglašal (Mal, 1931, str 7). Tako se je leta 1831 ustanovilo še Muzejsko društvo (Verein des Krainischen Landesmuseums), se je slovenski naravoslovni – tudi entomološki – diletantizem nekako osredotočil. Duša tega gibanja sta postala vodja muzeja grof Franc Hochenwart in muzejski kustos farmacevt, botanik in entomolog Henrik Freyer (1802–1866) iz Idrije, ki je bil na tem mestu, ki ga je leto prej kot prvi zasedel jurist Jan. pl. Schildenfeld, od leta 1832 do 1852.

Ferdinand Josef Schmidt (1791–1878) – „Nestor der Illyrischen Naturforscher", kot ga je nazval Rossmaessler v svoji ikonografiji evropskih mehkužcev (prim. S.A. Rossmaessler: Iconographie der Land und Süsswasser Mollusken), nemški trgovec iz Šopronja (Odenburg), ki je preživel večino svojega življenja v Ljubljani, je s svojo amatersko zbirateljsko dejavnostjo ustvaril temelje slovenske entomologije in speleobiologije – kot je njegovo delo pravilno ocenil tudi njegov sodelavec, takratni muzejski kustos Karel Dežman, ki je leta 1852 nasledil to mesto po odhodu svojega prednika Henrika Freyerja v Trst. Z dejavnostjo slednjega naj bi bil v zvezi naravoslovni nastop F. Schmidta (Slodnjak, 1934, str. 20)

Schmidt se je ukvarjal že od leta 1827 s študijem domače entomofavne, kot pričuje sam v svoji avtobiografiji (Schmidt, 1875, str 9). V pomoč si je organiziral gosto mrežo sodelavcev. Med njimi: trgovski pomočnik Henrik Hauffen, grof Franc Hochenwart, notar Nikolaj Hoffmann, Karel Dežman itd., od leta 1840 pa še takratni srednješolci Fran in Matija Erjavec, Ivan Tušek in Anton Stubec. Fran Erjavec je sam zbral zbirko kranjskih insektov, ki jo je leta 1854 podaril Deželnemu muzeju (Sbornik, 1934 str 31). Danes žal ni več ohranjena. S področja odonatologije ni pustil Schmidt ne v tisku ne v rokopisu nobenega dela, niti ni ohranjen kataloški seznam za ta oddelek njegove zbirke, kljub temu, da se seznami za skoro vse ostale oddelke še danes nahajajo v Schmidovi zapuščini v Prirodoslovnem muzeju v Ljubljani. Ohranjena pa je odonatna zbirka sama, ki predstavlja tako najstarejšo ohranjeno zbirko odonatov s slovenskega ozemlja. Poleg domačega materiala vsebuje zbirka tudi nekaj inozemskih primerkov, ki jih je zbral znani entomolog in odonatolog <u>Brittinger</u> leta 1851 v Steyru (Avstrija), nekaj pa je celo eksotov iz Sirije. ~~Po Schmidovi smrti je~~ l. 1877 je [illegible] ~~[illegible]~~ prešla ~~[illegible]~~ zbirka ~~[illegible]~~ prirodoslovnega ~~Kabineta~~ Državne realne gimnazije, danes pa se nahaja v Prirodoslovnem muzeju v Ljubljani.

(stanje)

Iz razumljivih vzrokov je material zbirke deloma v precej slabem stanju – od nekaterih osebkov so ohranjeni samo večji fragmenti, ali pa so celo preparati popolnoma uničeni. Skupno obsega danes Schmidova odonatna zbirka sto primerkov s štiridesetimi vrstami. Žal material v veliki večini ni opremljen s podatki o lokalitetah, tako da za večino vrst lahko le bolj ali manj pravilno sklepamo na slovensko nahajališče. Označbo iz Kranjske nosijo le vrste: <u>Calopteryx virgo L.</u>, <u>Sympecma fusca v.d. Lind.</u>, <u>Aeschna juncea L.</u>, <u>A. isosceles Müll.</u>,

in *Anax imperator* Leach. Nekatere vrste iz Schmidt=
tove zbirke, ki pa žal niso opremljene z lokalitetami=
mi, v Sloveniji niso bile pozneje več najdene.
Razen splošne oznake nosi le en primerek točno
lokacijo (Cerknica)

Mlajši sodobnik F. Schmidta je bil pravnik dr. Ju=
[illegible]
lius pl. Gspan (1848 – 1888), notar v Kostanjevici
in Krškem, oče znanega slovenskega koleopterolo=
ga Alfonza Gspana. Bil je v stalnih stikih
s takrat pri nas delujočimi entomologi in je tu=
di sam zbral znatno zbirko odonatnega materia=
la. Po njegovi smrti je zbirko odkupil neki
Radel iz Žužemberka, ljubitelj entomologije, takrat
dijak na meščanski šoli v Krškem. Ko se je
okrog leta 1903 zanj zanimal zbiratelja sin
Alfonz je bila zbirka že skoro popolnoma uni=
čena in je ~~do danes~~ kasneje žal neobdelana tudi
dokončno propadla.

Po Schmidtu je nadaljeval z zbiranjem Odona=
tov dr. Josip Staudacher (1876 – 1948), okrožni
zdravnik po rodu koroški Nemec, ki pa je stalno
živel v Ljubljani. Ta je podobno kot F. Schmidt
posvetil zbiranju odonatnega materiala le
manjšo pozornost, odonatologija mu predstav=
lja le stransko entomološko panogo. Mnogo več se
je ukvarjal z lepidopteri, koleopteri, hemipteri
(heteropteri in homopteri), ortopteri, himenopteri
in deloma celo dipteri. Zlasti na nekaterih pod=
ročjih domače entomološke favnistike predstav=
ljajo njegove zbirke – poleg Schmidtovih, ki je
bil prav tako polihistor v entomologiji – temelje
nadaljnjemu favnističnemu delu. Podobno kot

Schmidt je dr Staudacher uspel organizirati gosto mrežo zbiralcev materiala, tako da je zahvaljujoč se lastni delavnosti in številnim strokovnim prijateljem (med njimi znani koleopterolog Alfonz Gspan in lepidopterolog Rudolf Rakovec) zajel znaten del slovenskega ozemlja

Dr Staudacherjeva odonatna zbirka, ki se nahaja danes v Prirodoslovnem muzeju v Ljubljani, obsega 76 primerkov in 24 vrst. Material je bil zbiran nesistematično, zgolj priložnostno od leta 1884 do 1953. Zbirka je bila dopolnjevana tudi po dr Staudacherjevi smrti. Pri zbiranju so sodelovali: lekarnar Piccoli, poštni nadkontrolor J. Strussiner ter biologi M. Petričeva, S. Brelih in dr. Š. Michieli. Poleg domačega materiala vsebuje tudi nekaj dalmatinskih in eksotičnih primerkov iz Brazilije. Slednje je prispeval znani lepidopterolog Ciril Vrančič. Nekateri zoogeografski in fenološki podatki imajo še danes veliko vrednost (Kiauta, 1953/54, str. 221; 1954/55 str. 46).

V jugoslovanskem merilu prednjači Slovenija pri zbiranju in študiju lokalne favne odonatov skoro za celo stoletje pred ostalimi republikami. Šele leta 1856 je objavil Frauenfeld v Verhandlungen der Zoologisch-Botanischen Gesellschaft in Wien prve podatke o odonatni favni Dalmacije. Toda med tem ko je odonatno publicistično delo v Sloveniji zamrlo – znana mi je le študija dr Robert Latzel-a o koroški favni iz leta 1876 ki omenja le dve odonatni vrsti in sicer *Aeschna cyanea* Müll. iz okolice Celovca ter *Libellula depressa* L. (Latzel, 1876, str. 17) – material pa se je zbiral le pri-

[illegible] in nesistematično, so si sledile razprave o odonatih iz jugoslovanskega juga druga za drugo. Od leta 1856 pa do osvoboditve so prispevali z vidnejšimi prispevki k razjasnitvi jugoslovanskih odonatofavnističnih vprašanj: Stein (1863), Spagnolini (1877), Novak (1890), Kohaut (1896), Klapalek (1898), Mc Lachlan (1898), Rössler (1900), Galvagni (1902), Bartenev (1912), Campion (1918), Werner (1919), Doflein (1921), Fudakowski (1930, 1932), Pongratz (1944) in St. Quentin (1944). Po letu 1945 pa so objavili prispevke k favni odonatov Jugoslavije: Adamović (1948, 1949, 1956), Živojinović (1950), Kiauta (1953/54, 1954/55, 1958/59) in Filevsky (1954)

V seznamu navajam celoten pregled odonatnega materiala Schmidove in dr Staudacherjeve zbirke ne glede na to ali je bil ulov na ozemlju etnografske Slovenije ali drugod. Datume in lokalitete navajam povsod kjer so v zbirkah zabeleženi. Pri pisavi krajevnih imen sem se držal prvotne oblike, oziroma okrajšave (kot je zabeležena na etiketi) ne glede na poznejše preimenovanje kraja ali različno v pisavi. Prav tako navajam povsod kjer je na etiketi zabeleženo tudi ime zbiralca, ki je primerek za zbirko prispeval.

Schmidtovo zbirko je determiniral deloma *Schmid* sam, deloma pa *Brittinger*. Njuno sinonimiko navajam pri vseh dotičnih vrstah z navedbo determinatorja poleg danes veljavnega imena. Pri nekaterih primerih se je izkazalo, da je Schmidtova determinacija nepravilna. Dr. Staudacherjeva zbirka ni bila determinirana.

~~Nomenklatura, ki jo uporabljam je klasična, kot jo je postavil monograf *Selys-Longchamps*, sistem pa je povzet po dr *E. Schmidtu* (1929)~~

Nekaj primerkov v obeh zbirkah zaradi zelo slabe ohranjenosti ali pomanjkljive literature (ahati) ni bilo mogoče determinirati.

Upoštevajoč le z lokalitetami opremljene vrste, so po Scopolijevi Entomologiji v Schmidtovi zbirki za Slovenijo nove vrste: *Sympecma fusca v. d. Lind.*, *Aeschna juncea L.*, *A. isosceles Müll.* in *Anax imperator Leach*; v dr Staudacherjevi zbirki pa po Scopoliju in Schmidtu - upoštevajoč tudi današnja scopolieva „mutacija" in študija dr. Sotrela - so: Lestes viridis v. d. Lind., Platycnemis pennipes Pall., Agrion ornatum Selys, Gomphus vulgatissimus L., Onychogomphus forcipatus L., Cordulegaster bidentatus Selys, Somatochlora flavomaculata v. d. Lind., Crocothemis erythraea Brullé, Sympetrum vulgatum L. in S. sanguineum Müll.

I. Zbirka Ferdinanda Schmidta

1. fam.: Calopterygidae Burmeister

1. Calopteryx virgo L.
„Carn." 1♂ det. Sch.: Agrion vesta Hag., 1♀, 1 Kamnik ♂

2. Calopteryx splendens Harr.
1♂, 1♂ det. Sch.: Agrion parthenias var. ♀

2. fam.: Lestidae Leach (Brauer)

3. Sympecma fusca v. d. Lind.
1♂ „Steyer 1851" det. Britt.: Lestes fusca v. d. Lind., 1♂ „Carn" det. Sch.: Lestes fusca v. d. Lind., 1♀

4. Lestes virens Charp.
1♂ det. Sch.: Agrion virens Charp., 1♀

5. Lestes viridis v. d. Lind.
1♂ det. Sch.: Agrion leucopsalis Charp., 1♀

6. nedoločljiva vrsta

3. fam.: Agrionidae Steph. (Tillyard)

7. Platycnemis pennipes Pall.
1♂ juv. det. Sch.: Agrion lacteum Charp., 2♂♂ juv., 5♂♂, 1♀ juv.

8. Pyrrhosoma nymphula Sulz.
1♂ in 1♀ det. Sch.: Agrion minium Charp.

9. Ischnura elegans v. d. Lind.

1♂ det. Sch.: Agrion tuberculatum Charp.

10. Ischnura pumilio Charp.

2 primerka, pol nedoločljiv, det. Sch.: Agrion pumilio Charp.

11. Enallagma cyathigerum Charp.

1♂ det. Sch.: Agrion cyathigerum Charp.

12. Agrion hastulatum Charp.

1♀ det. Sch.: A. hastulatum F.

13. Agrion puella L

2♂♂; 1♀

14. en nedoločljiv primerek

4. fam.: Aeschnidae Jac. et Bianchi

15. Aeschna juncea L

„Corn." 1♀ det. Sch. A. juncea L

16. Aeschna grandis L

1♀ det. Sch.: Aeschna grandis L

17. Aeschna cyanea Müll

1♂ det. Sch.: Aeschna maculatissima Latr.; 2♂♂; 1♀ det. Sch.: Aeschna affinis v.d. Lind.

18. Aeschna mixta Latr.

1♂.

19. Aeschna isosceles Müll

„Zirknitz" 1♂; 1♀ det. Sch.: Aeschna chrysophthalma Charp.

20. Anax imperator Leach

„Corn." 1♂ det. Sch.: Aeschna azurea Charp.

5. fam.: Gomphidae Bks.

21. Gomphus vulgatissimus L

1♂ det. Sch.: Aeschna forcipata Charp.; 3♂♂; 1♀

22. Onychogomphus forcipatus L

1♀. det. Sch.: Aeschna hamata Charp.

6. fam.: Cordulegasteridae Bks.

23. Cordulegaster bidentatus Selys

Sch.: Aeschna lunulata Charp.; 1♀

7. fam.: Libellulidae (Steph.) Burm.

24. Cordulia aenea L.

1♂ det. Sch.: Libellula flavomaculata L.; 2♂♂

25. Somatochlora metallica v. d. Lind.

1♂ inv.

26. Libellula depressa L.

1♀ det. Sch.: Libellula depressa L.

27. Libellula fulva Müll.

1♂ (pri kopulaciji) det. Sch.: Libellula conspurcata Fab.; 1♀

28. Libellula quadrimaculata L.

3♂♂ det. Sch.: Libellula quadrimaculata L.

29. Orthetrum coerulescens Fbr.

1♀ (?)

30. Orthetrum brunneum Fonsc.

1♂ det. Sch.: Libellula coerulescens Fab.; 1♂ inv., 2♀♀ inv., 3♀♀

31. Orthetrum cancellatum

1♀ det. Sch.: Libellula cancellata L.; 1♀ inv.

32. Crocothemis erythraea Brullé

1♂ det. Sch.: Libellula coccinea Charp.; 1♂ det. Sch.: Libel-
lula ferruginea Ramb.; 1♀

33. Sympetrum flaveolum L.

1♂; 1♀ inv. det. Sch.: Libellula flaveola L.

34. Sympetrum vulgatum L.

1♂ det. Sch.: Libellula vulgata L.; 1♀ 10. XI. 1844

35. Sympetrum pedemontanum Allioni

1♂ det. Sch.: Libellula pedemontana Fab.; 1♂ (?) inv.

36. Sympetrum sanguineum Müll.

„Carn." 1♂ det. Sch.: Libellula rufescens; 2♂♂; 2♀♀

37. Sympetrum rubrinervis De Selys

„Syrien" 1♂, 1♀ det.: Libellula rubrinervis De Selys

38. Leucorrhinia albifrons Burm.

„Steyer" 1851 1♀ det. Smith.: Libellula albifrons Burm.

39. Leucorrhinia dubia v.d. Lind.
„Steyer" 1851 1♂, 1♀ det. Spitt.: Libellula dubia v.d. Lind.
40. Leucorrhinia pectoralis Charp.
1♂ det. Sch.: Libellula caudalis, 1♂, 1♀

II. Zbirka dr. Josipa Staudachera

1. fam.: Calopterygidae Burmeister

1. Calopteryx virgo L.
„Carn. Laibach" 1♂ (leg. Stussiner); „Slovenija, Ljubljana" 14.VI.1953 1♂ (leg. Michieli); „Carn. Laibach" 1♀ (leg. Stussiner); „Dr. Grinšiči" 15.VII.1938 1♀ (leg. Bertić)
2. Calopteryx splendens Harr.
„Carn. Veldes" 2.VIII.1931 1♂ (leg. dr. Staud.); „Carn. Veldes" 19.VII.1931 1♀ (leg. dr. Staud.)
3. Calopterygidae sp. = Hetaerina sp.
„Brasilia, Sao Paulo" V.1934 1♂, 1♀

2. fam.: Lestidae (Jac. et Calvert) Tillyard

4. Lestes viridis v.d. Lind.
„Carn. Veldes" 17.VIII.1932 1♀ (leg. dr. Staud.)
5. ~~Lestidae sp.~~ Mecistogaster sp.
„Sao Paulo" 1♂, 1♀

3. fam.: Agrionidae (Steph.) Tillyard

6. Platycnemis pennipes Pall.
„Dr. Grinšiči" 15.VI.1938 2♂♂ (leg. Bertić); „Črnuče" 17.V.1953 1♂ juv (leg. et det. Bralić); „Carn. Veldes" 2.VII.1931 1♂ (leg. dr. Staud.) „Carn. Laibach" 2.VIII.1929 1♂ (leg. dr. Staud.)
7. Pyrrhosoma nymphula Sulz.
„Črnuče" 17.V.1950 1♂, 1♀ (leg. Bralić); „Slovenija, Ljubljana" 14.VI.1953 1♀ (leg. dr. Michieli); „Salona" 1.VI.1923 1♂ (leg. dr. Staud.)
8. Agrion ornatum Selys

„Triest" 27. VI. 1944 1 ♂ (leg. H. Staud.)

9. Agrion puella L

„Carn. Laibach" 5 ♂♂, 1 ♀ (leg. Stussiner); „Slovenija, Ljubljana okolica" 14. VI. 1953 3 ♀♀ (leg. H. Michieli); „Triest" 27. VI. 1944 1 ♂ (leg. H. Staud.); 1 ♀ brez etikete

4. fam.: Aeschnidae Jac. et Bianchi

10. Aeschna cyanea Müll.

„Carn. Krain" 14. VII. 1926 1 ♂ (leg. H. Staud.); „Carn. Laibach" X. 1884 1 ♂ (leg. Piccoli); „Carn. Laibach" 24. IX. 1929 1 ♀ (leg. H. Staud.)

11. Aeschna mixta Latr.

„Ins. Silba, Dalm." 1934 1 ♂ (leg. H. Staud.)

12. Anax imperator Leach

„Slovenija, Ljubljana" 14. VI. 1953 1 ♂ (leg. H. Michieli)

5. fam.: Gomphidae Bks.

13. Gomphus vulgatissimus L

„Carn. Laibach" 9. VI. 1944 1 ♂ (leg. H. Staud.); „ob glinščici" 6. VI. 1938 1 ♂ (leg. Batrič); „Carn. Laibach" 2. VI. 1938 2 ♀♀ (leg. H. Staud.); „Carn" 18. V. 1926. 1 ♀ imm. (leg. H. Staud.), „ob glinščici" 6. VI. 1938 1 ♀ (leg. Batrič)

14. Onychogomphus forcipatus L

„ob glinščici" 15. VII. 1938 1 ♂ in 1 ♀ (leg. Batrič); „Laibach" 20. VIII. 1931 1 ♀ (leg. H. Staud.)

6. fam.: Cordulegasteridae Bks.

15. Cordulegaster bidentatus Selys

„Bad Vellach, Carinthia" 1 ♂ (leg. Stussiner)

7. fam.: Libellulidae (Steph.) Burm.

16. Somatochlora flavomaculata v.d. Lind

„Carn. Laibach" 10. VI. 1931 1 ♂ (leg. H. Staud.)

17. Libellula depressa L

„Carn. Babni dol" 5. VI. 1927 1 ♂ (leg. H. Staud.); „Carn. Laibach" 17. VII. 1913 1 ♂ (leg. H. Staud.); „Zagorod Dalm."

1934 1♂ (leg dr. Staud.?); „Carn. Laibach" 17. VII. 1934 1♀ (leg dr. Staud.); „Carn. Prasser" 13. V. 1926 1♀ (leg dr. Staud.)

18. Libellula quadrimaculata L.

„Slovenija, Ljubljana 14. VI. 1953 1♂ (leg dr. Michieli)

19. Orthetrum coerulescens F.

„Slovenija, Ljubljana" 14. VI. 1953 1♂ in 1♀ (leg dr. Michieli); 1♀ brez etikete

20. Orthetrum cancellatum L.

„Boccagnazzo, Zara, Dalm." 1. VII. 1932 1♀ (leg dr. Staud.)

21. Crocothemis erythraea Brullé

„Carn. Laibach" 1. VIII. 1931 1♂ (leg dr Staud.); „Croatia Krk-Čikes" 5. VII. 1953 2♂♂ (leg. Bralić); „Boccagnazzo, Zara, Dalm." 1. VII. 1931 1♀ (leg dr. Staud.)

22. Sympetrum vulgatum L.

„Carn. Prasser" 22. III. 1929 1♂ (leg. dr Staud.); „Ljubljana-Tivoli" 19. VI. 1926 1♀ (leg. Petrič)

23. Sympetrum sanguineum Müll.

„Carn. Laibach, Tivoliwald" 10. XI. 1881 1♂ (leg. Stussiner); „Carn. Laibach" 17. VII. 1943 1♂ (leg dr. Staud.); „Carn. Veldes" 3. VIII. 1930 2♀♀ imm. (leg dr. Staud.); „Boccagnazzo, Zara, Dalm." 1. VII. 1931 1♂ (leg dr. Staud.); 1♀ imm brez etikete

24. Leucorrhinia albifrons Burm.

1♂ brez etikete

S. fonscolombii

Literatura

1. Carnelutti J.: O dosedanjem in bodočem delu na favni lepidopterov Slovenije. Prir. vest. II. 1953
2. Dimitz A.: Geschichte Krains I, IV ~~([illegible])~~ Laibach ~~1874~~-1876
3. Ferdinand Josef Schmidt. Laibacher Tagblatt XI/1878 št. 43-44
4. Gogala M.: Prispevek k poznavanju favne stenic Slovenije Ljubljana 1958 (rokopis)
5. Grošelj P.: Prirodoznanska prizadevanja med Slovenci Proteus VI 1939
6. Gruden J.: Zgodovina slovenskega naroda. Celje 1911-1916
7. Hartig: Hacquets Autobiographie und Originalbriefe an Freiherrn v. Moll. „Wahrheit" München 1908
8. Heinrich F. X.: Johann Anton Scopoli. Ukh. Illyrisches Blatt XV 1833 Nr 12 str 47-49
9. Kiauta B.: Odonati v ljubljanski okolici. Proteus 1953/54 str 220-222
10. Kiauta B.: Tuji elementi med kačjimi pastirji Ljubljanskega Barja. Proteus 1954/55 str 44-46
11. Kiauta B.: Problematika in dosedanje delo na odonatni favni Slovenije. 1955 (rokopis predavanja v Entomološki sekciji v Ljubljani)
12. Kidrič F.: Zgodovina slovenskega slovstva od začetkov do Zoisove smrti. Ljubljana 1938
13. Kocbek F.: Dr. J. A. Scopoli – v spomin stoletnice smrti. Učiteljski tovariš XXVIII. 1888 str 18-23
14. Latzel R.: Beiträge zur Fauna Kärntens. Separ. des nat.-hist. Landesmus. von Kärnten. XII. 1876
15. Leitz W.: Scopolis Schicksale in Idria. Laibacher Schulzeitung XVI./1888 št 13-14 str 198-202
16. Mal J.: Zgodovina slovenskega naroda. Najnovejša doba

Celje 1928 – 1939

17. Mal J.: Zgodovinski pregled. Vodnik po zbirkah Narodnega muzeja v Ljubljani. Kulturnozgodovinski del. Ljubljana 1931

18. May E.: Libellen oder Wasserjungfern (Odonata). Die Tierwelt Deutschlands. 27 Teil. Jena 1933

19. Pintar L.: Balthasar Hacquet. Planinski Vestnik XVIII 1912

20. Pretner E.: O ohranjevanju krasa. V knjigi: V. Bohinec, F. Planina, J. Sottler: Slovensko Primorje v luči turizma. Ljubljana 1952 str 135 – 139

21. Prijatelj I.: Duševni profili slovenskih preporoditeljev. Ljubljana 1935

22. Rupel M.: Reformacija. V knjigi: Legiša L.: Zgodovina slovenskega slovstva I. Ljubljana 1956 str 185 – 323

23. Savnik R.: Razvoj domače speleologije in nekatere njene aktualne naloge SAZU. Acta carsologica I 1955

24. Scopoli J. A.: Entomologia carniolica exhibens insecta Carnioliae indigena et distributa ..... methodo linnaeana. Vindobonae 1763

25. Schmidt E.: Libellen. Odonata. Die Tierwelt Mitteleuropas IV. Lief. 16. Leipzig 1929

26. Schmidt F.: Leben und Streben des Ferd. J. Schmidt. Laibach 1875 (rokopis v Državnem arhivu LRS)

27. Schönleben L.: Carniola antiqua et nova. Labaci 1681

28. Slodnjak A.: Uvod v Frana Erjavca zbrano delo. Uredil Anton Slodnjak. Ljubljana 1934

29. Valvasor J. W.: Die Ehre des Herzogthums Crain. Laibach 1689

30. Voss W.: Joannes Antonius Scopoli (Lebensbild eines Österreichischen Naturforschers) Laibacher Schulzeitung VIII. 1880 št 23 str 353-360

31. Voss W.: Franz Xaver Freiherr von Wulfen (Lebenslauf und botanische Tätigkeit in Krain) Laibacher Schulzeitung XI. 1883 št 19 str 283-285

32. Wester J.: Baltasar Hacquet. Njegovo življenje in delo. Življenje in svet V. 1931 knj 9 št 2 str 30-33

33. Wester J.: Baltasar Hacquet, prvi raziskovalec naših Alp. Ljubljana 1954.

34. Ullepitsch K.A.: Johann Anton Scopoli. Illyrisches Blatt XXI. 1839 št 28 str 113-114

35. Ullepitsch K.A.: Hacquet. Illyrisches Blatt XXI. 1839 št 39 str 157-158

36. Ullepitsch K.A.: Franz Freiherr von Wulfen Illyrisches Blatt XXI. 1839 št 52 str 209-210

37. Gspan A.: Razsvetljenstvo. V knjigi: Legiša L.: Zgodovina slovenskega slovstva I. Ljubljana 1952 str 327-440

38.) Dolšak F.: 1925 J. A. Scopoli. Odličnemu botaniku ob dvestoletnici. Glas. Muz. dr. za Slovenijo 1924-25 str 64-69

Related Books

ISBN: 9789388043304

ISBN: 9788172336417

ISBN: 9788172335090

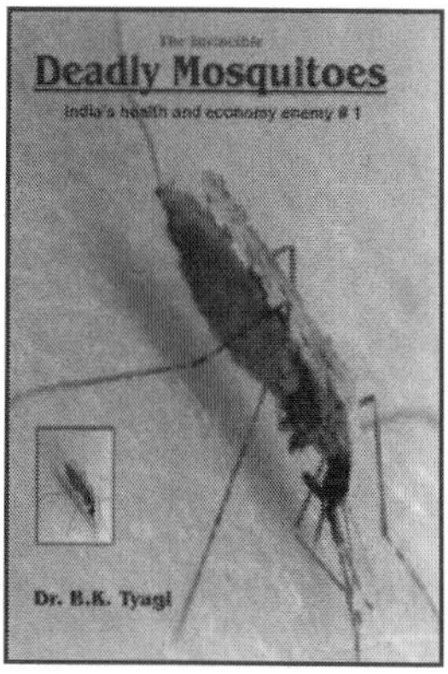

ISBN: 9789388043502

ISBN: 9789388043434

ISBN: 9789386102751

ISBN: 9788172335328

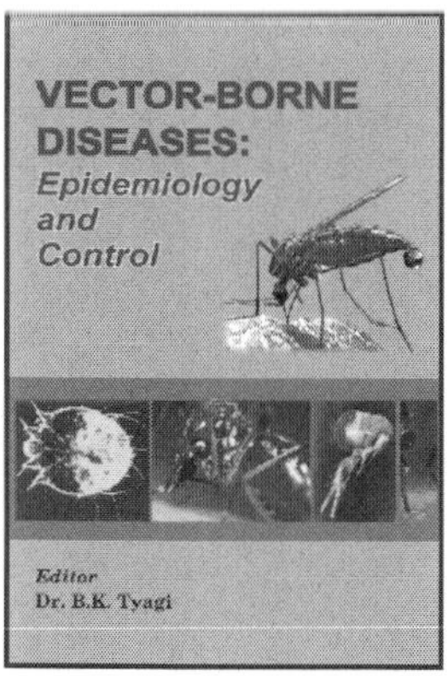

ISBN: 9788172335410

ISBN: 9788172335991

Other Related Books

ISBN: 9789388449076

ISBN: 9789388172752

ISBN: 9789388172981

ISBN: 9789388172950

ISBN: 9789388043311

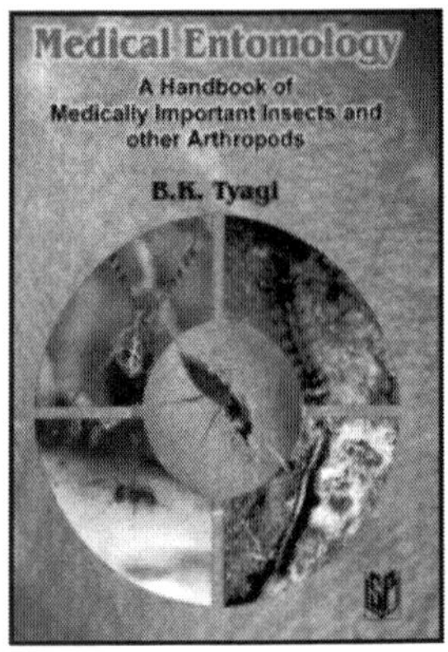

ISBN: 9788172333515

ISBN: 9788172333652

ISBN: 9788172335410

ISBN: 9789386652355

for Full Content and Sample Pages visit www.scientificpub.com